CONTEMPORARY MATHEMATICS

ALGEBRAISTS' HOMAGE:
Papers in Ring Theory and Related Topics

AMERICAN MATHEMATICAL SOCIETY

VOLUME 13

Titles in this Series

VOLUME 1 **Markov random fields and their applications**
Ross Kindermann and J. Laurie Snell

VOLUME 2 **Proceedings of the conference on integration, topology, and geometry in linear spaces**
William H. Graves, Editor

VOLUME 3 **The closed graph and P-closed graph properties in general topology**
T. R. Hamlett and L. L. Herrington

VOLUME 4 **Problems of elastic stability and vibrations**
Vadim Komkov, Editor

VOLUME 5 **Rational constructions of modules for simple Lie algebras**
George B. Seligman

VOLUME 6 **Umbral calculus and Hopf algebras**
Robert Morris, Editor

VOLUME 7 **Complex contour integral representation of cardinal spline functions**
Walter Schempp

VOLUME 8 **Ordered fields and real algebraic geometry**
D. W. Dubois and T. Recio, Editors

VOLUME 9 **Papers in algebra, analysis and statistics**
R. Lidl, Editor

VOLUME 10 **Operator algebras and K-theory**
Ronald G. Douglas and Claude Schochet, Editors

VOLUME 11 **Plane ellipticity and related problems**
Robert P. Gilbert, Editor

VOLUME 12 **Symposium on algebraic topology in honor of José Adem**
Samuel Gitler, Editor

Titles in this series

VOLUME 13 **Algebraists' homage: Papers in ring theory and related topics** Edited by S. A. Amitsur, D. J. Saltman and G. B. Seligman

VOLUME 14 **Lectures on Nielsen fixed point theory** Boju Jiang

VOLUME 15 **Advanced analytic number theory. Part I: Ramification theoretic methods** Carlos J. Moreno

ALGEBRAISTS' HOMAGE:
Papers in Ring Theory and Related Topics

NATHAN JACOBSON

CONTEMPORARY MATHEMATICS

Volume 13

ALGEBRAISTS' HOMAGE:

Papers in Ring Theory and Related Topics

AMERICAN MATHEMATICAL SOCIETY

Providence · Rhode Island

PROCEEDINGS OF THE CONFERENCE ON ALGEBRA IN HONOR OF NATHAN JACOBSON

HELD IN NEW HAVEN, CONNECTICUT

JUNE 2–5, 1981

EDITED BY

S. A. AMITSUR, D. J. SALTMAN, AND G. B. SELIGMAN

1980 *Mathematics Subject Classification.* Primary 16-06; Secondary 17-06.

Library of Congress Cataloging in Publication Data

Main entry under title:

Algebraists' homage.

(Contemporary mathematics, ISSN 0271-4132; v. 13)

"Proceedings of the Jacobson conference held at Yale University, New Haven, Connecticut, June 2–5, 1981, edited by Shimshon A. Amitsur, David J. Saltman, and George B. Seligman"—Verso of t.p.

Includes bibliographies.

1. Associative rings—Congresses. 2. Nonassociative rings—Congresses. 3. Associative algebras—Congresses. 4. Nonassociative algebras—Congresses. 5. Galois theory—Congresses. I. Amitsur, Shimshon A., 1921– . II. Saltman, David J., 1951– . III. Seligman, George B., 1927– . IV. Series: Contemporary mathematics (American Mathematical Society); v. 13.

QA251.5.A44 1982 512'.4 82-18934

ISBN 0-8218-5013-X

CONTENTS

Introduction.		1
Surveys:		
S. A. Amitsur,	Division algebras, a survey	3
M. Auslander,	Representation theory of finite dimensional algebras	27
E. Formanek,	The polynomial identities of matrices	41
T. A. Springer,	The order of a finite group of Lie type	81
J. E. Humphreys,	Restricted Lie algebras (and beyond)	91
Addresses:		
P. M. Cohn,	Determinants on free fields	99
M. E. Sweedler,	Weak cohomology	109
Papers in Ring Theory:		
M. Schacher,	Applications of the classification of finite simple groups to Brauer groups	121
D. Saltman,	Generic structures and field theory	127
D. Haile,	Balanced spaces of forms and skew cyclic transformations	135
A. Klein,	Rings with bounded index of nilpotence	151
J. Zelmanowitz,	On the Jacobson density theorem	155
I. N. Herstein,	Derivations of prime rings having power-central values	163
W. S. Martindale,	Lie and Jordan mappings	173
B. Osofsky,	Some properties of rings reflected in endomorphism rings of free modules	179
C. C. Faith,	On the Galois theory of commutative rings I: Dedekind's theorem on the independence of automorphisms revisited	183
M. Gerstenhaber and S. D. Schack	On the deformation of diagrams of algebras	193
M. Lorenz and L. W. Small	On the Gelfand-Kirillov dimensions of noetherian PI-algebras	199
L. H. Rowen,	A simple proof of Kostant's theorem, and an analogue for the symplectic involution	207
W. Schelter,	Affine rings satisfying a polynomial identity	217
M. Artin,	Algebraic structure of power series rings	223

C. Huneke, Linkage and the symmetric algebra of ideals 229

S. Steinberg, Polynomial constraints in lattice-ordered rings 237

Lie, Jordan and Nonassociative Systems:

W. Feit and G. J. Zuckerman, Reality properties of conjugacy classes in spin groups and symplectic groups 239

G. B. Seligman, Variations on even Clifford algebras 255

G. M. Benkart and J. M. Osborn, On the determination of rank one Lie algebras of prime characteristic 263

R. E. Block, Restricted simple Lie algebras of rank 2 or of toral rank 2 267

R. L. Wilson, Restricted simple Lie algebras with toral Cartan subalgebras 273

K. McCrimmon, Composition triples 279

B. Harris, Triple products, modular forms and harmonic volumes 287

W. G. Lister, Bivariants of ternary rings 297

R. Bix, Separable alternative algebras over commutative rings 305

H. Petersson and M. Racine Exceptional Jordan division algebras 307

J. K. Faulkner, An apology for Jordan algebras in quantum theory 317

L. Hogben, Identities of non-associative algebras studied by computer 321

Galois Theory and Other Topics:

T. Tamagawa, On regularly closed fields 325

P. Blass, Some geometric applications of a differential equation in characteristic $p > 0$ to the theory of algebraic surfaces 335

R. P. Infante, On the inverse problem in difference Galois theory 349

H. F. Kreimer, Quadratic Hopf algebras and Galois extensions 353

W. D. Nichols and E. J. Taft, The left antipodes of a left Hopf algebra 363

D. Zelinsky, Similarity of nilpotent integer matrices, or, four elementary categories 369

I. Kaplansky, Normal antilinear operators on a Hilbert space 379

S. MacLane, Why commutative diagrams coincide with equivalent proofs 387

D. J. Winter, Axiomatic game theory 403

INTRODUCTION

These Proceedings contain papers presented in person or by title at the Conference in Algebra held at Yale University, June 2-5, 1981, on the occasion of the retirement of Professor Nathan Jacobson. Support was generously provided by the National Science Foundation[1]. About seventy mathematicians were visitors in residence at Yale for all or part of the conference. In addition, numbers of people within commuting distance participated on a daily or occasional basis.

The theme of the conference was to discuss the current status and to suggest directions for the future in those areas that have been decisively influenced by Nathan Jacobson. The organizers of the conference apologize for the fact that Jordan rings were not so prominent on the agenda as would befit the contributions of the man we honor. Ideally, E. I. Zelmanov would have been on hand to present the dramatic developments he has recently brought about. This omission is being remedied, however, because of the publication, in September, 1981, of an extensive report on the work of Zelmanov, along with other results of Jacobson and McCrimmon. This volume by Jacobson, "Structure Theory of Jordan Algebras", is in the University of Arkansas Lecture Notes in Mathematics (volume 5), published by that university in Fayetteville, Ark.

The Organizing Committee for the conference was made up of Charles Curtis, Kevin McCrimmon, George Seligman, Earl Taft and Maria Wonenburger. Editorial assistance with the Proceedings has been graciously rendered by Walter Feit, Angus Macintyre, Charles Rickart and a few external referees whom we thank while respecting their anonymity. The secretarial staff of the Yale Mathematics Department prepared the copy smoothly and cheerfully.

Some of the papers bear dedications setting forth the authors' own affection and scientific indebtedness. In the other cases, contributors requested to be included in a general expression of these sentiments. On behalf of all the participants, then, we present Jake with these efforts. We hope they are worthy of him. We thank him for the standards he has set for us. We rejoice that we continue to enjoy and to cherish his advice and his friendship.

---

[1]Grant No. MCS80-19614

Contemporary Mathematics
Volume 13, 1982

DIVISION ALGEBRAS. A SURVEY

S. A. AMITSUR

Division algebras in this survey will always mean division algebras which are of finite dimension over their centers. The research in this area reached its peak in the famous result of Brauer-Hasse-Noether and Albert (1930), who proved that all division algebras over the rationals are cyclic. Their method was later polished up to cover division algebras over global fields. For many years the activity in this field was kept low, or rather its success was unnoticable though almost every algebraist has, at one stage or another, tried his hand in the field of division algebras. Various results which were obtained indicated that there is no hope for an extension of the results of 1930 to arbitrary fields. Nevertheless, important generalizations and extensions of the theory of central simple algebras were introduced, e.g. Azumaya algebras and various forms of Brauer groups and algebraic cohomologies adapted to other fields.

Recently, some of these methods together with new approaches have led to the solutions of old outstanding problems. In fact, it became apparent that some of the structure questions of division algebras over a field C depend heavily on the arithmetic of C, and others are of a general nature independent of the base field. The tools to attack these properties of division algebras were the generic constructions independent of any base field; but then, the structure of the specific centers became interesting and important.

The present survey will cover some aspects of the old and new approaches which led to the solution of some outstanding problems. We also shall try to raise some forgotten methods. No attempt is being made to cover all the significant results like the solution of the Tanaka-Artin problem by Platnov, the Schur subgroups of the Brauer group, the finite subgroups of division rings, etc.

The topics which are at least partially, dealt with in this survey are:

A. Methods of construction of division algebras.

B. Division algebras over specific fields.

C. Splitting fields; generation of the Brauer group.

Generic methods

D. Generic splitting fields.

0271-4132/82/0510/$07.00

E. Universal division algebras.

F. Generic crossed-products; involutions.

The bibliography is somewhat selective (according to the taste of the author) and arranged in corresponding sections, with older papers mentioned in so far as they influenced the developments discussed in this survey.

As a final remark, note that we have tried to deal with the structure of a single division ring, rather than the Brauer group of all division rings over a field, and so the survey is lacking in detailed information on Brauer groups.

A. METHODS OF CONSTRUCTION OF DIVISION ALGEBRAS

Let C be a field and F an extension of C. The classical way of constructing division algebras is to consider normal separable extensions F/C with Galois group G, and cocycles f of $H^2(G,F^*)$. The algebra $(F/C,G,f) = A$ is the crossed product of F with G; with the aid of the cocycle f.

A closely related representation of division algebras is the use of cocycles of the Amitsur-cohomology, ([A1] [A4]). Sweedler in ([A14]) gave a rather simple method of constructing these algebras.

A cocycle $f \in H^2(F/C)$ is represented by an invertible element $f \in F \otimes_C F \otimes_C F$ satisfying: $\varepsilon_0(f)\varepsilon_1(f)^{-1}\varepsilon_2(f)\varepsilon_3(f)^{-1} = 1 \otimes 1 \otimes 1 \otimes 1$ where

$$\varepsilon_0(x_1 \otimes x_2 \otimes x_3) = 1 \otimes x_1 \otimes x_2 \otimes x_3, \ \varepsilon_1(x_1 \otimes x_2 \otimes x_3) = x_1 \otimes 1 \otimes x_2 \otimes x_3,$$

etc. Sweedler defines the central simple algebra U_f as follows: As an additive group, it is the additive group $U_1 = \mathrm{Hom}_C(F,F)$. The elements of F are embedded as subring of U_1 by considering their right multiplication. The multiplication is defined in U_f by:

$$\text{for } T,S \in U_1, \quad T * S = \Sigma a_i T b_i S c_i \quad \text{for } f = \Sigma a_i \otimes b_i \otimes c_i.$$

Every central simple algebra is obtained in this form. It is not required as previous cases that F be a normal extension, or even separable.

We do not know of any application or use of this construction, though the Amitsur-cohomology has been used and generalized (e.g. [A15], [F13] etc.).

An old paper of Jacobson ([A9]), contains a method of describing central simple algebras split by F as F-bimodules, for arbitrary F, with a kind of generalized crossed product. The method of Sweedler is more straightforward.

Altogether a different method of constructing division algebras is by obtaining the algebras as centralizers of simple modules. That is; let R be a ring and ${}_RV$ be a simple R module. Then $\mathrm{Hom}_R(V,V) = D$ is a "general" division ring, but for special rings and modules D will be of finite dimension over its center; and with further restriction on V and K it will have a given center.

A first example of this type (though not in this language) is in Jacobson's dissertation ([A,8]): Let F/C be a cyclic extension and $R = F[x,\sigma]$ the ring of non commutative polynomials defined by the relation $xa = \sigma(a)x$, $a \in F$. The centralizer of some irreducible modules of R are the cyclic division algebras over C split by F.

In the same direction, the author ([A2]) gave a description of all central simple algebras over C (of characteristic 0) split by a field extension K which is transcendental over C, as follows:

Let $d:a \to a'$ be a derivation in K with the field of constants C, then any division algebra D over C split by K is the centralizer of a $K[t]$-simple module $V = K[t]/K[t]f[t]$, where $K[t]$ is the ring of differential polynomials defined by: $ta = at + a'$, and $f[t]$ is an irreducible polynomial of a certain type. This is similar to the construction of commutative field extensions, since D is isomorphic with the set of all polynomials $g[t] \in K[t]$ taken modulo $K[t]f[t]$ satisfying $f[t]g[t] = g_1[t]f[t]$, for some $g_1 \in K[t]$.

Recently, Quebbemann [B12] has shown the following surprising fact. In characteristic 0, take $R = A(C)$ to be the Weyl algebra $A(C) = C[x,y]$ where $yx - xy = 1$ is the only relation between x and y. Then every division algebra D with center C can be obtained as the centralizer of an irreducible module V. His method of constructing V is by using D itself in the definition. Namely, $V = D[x]$ where V is turned into an $A(C)$ module by left multiplication by $C[x]$, and by defining:

$$yf[x] = f[x]p[x] + \frac{df}{dx}$$

for a fixed $p \in D[x]$. If the coefficients of $p[x]$ generate D, then the centralizer $\mathrm{Hom}_R(V,V)$ is exactly D. Farkas and Snider have extended this result to D finite dimensional over C ([A6]).

This result of Quebbemann indicates that structure of the left ideals of $A(C)$ contains information on arithmetical properties of the field C, since the simple left $A(C)$-modules are quotients $A(C)/L$ for maximal left ideals of $A(C)$.

One finds in the literature various constructions of special division algebras for special purposes, e.g. noncyclic algebras and division algebras of fixed exponent ([A3]). Most of these algebras are tensor products of cyclic algebras. Generic methods of construction will be discussed in part D, and here we only mention a method introduced by Rosset [A13]: Let Φ be a finite group, and G be a group without torsion which is an extension of a finitely generated abelian group A by Φ, and assume Φ acts faithfully on A. In other words, there is an exact sequence:

$$\alpha: 1 \to A \to G \to \Phi \to 1.$$

The group ring $C[G]$ is then a domain and its ring of quotients $C(G)$ is a crossed product $(C(A),\Phi,\alpha)$. Here α, as a short exact sequence, is an element of $H^2(\Phi,A)$. He proves that $C(G)$ is a division algebra of dimension $|\Phi|^2$ and exponent equal to the order of α in $H^2(\Phi,A)$. Furthermore, given Φ one can choose G and A so that $C(G)$ has a fixed order m, provided that m and n have the same set of prime divisors.

Finally, constructions of p-algebras, that is central simple algebras of degree p^{2n} over fields of characteristic p, are of a different type. One replaces the automorphisms of the Galois group by derivations, and then the norm condition for cyclic algebras is replaced by a different form. This has been done first by Jacobson [A11] and later by Hoechsman [A7] and Yuan [A15] and [B14].

B. DIVISION ALGEBRAS OVER FIXED FIELDS

After the success of determining the division algebras over global fields, attempts were made to generalize the methods used for global fields. One of the main tools was the Hasse principle, which states that a division algebra D splits if $D \otimes C_p \sim C_p$ for all primes of C. Witt in [D11a] gave a counter-example to Hasse's principle for fields of genus zero, and for genus 1 an example is given in [B8].

A positive and surprising result was Tsen's theorem [B11] that there are no division algebras over function fields of transcendence degree 1 over an algebraically closed field. The first non-global fields to be dealt with are the fields of rational functions $C(t)$ with C not algebraically closed. Combining the results of Faddeev (B3], Auslander and Brumer [B22] and Fein and Schacher [B5], one has the following:

Let C be a global field of characteristic $q \geq 0$, and let $p \neq q$ be a prime. Then every class of order p^n in $B_p(K(t))$ can be represented in the form

$$[A \otimes K(t)] \otimes \mathrm{cor}[L_1(t)/F_1(t),\sigma_1,f_1] \otimes \cdots \otimes \mathrm{cor}[L_n(t)/F_n(t),\sigma_n,f_n]$$

where $[A] \in \mathrm{Br}(K)$ is of order dividing p^n; F_i is a finite separable extension of C, L_i/F_i is a cyclic extension of order dividing p^n with a generating automorphism σ_i, $f_i \in F_i[t]$ and cor is the corestriction from $B(F_i(t))$ to $B(K(t))$. This together with a recent result of Rosset ([C6]) proves a result of Bloch's [C1], namely, that any algebra over $C(t)$ is similar to a product of cyclic algebras.

Work of Fein and Schacher [B5], together with a result of Sonn [B10], give a detailed analysis of the Brauer group of $C(t)$ and it's Ulm invariants. This led to a complete description of the Brauer groups of the fields of rational functions $C(t_1,\ldots,t_n)$ over a global field C ([B4]). An interesting result of this analysis is that the groups are independent of the transcendence degree. Namely, $Br(C(t_1,\ldots,t_n)) \cong Br(C(t))$, for global fields C. A detailed and interesting account of this important result is given in [B5], where an example is given of a field C for which $Br[C(t_1,t_2)] \ncong Br(C(t_1))$.

Other types of fields which have been studied are the local fields. Witt in [B12] describes the Brauer group of complete fields with respect to a discrete valuation whose residue field is perfect. The case of an imperfect residue field $\bar{C}$ is far more complicated, in particular, when C and $\bar{C}$ are of different characteristics. A first attempt in this direction was made by Saltman in [B9] for algebras of dimension p^2. A cohomological description of the equal characteristic case is given in [B13] by Yuan.

For fields C of characteristic zero which are not number fields, another factor appears and this is the derivation of C. It is an interesting fact that all these derivations can be extended uniquely, up to an inner derivation, to any central simple algebra over C, as it can be done for separable commutative field extensions. This is generally not true for automorphisms of C. ([B1a]).

In the finite characteristic case, the field C has isomorphisms $\sigma^r : x \to x^{p^r}$, which can be extended in a natural way to a matrix ring $M_t(C)$. But, for arbitrary central simple algebras D, σ^r can be extended from C to D if and only if $p^r \equiv 1$ modulo the exponent of D. Thus σ can be extended for every r if and only if D is a matrix ring over C.

C. SPLITTING FIELDS, GENERATION OF BRAUER GROUPS

Let D be a division algebra of dimension n^2 over its center C. One of the major questions is to determine the splitting fields $F \supset C$, of D. That is, the fields F such that $D \otimes_C F \cong M_n(F)$. It is easily verified that D has a splitting field which is separable normal and of finite dimension over C. One can ask what types of Galois groups can be expected. If C is a local or a global field, a cyclic extension F/C can be found which splits D, and actually F can be taken to be a maximal subfield of D. It is known that for arbitrary fields one cannot always expect a cyclic splitting field, and so a major problem in the theory of division algebras is:

QUESTION 1: *Does every division algebra D over an arbitrary field C have an abelian splitting field?*

This question is closely related with another unsolved problem in the theory of Brauer groups Br(c).

QUESTION 2: *Is every class* $[D] \in B(K)$ *a class of a tensor product of cyclic algebras?*

If the second question has a positive answer, then clearly the answer to question 1 will also be positive.

In the next section we shall consider the generic approach to the problem of splitting fields. Here we survey only some of the new approaches and results related to the above questions, in particular - the use of K-theory.

Division algebras D od degree 2,3 are well known to be cyclic, p-algbras also have a cyclic splitting field. If D is a division algebra of exponent 2 (i.e., division algebras with involution) over a field of characteristic $\neq 2$, there is an old result of Albert which states that such a division algebra of degree $n = 4$ is a tensor product of two quaternion (and so cyclic) algebras. Rowen in [F10] shows that algebras of exponent 2 and degree 8 have a maximal subfield with a Galois group $Z_2 \oplus Z_2 \oplus Z_2$, and Tignol (F17]) has shown that a matrix algebra $M_2(D)$ over such algebras D is a product of quaternions. We shall return in the last section to the major question connected with these results, solved in ([F3]), as well as to the problem related to the splitting fields of the universal division algebras.

Snider in [C15] has shown that if D is of degree 4 and $\sqrt[4]{1} \in C$, then D is similar to a product of cyclic algebras; and also for D of characteristic prime to 2n, $\sqrt[n]{1} \in C$, and if D is split by a Galois extension F/C with a dihedral group D_{2n} of automorphisms n odd, then D is similar to a product of cyclic algebras. His method is a combination of a generic approach to be discussed later and a method of K-theory (used first incorrectly in [C7]).

Rowen and Saltman [C10] improved this last result by showing that if $(D:E) = n^2$ and C contains an n-primitive root of one, n is odd, and D has a splitting field F/C with the dihedral group D_{2n}, then D is also split by a cyclic extension of dimension n.

Before discussing the K_2-method, we mention that the question of algebras D of a prime degree seems to be the harder part of the problem and the general questions can be reduced to this case. An interesting result in this direction was obtained by Brauer ([C27], who showed that division algebras of degree 5 have a solvable splitting field, this leads us to our next question:

QUESTION 3: *Does every division algebra have a solvable splitting field*

This problem, which is milder than the previous question, can be, sometimes, answered positively. Rosset in [C8] and [C6] proves that every division algebra of prime degree has, in fact, a splitting group G with the following description. G has normal abelian subgroup H which is a p-group and has $(p-1)!$ generators. G/H is cyclic of order dividing $p-1$. G is of course solvable. We note that $G \neq H$ only if the base field C does not contain the group μ_p of p-roots of 1, since Rosset adds them in order to apply properties of $K_2(C)$.

For an arbitrary integer n, if the field C contains μ_n then the following is a description of the relationship between $K_2(C)$ and $Br(C)$, along with some of its useful properties ([C5],[C1]).

$K_2(C)$ is the group (written multiplicatively) generated by elements $\{a,b\}$, a,b non zero elements of C, satisfying the relations

$$\{a,1-a\} = 1 \quad \text{for} \quad a \neq 1: \{a_1a_2,b\} = \{a_1b\}\{a_2b\},$$
$$\{a,b_1b_2\} = \{ab_1\}\{ab_2\}.$$

To each element $\{a,b\}$ we define an algebra in $Br(C)$, $A_\omega(a,b)$, generated by two elements x,y over C with the relations $x^n = a$, $y^n = b$, and $yx = \omega xy$ where ω is a fixed primitive n-th root of 1.

The correspondence $\{a,b\} \to A_\omega(a,b)$ defines a homomorphism of $K_2(C)$ into $Br(C)$. If dependence on ω is to be avoided, one may correspond to $\{a,b\}$ the element $A_\omega(a,b) \otimes \omega \in Br(C) \otimes \mu_n$, which is independent of the choice of ω. The corresponding image will be denoted by (a,b) (called - a symbol). The mapping $\{a,b\} \to (a,b)$ induces a homomorphism

$$R_{n,F}: K_2(C)/nK_n(C) \to Br_n(C) \qquad \text{(actually } Br_n(C) \otimes \mu_n)$$

where $nK_2(C)$ is the group generated by all n powers of $K_2(C)$ and $Br_n(C)$ is all elements of $Br(C)$ of exponent divisible by n.

Let F/C be a separable extension, then there are two maps; the "transfer" $tr_{F/C}: K_2(F) \to K_2(C)$, and the corestriction: $cor_{C/F}: Br(F) \to Br(C)$, and we have the relation:

$$(*) \qquad R_{n,F}tr_{F/C} = cor_{F/C}R_{n,F} \qquad ([C8]).$$

Furthermore, it is shown in [C6] that $tr_{F/C}\{u,v\}$, $\{u,v\} \in K_2(F)$, is a product of at most $(F:C)$ symbols $\{a,b\}$. This together with the fact that $cor_{F/C}res_{C/F} = (F:C)$ readily implies the existence of solvable (abelian if $\mu_n \in C$) fields splitting algebras of prime degree ([C8]).

Indeed, if $(D:C) = p^2$, then there exists a Galois splitting field L/C of degree pr where $r|(p-1)!$. Let F be the subfield of L invariant under its p-Sylow subgroup. Then $R = D \otimes_C F$ is a symbol and by taking the

$\mathrm{cor}_{F/C}\mathrm{res}_{C/F}(D)$, one readily obtains the above results. Note also that the same methods yield the result stated in section B that all algebras with center the rational functions over a global field are similar to a tensor product of cyclic algebras.

Another interesting result of K-theory which can be useful to our questions is the theorem of S. Bloch ([C1]):

Let $\mu_n \subseteq C$, $C = k(t_1,\ldots,t_n)$ be a field of rational functions in n commutative indeterminates, then

$$(**) \qquad \mathrm{Ker}(R_{n,C}) \cong \mathrm{Ker}(R_{n,k}) \ ; \ \mathrm{Coker}(R_{n,C}) \cong \mathrm{Coker}(R_{n,k})$$

Thus, if $\mathrm{Coker}(R_{n,k}) = 0$ i.e., every algebra over k is a product of cyclic algebras then $C = k(t_1,\ldots,t_n)$ also has this property.

This result is very useful when combined with generic methods as will be discussed in the next part of the survey.

In part A we have seen that there is a method of constructing division algebras over C with the aid of splitting fields K/C which are transcendental extensions. This raises the project of:

<u>Problem 4: Describe the transcendental splitting fields</u> K/C <u>of a division algebra</u> D_C. <u>Also, for a given</u> K/C, <u>find the division algebras over</u> C <u>split by</u> K, <u>i.e., the elements of</u> $\mathrm{Br}(K/C)$.

Beside the existence of such K, very little is known about the first part of our problem.

For some fields K, namely - the generic splitting fields, we shall see that the group of algebras split by K is just a cyclic group generated by a single algebra.

Roquette in [C9] made some progress on this problem. From the fact that K splits D_C, then if K_p is a residue field of a C-place on K/C, then also K_p splits D_C and so $\mathrm{Br}(K/C) \subseteq \cap \mathrm{Br}(K_p/C) = \mathrm{Br}_0(K/C)$. This implies that there is a bound on the degree of the elements of $\mathrm{Br}(K/C)$. In particular if K has a rational point in C then $\mathrm{Br}(K/C) = \mathrm{Br}(C/C) = 1$. Generally, one cannot expect that $\mathrm{Br}(K/C) = \mathrm{Br}_0(K/C)$, but Roquette has shown in [C9] the following. For a local field C, let $d(K/C)$ be the minimal positive degree of the divisors of K, which is also the greatest common multiple of the degrees (K_p/C), then $\mathrm{Br}(K/C) = \mathrm{Br}_0(K/C)$ is a cyclic group of order $d(K/C)$. He also gives examples where $\mathrm{Br}(K/C) \neq \mathrm{Br}_0(K/C)$.

Given a finite group G, and a field C, one can ask:

<u>Problem 5: Are there division algebras</u> D <u>over</u> C <u>which are split by (or contain) an extension</u> L/C <u>which is Galois with the group</u> G?

We shall refer to G as "splitting" D.

In other words, does there exist division algebras over C which are a crossed product with G (known as a C-adequate group G) or similar to a G-crossed product. There are various results in this direction, mainly by Fein and Schacher e.g. [C3], [C13]. We quote here interesting results of Schacher.

Given a group G there is always a number field C, over which there is a division algebra "split" by G ([C13]). More surprising is the fact that there is a number field C, and a division algebra D over C of degree n which is split by <u>all</u> groups G of order n. This means that no uniqueness can be expected in the representation of algebras as crossed products. The construction of arbitrary G-crossed products which are division algebras over some field C will be given by generic methods.

We close this section with an interesting result of Saltman ([E11]) about groups splitting p-algebras:

Let D be a p-algebra of degree $n = p^e$. If D is cyclic then every p-group of order n splits D.

GENERIC METHODS

As we pointed out before, recent advances in the theory of division algebras indicate that a large class of properties of division algebras are independent of the properties of the base field. We shall refer to the methods of dealing with division algebras independent of the properties of the center as "generic methods". The first to be surveyed is the following.

D. GENERIC SPLITTING FIELDS

A Brauer-Severi variety V over a ground field C is a scheme over C which becomes isomorphic to a projective space P_{n-1} over a separable extension F/C. Chatelet in [D3], who introduced this notion, finds a correspondence between central simple algebras over C and the Brauer-Severi varieties. In [D1] the author has found for each central simple algebra D of dimension n^2 over C, a field $F(D) \supseteq C$ which happens to be the field of functions on the V which corresponds to D.

The field F(D) is a finitely generated regular extension of C of transcendence degree n-1, and C is algebraically closed in F(D). This field splits D and actually F(D) is a generic splitting field of D in the following sense:

(i) A field $K \supseteq C$ splits D if and only if there is a place $p: F \to F$.

Fields with this property will be called "generic splitting fields". The generic splitting fields defined below have the following stronger property.

(ii) K splits D if and only if the composite $KF(D)$ is a field of rational functions in $n-1$ indeterminates.

The case $n = 2$ was first noted by Witt ([B12]).

Other properties of $F(D)$ are that it has a group of automorphisms isomorphic with D^*/C^* where D^* denotes the multiplicative of the regular elements of D; and that an algebra A is split by $F(D)$ if and only if $[A] = [D]^q$ in the Brauer group $Br(C)$.

An algebra D has other generic splitting fields, namely the fields $F_m(D) = F(M_r(D))$ which are of transcendence degree $m-1$ with $m = nr$. This follows from the fact that D and $M_r(D)$ belong to the same class in $Br(C)$ and have, therefore, the same splitting fields. In fact, Roquette defines in [D9] this sequence of fields, using the representation of $[D]$ by cocycles of $H^1(G, PGL(n,E))$ where E is a normal splitting field of D and $PGL(n,E)$ is the projective linear group.

Let A,B be two central simple algebras. If $[B] = [A]^{\ell}$ then there exists a C-homomorphism of $F_m(B)$ into $F_m(A)$. This is an immediate consequence of the generic property (ii) and the following interesting lemma of Roquette ([D9]):

Let F,K be two extensions of an infinite field C, of finite transcendence degree over C, with C algebraically closed in both. If F can be embedded isomorphically (over C) into a pure transcendental extension of K, and $\text{tr deg}(K/C) \geq \text{Tr deg}(F/C)$, then F can in fact be embedded isomorphically in K itself.

In our case take $K = F_m(A)$ and $F = F_m(B)$. Another result is that all $F_m(A)$ are pure transcendental extensions of the lower field $F_n(D) = F(D)$ of [D1], where D is the division algebra of the class $[A] \in Br(C)$, of dim. n^2 over C. Actually, every splitting field K, of A, in which C is algebraically closed, and of transcendence degree $\geq n-1$, will contain $F(D)$ since $F(D) \subseteq F(D)K = K(t_1,\ldots,t_s)$ and the previous lemma can be applied.

The following problem, which was raised in [D1], is still unanswered.

<u>QUESTION 6</u>: <u>Let</u> $[A] = [B]^q$, <u>where</u> q <u>is relatively prime to the exponent of</u> $[B]$, <u>and</u> A <u>and</u> B <u>have the same dimension</u>. <u>Is</u> $F_m(A)$ <u>isomorphic with</u> $F_m(B)$ <u>for all</u> m?

Our previous remark shows that $F_m(A)$ can be embedded in $F_m(B)$, and conversely, but it seems difficult to provide an isomorphism between them. The case of cyclic division algebras was proved in [D1], and Roquette in [D10] proved this result in the following cases:

(1) $m > s$, s the Schur index.

(2) $q \equiv -1$ mod the exponent.

(3) A is division algebra and a crossed product of a solvable group.

Two new generic splitting fields defined in a rather natural way appeared recently, in the literature ([D4], [D5],[D8]).

In [D4], Heuser considered the algebraic variety defined by the Norm form of an algebra $A:f(x_1,\ldots,x_{n^2}) = \mathrm{Norm}(\Sigma x_i a_i)$ where $\{a_i\}$ is a C-base of A and the Norm is the reduced norm of the algebra A. The field of functions $N(A)$ of the variety V defined by the relation $f[x_1,\ldots,x_{n^2}] = 0$ is a generic splitting of A [D8]. (The case of a crossed product was proved in [D4] by Heuser).

The extension to Azumaya algebras is given in [D8], where Saltman also shows that this field is a purely transcendental extension of $F_n(D)$, where $[A] = [D]$ and D is a division algebra of degree n. Using the argument of Roquette, algebras A, B of the same degree generate the same subgroup of the Brauer group if, and only if the corresponding Norm fields $N(A)$ and $N(B)$ are isomorphic.

A second definition of a generic splitting field (of the same transcendence degree) was introduced by Kovacs in [D5], using the idea of a universal domain of the embeddings of an algebra A of degree n in matrix rings ([D2], [D7]).

To an algebra A over C, there corresponds a domain $V_C^n(A) = S$ and a C-homomorphism $\varphi:A \to M_n(S)$ such that for every C-homomorphism $\psi:A \to M_n(R)$ into a matrix ring over a commutative C-algebra R, there exists a unique C-homomorphism $\eta:S \to R$ such that the following diagram is commutative:

$$\begin{array}{ccc} A & \xrightarrow{\varphi} & M_n(S) \\ & \searrow^{\varphi} & \downarrow M_n(\eta) \\ & & M_n(R) \end{array}$$

The ring S, which solves this universal problem for a central simple algebra A of degree n, is shown to be a domain and its ring of quotients $Q_n(A)$ is a generic splitting field. In [D8] it is shown that $Q_n(A)$ is isomorphic with $F_n(t_1,\ldots,t_{n^2-n})$ and so with $N(A)$.

E. UNIVERSAL DIVISION ALGEBRAS

A division algebra of a very different nature came out of the theory of polynomial identities. This is the universal division algebra $UD(n,C)$ defined as follows

Let $\{x_{ik}^{\lambda}; 1 \leq i,k \leq n, \lambda = 1,2,\ldots\}$ be commutative independent variables over C and let $X_{\lambda} = (x_{ik}^{\lambda})$ be $n \times n$ generic matrices. The ring $C[X_1,X_2\ldots]$ is an Ore domain and its ring of quotients $C(X)$ is a division algebra of dimension n over a commutative field, whose structure is unknown. The division algebra $UD(n,C)$ received its name - the universal division algebra - from the following property ([E1]):

(P_n) Given a central simple algebra A over a field $F \supseteq C$ of $\dim n$, and an element $0 \neq r \in C(X)$ (which can be assumed to be central) - then there exists a place $\varphi: C(X) \to A$ such that $\varphi(r) \neq 0$.

A place means, in our context, a subring $R \subseteq C(X)$ whose ring of quotients is $C(X)$ and a homomorphism φ of R into A_F such that $\varphi(R)F = A$. One can readily show that R can be chosen to be a local ring with a unique maximal ideal $M = \operatorname{Ker} \varphi$.

The importance of the universal property (P_n) is that many of the classical questions about division algebras, like the crossed-product representation of any division algebra or the decomposibility into products of algebras, can be formulated in a form which is preserved under taking places, as long as some fixed element r, depending on the problem, remains non zero. Hence, if the question can be answered positively for $UD(C;n)$ then it can be shown to hold in any algebra over any field $F \supseteq C$. Actually, the application of (P_n) is in the opposite direction - namely, by providing different division algebras over different fields $F \supseteq C$ with contradictory forms of a fixed property - one proves that $UD(C,n)$ does not have this property. The first application of this method was in [E1], where this method was applied to show that:

(a) If $UD(C,n)$ is a crossed product of a group G, then any division algebra of dimension n^2 over a field $F \supseteq C$ is also a crossed product of the group G.

If $8|n$ or $p^2|n$, p odd, it was shown that for different fields $F \supseteq C$ and different division algebras D_F, the group G had to satisfy contradictory conditions, and therefore in this case;

$$UD(C,n) \text{ is not a crossed product}$$

The original theorem was proved for $C = Q$. Since then it has been extended to other fields, including fields of characteristic p with $(n,p) = 1$ see e.g. [E2] and [E10]. A new method of constructing division algebras was used by Saltman ([D12]) to prove that $UD(C,p^r)$ is not a crossed product if C is of characteristic p and $r \geq 3$.

These major results did not solve the old problem:

Problem 7: Does there exist a division algebra of prime degree which is not a crossed product (i.e. cyclic)?

Another generic property proved by the author which has not yet appeared, is the following:

(b) If UD(C,n) has a Galois splitting field K/C with a group G = Gal(K/C), then every division algebra of dimension n over a field $F \supseteq C$, has a splitting group H which is a subgroup of G.

Again by providing different fields F/C with division algebras having different splitting groups one can find conditions on possible G's satisfying (b). These are first steps in attacking a generalization of problem 5 which will give information on splitting groups of division algebras.

Problem 8: What are the splitting groups of UD(C,n)?

Of course the symmetric group S_n is one, but there are also other groups if for example n is a composite number. Indeed, UD(C,n) is then a product of prime power division algebras, and the composite of their splitting fields will not have S_n as a Galois group. In fact we can show that if $n = n_1, \dots, n_k$ a product of relatively prime numbers, then $S_{n_1} \times \cdots \times S_{n_k}$ splits UD(C,n).

The universality property of UD(C,n) raises interest in two directions of research in division algebras: (U) Construction of universal division algebras with respect to other fixed properties. (M) New methods of constructing division algebra - generic constructions.

Some of the achievements in (M) will be discussed in the next section. We devote this section to results of the type (U) - universal algebras.

The algebra UD(C,n) is universal in the class of all division algebras of dimension n^2 over a field containing C. It is an algebra of exponent n. So, if we restrict ourselves to a class $T_{n,m}$ of all division algebras of degree n and exponent $m < n$ one naturally asks if this class also has a universal algebra UD(C;n,m) of degree n and exponent m satisfying a parallel property $(P_{n,m})$, namely:

$(P_{n,m})$ if A is a central simple algebra of degree n and exponent n over a field $F \supseteq C$, and $0 \neq r \in UD(C;n,m)$ is a fixed element, then there is a place $\varphi: UD(C;n,m) \to A$ such that $\varphi(r) \neq 0$.

A natural candidate is the algebra of degree n in the Brauer class of $[UD(C;n)]^{n/m}$, which is of exponent m, but fails to be a division algebra. This is because it can be shown that:

If a division algebra D has degree n and equal exponent n, then D^r has exponent $\frac{n}{r}$ and Schur index $\frac{n}{r}$, for $r|n$.

This has been shown in the special case $r = p$ in [A1], but it can be extended to arbitrary r.

In [E7], Saltman gives a method of constructing the universal algebras $UD(C;n,m)$. He uses the domains $V_C^n(A)$ of [D2] described in the previous section, in the following way:

Let n,m have the same set of prime divisors and let $m|n$. Set R to be $C[X_1,X_2,\dots]$, the ring of generic matrices (whose quotient is $UD(C;n)$). Let $d = d[X]$ be a fixed non zero central polynomial in R and R_d the localization of R with respect to d. Next, we define a commutative ring $S = S_{n,m}$.

Consider the algebra $R_d^s = R_d \otimes \cdots \otimes R_d$, where $s = \frac{n}{m}$ and where the tensor product is taken with respect to the center Z_d of R_d. Let $S = V_C^t(R_d)$, for $t = n^m$, be the universal domain of the embedding problem of representations of R_d^m in $t \times t$ matrices over a commutative ring. ([D2]-described in the previous section). Then S is a domain and so is $T_{n,m} = R \otimes_Z S$, where Z is the center of R. Saltman shows that the ring of quotients of $T_{n,m}$ is the required universal division algebra $UD(C;n,m)$. That is, it is of degree n and exponent m, and satisfies $(P_{n,m})$.

Now, by constructing various algebras of small exponent m, and following the now classical use of property $(P_{n,m})$, he shows ([D6],[D7]):

If p is a prime, $p^3|m$, $m|n$ then $UD(C;n,m)$ is not a crossed product division algebra. If C is an algebraic number field not containing a primitive p-th root of 1 - then it suffices to assume $p^2|m$.

Another problem which can be attacked in this manner is the question of whether a division algebra of low exponent m can always be decomposed as a product of algebras of degree = exponent. For a long time, this was the only way to construct algebras of low exponent. Again, by proving that the property of having a prefixed decomposition is inherited by places, and providing different decomposable algebras, Saltman shows:

$UD(C;p^r,p^s)$ is indecomposable if $r \geq 4$, $s \geq 1 + \frac{r}{2}$, and the characteristic of $C \neq p$. If the characteristic of C is p - this is shown only for $s = r-1$.

Algebras of exponent 2 are the division algebras with involutions (which always mean of the first kind), and their universal algebras had been found long before ([D5]) by Rowen, in a method similar to the construction of generic matrices.

Let $C[x]$ be the ring of polynomials generated by commutative indeterminates $\{x_{ik};| \leq i,k \leq n,\ \lambda \geq 1\}$. The $n \times n$ matrix ring $M_n(C[x])$ has two involutions: the ordinary transpose $X_\lambda^t = (x_{ki}^\lambda)$ of $X_\lambda = (x_{ik}^\lambda)$, and the symplectic involution $X_\lambda^s = w^{-1}X^t w$, where $w = \begin{pmatrix} 0 & -I \\ I & 0 \end{pmatrix}$ and I denotes the $\frac{n}{2} \times \frac{n}{2}$ unit matrix. For $n = 2^m$ (which is a natural restriction) the rings $C[X,X^t]$, and $C[X,X^s]$ are domains and their quotients are universal division algebras of degree n with involution (symplectic) for all algebras of this type over fields containing C.

The problem of applying these universal algebras is awaiting new methods of constructing division algebras with involution. One such method will be discussed in the last section.

A better understanding of the universal algebras UD(C;n) of generic matrices may also lead to information on general division algebras. For example, if one could show that the center is a pure transcendental extension of the field C, then Bloch's theorem mentioned in section B is applicable, and we would have on hand a method for answering the question about abelian splitting fields. Unfortunately, this has been shown only for $n = 2,3,4$ (Formanek [D3], [D4] and Procesi for $n = 2$). There are reasons to believe that the center of $C(X) = UD(C,n)$ is not a pure transcendental extension of C for all $n \geq 5$.

Formanek's papers also contain an interesting description of the center Z(n) of UD(C;n) (due to Procesi) as a field invariant under a symmetric group. First, we point out that in the construction of UD(C;n) we have not limited the number of generic matrices X_λ, but for all our considerations it suffices to take two generic matrices $X_1 = (x_{ik})$ and $X_2 = Y = (y_{ik})$ ([D3]) and the corresponding algebra $UD_2(C,n)$.

Let $U = \langle u_1,\dots,u_n\rangle$, $B = \langle x_i, y_{ij}; 1 \leq i,j \leq n\rangle$ free abelian groups on the indicated generators. Define an action of the symmetric group S_n on these abelian groups by setting $\sigma(x_i) = x_{\sigma(i)}$, $\sigma(y_{ij}) = y_{\sigma(i)\sigma(j)}$ and $\sigma(u_i) = u_{\sigma(i)}$. Define an S_n-homomorphism $\alpha: B \to U$ by $\alpha(x_i) = 1$, $\alpha(y_{ij}) = u_i u_j^{-1}$. Let $A = \ker \alpha$. Then A is generated by $\{x_i\}$ and all monomials of the form $y_{i_1 i_2} y_{i_2 i_3} \cdots y_{i_q i_1}$ $(q \geq 1)$. S_n acts as a group of automorphisms on the group ring C[A] and on its field of quotients C(A). The description of the center $Z_2(n)$ is as follows.

$Z_2(n)$ is C-isomorphic with the fixed field $C(A)^{S_n}$ under the group S_n. If UD(C,n) is generated by m generic matrices, then its center is a pure transcendental extension of $Z_2(n)$ in $(m-2)n^2$ indeterminates.

A final topic of a "generic" nature connected with the division rings of generic matrices is the relationship of central polynomials to maximal subfields ([E12]).

Maximal subfields F of a division algebra D splits D, and so D can be embedded in $D \otimes F \cong M_n(F)$, the matrix ring over F. One would think that this representation depends on the nature of F, but it was proved in [E12] that the representation of D in $M_n(F)$ can be obtained in a generic way, with the aid of fixed central polynomials of matrix rings, and is dependent neither on D nor on F, but only on n.

There exists a central polynomial: $\delta[x,y] = \delta[x_1,\ldots,x_{n^2};y_1,\ldots,y_{n-1}]$, and $\delta_{jk}[x_1,\ldots,x_n;y_1,\ldots,y_{n-1}]$ such that for a generic matrix u:

$$(***) \qquad \delta[x_1z^k;y]ux_i = \sum_j x_j\tau_{ij}[u,x;z]$$

where $\tau_{ij}[u,x,z] = \sum_k \delta_{jk}[x_1,\ldots,ux_i,x_{i+1},\ldots,]z^k$. This can be interpreted as a generic representation of u as a matrix $(\tau_{ij}[u,x,z]) = T_u$ with entries which are polynomials in z with coefficients which are central elements of UD(C,n). Thus, if $F = C(\xi)$, the representation (***) gives, by substituting $z = \xi$ and other elements for the x's and y's, a representation of D in $M_n(F)$.

F. GENERIC CROSSED-PRODUCTS AND INVOLUTIONS

The ideas involved in the study of the universal division algebras indicate that some of the methods to construct algebras should be independent of a fixed center. To this end, the author and Saltman [F1] defined a generic crossed-product of an abelian group G, by using the generators and the relations of G. On one hand, this idea is old and goes back to Dickson ([F4]), and on the other hand the corresponding cocycle can be described in a modern language by using a different complex for G (e.g. [F16]). Our construction is as follows.

Let F/C be an abelian extension with an abelian Galois group $G = S_1\ S_2 \times\cdots\times S_r$; each S_i being a cyclic group of order q_i generated by an automorphism σ_i, and $n = q_1q_2\cdots q_r$. Denote by F_i the field invariant by σ_i, and $N_i(x) = \mathrm{Norm}(F/F_ix)$. Any central simple algebra A_C having the maximal field F is a crossed-product (F/C,G,c) where the cocycle can be determined by the following parameters.

If $z_i \in A$ satisfying $z_ia = \sigma_i(a)z_i$ for $a \in F$, then

(1) $z_iz_k = u_{ik}z_kz_i$

(2) $z_i^{q_i} = b_i$, i,k=1,...,2

and the elements u_{ik}, $b_i \in F$ satisfy the conditions:

(3) $u_{ij} = 1$, $u_{ij}^{-1} = u_{ji}$;

(4) $\sigma_k(b_i) = b_iN_i(u_{ki})$ and hence $N_iN_k(u_{ki}) = 1$

(5) $\sigma_i(u_{jk})\sigma_j(u_{ki})\sigma_k(u_{ij}) = u_{jk}u_{ki}u_{ij}$.

The set $\{u_{ik},b_i\}$ uniquely determines the cocycle $c(\cdot,\cdot)$ of A. Conversely if a set of elements u_{ij},b_i satisfy (3), (4) and (5) one can construct, with the aid of (1) and (2), an abelian crossed-product A, which we shall denote by $A = (F/K,G,u)$.

Given a set $\{u_{ik}\}$ which satisfies (3) and (5), we define the generic crossed-product $D(F/K,G,u)$ as the ring of quotient $F(x_1,\ldots,x_r)$ of the non-commutative polynomial ring $F[x_1,\ldots,x_r]$ which is defined by the relations $x_i a = \sigma_i(a)x_i$, $x_i x_k = u_{ik}x_k x_i$ for $a \in F$ and $1 \leq r,k \leq r$.

A generic algebra of this type, with $u_{ij} = 1$ for all i,j, was defined by Kuyk ([F6]) in order to construct a non cyclic algebra. In this case $D(F/K,G,1)$ is a tensor product of r cyclic algebras.

A basic property of the algebra $D(F/C,G,u)$ is that it is a division algebra of degree n over a center which is a field of rational functions $C(y_1,y_2,\ldots,y_n)$, where $y_i = b_i x_i^{q_i}$ for some elements $b_i \in F$ which satisfy (4).

This construction was used in [F1] to show that there are p-algebras which are not cyclic, though they are always similar to cyclic ones. Saltman in [E8] determined the splitting groups of the generic algebra for characteristic p, and for $n = p^r$, and applied it to settle the existence of non crossed-product p-algebras (see section D).

One can prove a "generic property" for $D(F/C,G,u)$;

(Ab): **For any abelian crossed-product $A = (F/C,G\ u,\bar{b})$ with the same set $\{u_{ij}\}$; and $0 \neq r \in D(F,C,G,u)$ there exists an F-place $\varphi:D(F/C,G,u) \to A$ such that $\varphi(r) \neq 0$.**

For it can be proved that $\bar{b}_k = \gamma_k b_k$ for $\gamma_k \in C$, and by mapping $x_i \to \lambda_i z_i$, for some λ_i, one can obtain a homomorphism φ for which $\varphi(r) \neq 0$.

A cohomological representation of the sets $\{u_{ij}\}$ is given in [D16] by Tignol.

The generic abelian crossed-product has proved to be very helpful in solving an old standing problem in the theory of algebras with involutions. A quaternion algebra always has an involution. Hence every tensor product $D = D_1 \otimes\cdots\otimes D_n$ of quaternion's has an involution induced from the factors. But are these the only involutions of D?

An older and more important problem was the question of Albert: Are the tensor products of quaternions the only division algebras with involutions?

A negative answer was given to both questions by Tignol-Rowen-Amitsur in [F3]. There is a generic abelian crossed-product of degree 4 with an involution such that the involution has no invariant quaternion subalgebra. Also, an abelian crossed-product of degree 8 is constructed which is not a product of quaternions.

We outline the construction of this algebra. Let $A = (F/C,G,u,b)$ be an abelian crossed-product of a group $G = S_1 \times \cdots \times S_r$, where $S_i = <\sigma_i>$ is cyclic of order 2. Let $\tau \in G$ and assume A is an algebra with involution. Then by modifying the u's and the b's one can satisfy, in addition to the previous conditions (3) - (5), the relations;

$$\text{(6)} \qquad \tau(u_{ij})\sigma_i\sigma_j(u_{ij}) = 1, \ \tau(b_i) = b_i$$

and the involution restricted to F will be exactly τ. The converse also holds as (6) will imply the existence of an involution.

In the special case $G = S_1 \times S_2 \times S_3$ is of order 8, there is a straightforward method to find u's and b's which satisfy all requirements. Let $F = C(\xi_1, \xi_2, \xi_3)$, $\xi_i^2 \in C$, $\sigma_i(\xi_i) = -\xi_i$, and $\sigma_i(\xi_j) = \xi_j$ for $j \neq i$. Choose an element b satisfying;

$$\text{(7)} \qquad 0 \neq b \in C(\xi_2\xi_3) \cap C(\xi_2)N_1(F) \cap C(\xi_3)N_1(F)$$

where $N_1(x) = x\sigma_1(x)$ and $N_1(F) = \{N_1(x) : x \in F\}$. Then we can define the u's and b's as follows. Set $-_1 = b$ and write $b = a_i N_1(y_3^{-1}) = a_3 N_1(y_2^{-1})$ where $a_i \in C(\xi_i)$ and $y_i \in F$. Put $u_{31} = \sigma_2(y_3)^{-1}y_3$, $u_{12} = \sigma_3(y_2)^{-1}y_2$ and $u_{23} = (u_{12}u_{31})^{-1}$. Then the u's and b's will satisfy (5) and the respective part of (6). One then can find the additional b_2, b_3 to satisfy all requirements.

An equivalent condition to (7) for $b \in C(\xi_2\xi_3)$ can be shown to be

$$\text{(7')} \qquad N_2(b) = N_3(b) \in \mathrm{Norm}[C(\xi_1,\xi_2)/C(\xi_2)] \cap \mathrm{Norm}[C(\xi_1,\xi_3)/C(\xi_3)].$$

Now, given b satisfying (7'), one finds the u's and defines the generic algebra $D = (F/C, G, u)$. It is then proved that D is not a product of quaternion's if and only if,

$$\text{(8)} \qquad b \notin N_1(F)C, \quad F = C(\xi_1, \xi_2, \xi_3).$$

Thus, to complete our construction of an algebra of degree 8 which is not a product of quaternion's, we have to find an element b which satisfies (7) and (8). The following field will have such a b:

Let $\xi_1^2 = -1$, $\xi_2^2 = -(1 + \lambda^2)$, $\xi_3^2 = \lambda$, where $C = k(\lambda)$ is the field of rational functions in λ over a field of characteristic $\neq 2$.

Finally, let $b = \xi_2\xi_3$. Then (7') holds, since

$$\begin{aligned} N_2(\xi_2\xi_3) &= N_3(\xi_2\xi_3) = \lambda(\lambda^2 + 1) \\ &= N_1[\tfrac{1}{2}\xi_2[(\lambda - 1 - \xi_1) + F_1(\lambda - 1 + \xi_1)] \\ &= N_1[(\xi_1\xi_3)(\lambda\xi_1 - 1)]. \end{aligned}$$

The negative answer to the first problem is the algebra $(F/C, S_1 \times S_2, u)$ where $F = C(\xi_1, \xi_2)$; $\xi_1^2 = 2, \xi_2^2 = \lambda$ and $b = N_1(\xi_2 + 1 - \xi_1)$. From these the other constants can be computed.

A remark on the merit of using the generic product is now appropriate. Tignol ([F17]),[F18]) had been working on the problem of decomposing a product $(F/C,G,u,b)$ as a product of algebras whose maximal fields are subfields of F. It is interesting that when one turns to generic products, an arbitrary decompositor will yield another restricted decomposition which uses subfields of a maximal field. Thus ideas can be combined to give the final result.

The problem of decomposition of this restricted type seems to be closed related with quadratic forms and Witt rings of the field C. This relationship is analysed in the papers [F14] and [F15].

Rowen has further studied properties of algebras with involution ([F10]-[F13]). In particular, he has shown in [F11], that any such algebra of degree 8 is a crossed-product of the group $Z_2 \oplus Z_2 \oplus Z_2$. Tignol in [F17] showed that algebras of degree 8 with an involution are similar to a product of quaternions. This is significant in view of the fact that the algebra itself may not be actually a product of quaternions. Rowen gave in [F12] some counter-examples of algebras with involutions.

Finally, we describe the construction of a generic division algebra depending on a ground field C and a finite group G.

To each pair $(\sigma,\tau) \in G \times G$, where $\sigma \neq e$ and $\tau \neq e$, (e the identity of G), we choose a commutative indeterminate $t(\sigma,\tau)$. We consider the field $F = C(t(\sigma,\tau):(\sigma,\tau) \in G \times G)$ which is of transcendence degree $(n-1)^2$ over C. Complete the definition by putting $t(\varepsilon,\sigma) = t(\sigma,\varepsilon) = 1$ for all $\sigma \in G$. We make G act on F/C, using the cocycle condition. That is,

$$\rho(t(\sigma,\tau)) = t(\rho\sigma,\tau)t(\rho,\sigma\tau)^{-1}t(\rho,\sigma).$$

Let $K = F^G$ be the fixed subfield of F. Now construct the crossed-product $D_G = (F/K,G,t)$ with the cocycle $t(\ ,\ )$. i.e. $z_\sigma z_\tau = t(\sigma,\tau)z_{\sigma\tau}$; $z_\sigma a = \sigma(a)z_\sigma$. Clearly D_G is a central simple algebra.

D_G can be defined in another way. Let $X = \langle x_\sigma ; \sigma \neq \varepsilon \rangle$ be a free group with generators $\{x_\sigma\}$, let $\varphi: X \to G$ be the homomorphism $\varphi(x_\sigma) = \sigma$, and let $Y = \operatorname{Ker} \varphi$. Next abelianize the sequence $1 \to Y \to X \to G \to 1$ by taking the groups modulo the commutator subgroup $[Y,Y] = Y'$, and set $A = Y/Y'$, $\bar{G} = X/Y'$ and form the exact sequence: $1 \to A \to \bar{G} \to G \to 1$. D_G is then isomorphic to the ring of quotients of the group ring $C(\bar{G})$. This type of construction was mentioned in part A. (See also [A13]).

It is now also possible to construct the generic splitting field of D_G: add new $n-1$ commutative variables y_σ, $\sigma \in G$ $\sigma \neq \varepsilon$ and put $y_\varepsilon = 1$. Let

$L = F(y_\sigma; \sigma \in G)$, and extend the action of G on L, by adding for $\tau \in G$ the additional definition: $\tau(y_\sigma) = y_{\tau\sigma} y_\sigma^{-1} t(\tau,\sigma)$. Hence the set (y_σ) is a generic splitting of the cocycle $t(\tau,\sigma)$. The generic splitting field $F(D_G)$ of D_G is the fixed field L^G.

The division algebra D_G and its splitting field $F(D_G) = F_G$ own their title "generic" because of the following property.

Let E be an algebra which is a crossed-product of the group G over a field $C' \supseteq C$, and $0 \neq r \in D_G$. Then there is a place $\varphi: D_G \to E$ such that $\varphi(E)C' = E$ and $\varphi(r) \neq 0$. Furthermore, φ can be extended to a place $\varphi: F_G \to F(E)$ into the generic splitting field of E, such that $\varphi(F_G)C' = F(E)$.

A construction of this type was used by Snider in [D15]. A special case for abelian groups and symmetric and invariant cocycles $(t(\rho,\sigma) = t(\sigma,\rho)$ is given in [F7].

One can extend this construction to cover non crossed-product algebras, by adding the condition $t(\sigma,\tau) = 1$ when σ or τ belong to a fixed subgroup H of G. This construction will be generic for algebras which are similar to crossed-product of G, and which are split by the field fixed by H.

Hopefully, one will be able to find the possible groups which split D_G, which will yield an interesting correspondence between groups through division algebras (probably dependent on C).

SELECTED BIBLIOGRAPHY

A. CONSTRUCTIONS

[1] S. A. Amitsur, Simple algebras and cohomology groups of arbitrary fields. Trans. Am. Math. Soc. 90 (1959), 73-92.

[2] S. A. Amitsur, Differential polynomials and division algebras. Ann. Math. 59 (1954), 245-278.

[3] R. Brauer, Über der Index und der Exponenten von Division-algebren. Tōhoku Math. J. 37 (1933), 77-87.

[4] S. U. Chase and A. Rosenberg, Amitsur cohomology and the Brauer group. Mem. Am. Math. Soc. 52 (1968), 34-79.

[5] J. Dauns, Non cyclic algebras. Math. Zeit. 69 (1979), 195-204.

[6] D. R. Farkas and R. L. Snider, Commuting rings of simple A(k) modules. To appear.

[7] K. Hoechsmann, Algebras split by a given purely inseparable field. Proc. Am. Math. Soc. 14 (1963), 768-776.

[8] N. Jacobson, Generation of separable and central simple algebras. J. Math. Pures. App. 36 (1957), 217-227.

[9] N. Jacobson, Construction of central simple associative algebras. Ann. Math. 45 (1944), 658-666.

[10] N. Jacobson, Non commutative polynomials and cyclic algebras. Ann. Math. 35 (1933), 197-208.

[11] N. Jacobson, p-algebras of exponent p. Bull. Am. Math. Soc. 43 (1937), 667-670.

[12] H. G. Quebbemann, Schiefkörper als Endomorphismenringe einfache Moduln einer Weyl-Algebra. J. Alg. 59 (1979), 311-312.

[13] S. Rosset, Group extensions and division algebras. J. Alg. 53 (1978), 297-303.

[14] M. E. Sweedler, Multiplication alteration by two cocycle. Ill. J. Math. 15 (1971), 302-323.

[15] Yuan, S., Central separable algebras with purely inseparable splitting fields of exponent 1. Trans. Am. Math. Soc. 62 (1971), 427-450.

B. DIVISION ALGEBRAS OVER FIXED FIELDS

[1] S. A. Amitsur, Some results on central simple algebras. Ann. Math. 63 (1956), 285-293.

[1a] S. A. Amitsur, Extension of derivations and automorphism. To appear.

[2] M. Auslander, A. Brumer, Brauer groups of discrete valuation rings. Neder. Akad. Wetensch. Proc. Ser. A. 71 (1968), 286-296.

[3] D. K. Faddeev, Simple algebras over a field of algebraic functions of one variable. Trudy. Mat. Steklov, 38 (1951), 321-344. A.M.S. Translations Ser. II 3 (1956), 15-38.

[4] B. Fein, M. Schacher, J. Sonn, Brauer groups of rational function fields. Bull. Am. Math. Soc. (New series) 1 (1979), 766-768.

[5] B. Fein, M. Schacher, Brauer groups of rational function fields over global fields. "Groupe de Brauer" Lecture notes 844, Springer-Verlag 1981.

[6] B. Fein, M. Schacher, A. Wadsworth, Division algebras and the square root of -1. J. Alg. 65 (1980), 340-346.

[7] V. I. Jancevkii, Division algebra over Henselian discrete valued field k. Doklady Akad. Nach. USSR 20 (76), 971-974.

[8] T. Nyman, G. Whaples, Hasse's principle for simple algebras over function fields of curves I. J. reine u. ang. Math. 299/300 (1978), 396-405.

[9] D. J. Saltman, Division algebras over discrete valued fields. Comm. Alg. 8 (1980), 1749-1774.

[10] J. Sonn, Classes groups and Brauer groups. Israel J. Math. 34 (1979), 97-105.

[11] C. C. Tsen, Divisionalgebren über Funktionenkörper. Göttingen Nach. II 48 (1933), 335-339.

[12] E. Witt, Schiefkörper über diskret bewerteten Körper. J. reine u. ang. Math. 176 (1937), 31-44.

[13] S. Yuan, On the Brauer group of a local field. Ann. Math. 82 (1965), 434-444.

[14] S. Yuan, On the theory of p-algebras ... . J. Alg. 5 (1967), 280-304.

C. SPLITTING FIELDS, GENERATION OF BRAUER GROUPS

[1] S. Bloch, Torsion algebraic cycles, K_2, and the Brauer groups of function fields. Groupe de Brauer, Lecture notes in Math. 844 (1980), 75-102.

[2] R. Brauer, On normal division algebras of index 5. Proc. Nat. Acad. Sci. USA. 24 (1958), 243-246.

[3] B. Fein, M. Schacher, Galois groups and division algebras. J. Alg. 38 (1976), 182-191.

[4] Kersten, Ina, Konstruktion von zusammenhangenden Galoisschen Algebren de characteristic p^k., J. reine u. ang. Math. 306 (1979), 177.

[5] J. Milnor, Introduction to algebraic K-theory. Ann. Math. Studies 72, Princeton 1971.

[6] S. Rosset and J. Tate, The corestriction preserves sums of symbols. To appear.

[7] S. Rosset, A remark on the generation of the Brauer group... . Bull. Am. Math. Soc. 81 (1975), 707-708.

[8] S. Rosset, Abelian splitting fields of division algebras of prime degree. Comm. Math. Helv. 52 (1977), 519-523.

[9] P. Roquette, Splitting of algebras by function fields of one variable. Nagoya Math. J. 27 (1966), 625-642.

[10] L. H. Rowen, D. J. Saltman, Dihedral algebras are cyclic. To appear Proc. Am. Math. Soc.

GENERIC METHODS

D. GENERIC SPLITTING FIELDS

[1] S. A. Amitsur, Generic splitting fields of central simple algebras. Ann. Math. 62 (1955), 8-43.

[2] S. A. Amitsur, Embedding in Matrix rings. Pacific J. 36 (1971), 21-39.

[3] F. Chatelet, Variations sur un theme de H. Poincaré. Ann. E.N.S. 61 (1944), 249-300.

[4] A. Heuser, Über der Funktionenkörper der Normfläche. J. reine ang. Math. 301 (1978), 105-113.

[5] A. Kovacs, Generic splitting fields. Comm. Alg. 6 (1978), 1017-1035.

[6] M. Ojanguren, A remark on generic splitting fields. Arch. Math. (Basel) 27 (1976), 369-371.

[7] C. Procesi, Finite dim. representations of algebras. Israel J. Math. 19 (1974), 109-182.

[8] D. J. Saltman, Norm polynomials and algebras. J. Alg. 62 (1980), 333-348.

[9] P. Roquette, On the Galois cohomology of the projective linear group and its applications to the construction of generic splitting fields of algebras. Math. Ann. 150 (1963), 411-439.

[10] P. Roquette, Isomorphisms of generic splitting fields of simple algebras. J. reine u. ang. Math. 214/215 (1964), 207-226.

[11ä] E. Witt, Gegenbeispiel zum Normensatz. Math. Zeit. 39 (1934), 462-467.

[11] D. J. Saltman, Splitting of cyclic p-algebras. Proc. Am. Math. Soc. 62 (1977), 223-228.

[12] M. Schacher, Cyclotomic splitting fields. To appear.

[13] M. Schacher, Subfields of division rings I. J. Alg. 9 (1968), 451-477.

[14] M. Schacher, Non-uniqueness in crossed-products. Rep. Theory of Alg. (Temple Conf.) Marcel Dekker 1976.

[15] R. Snider, Is the Brauer group generated by cyclics? Lecture notes 734 (Waterloo conference, 1978). Springer-Verlag, 1979.

E. UNIVERSAL DIVISION ALGEBRAS

[1] S. A. Amitsur, On central division algebras. Israel J. Math. 12 (1972), 408-420.

[2] N. Jacobson, P.I. algebras, An introduction. Lecture notes in Math. 441. Springer-Verlag 1975.

[3] E. Formanek, The center of the ring of 3 × 3 generic matrices. Linear and Mult. Alg. 7 (1979), 203-212.

[4] E. Formanek, The center of the ring of 4 × 4 generic matrices. J. Alg. v. 62 (1980), 304-320.

[5] L. H. Rowen, Identities in algebras with involutions. Israel J. Math. 20 (1975), 70-95.

[6] D. J. Saltman, Non crossed-products of small exponent. Proc. Am. Math. Soc. 68 (1978), 165-168.

[7] D. J. Saltman, Indecomposable division algebras. Comm. Alg. 7 (1979), 791-817.

[8] D. J. Saltman, Non crossed-product algebras and Galois p-extensions. J. Alg. 52 (1978), 302-314.

[9] D. J. Saltman, On p-power central polynomial. Proc. Am. Math. Soc. 78 (1980), 11-13.

[10] M. Schacher, The crossed-product problem. Ring Theory II, Ed. M. McDonald, R. Morris, Marcel Dekker, 1977.

[11] D. J. Saltman, Abelian crossed-product and p-algebras II.

[12] S. A. Amitsur, L. W. Small, Prime ideals in PI-rings. J. Alg. 62 (1980), 363.

F. GENERIC CROSSED-PRODUCTS AND INVOLUTIONS

[1] S. A. Amitsur, D. J. Saltman, Abelian crossed-products and p-algebras I. Israel. J. Math. v. 51 (1978), 76-87.

[2] S. A. Amitsur, Generic crossed-product for arbitrary group, and their generic splitting fields. In preparation.

[3] S. A. Amitsur, L. H. Rowen, J. P. Tignol, Division algebras of degree 4 and 8 with involution. Israel J. Math. 33 (1979), 133-148.

[4] E. Dickson, Construction of division algebras. Tran. Am. Math. Soc. 32 (1930), 319-334.

[5] D. E. Haile, On central simple algebras of given exponent. J. Algebra 57 (1979), 449-465.

[5a] D. E. Haile, On central simple algebras of given exponent, II. J. Algebra 66 (1980), 205-219.

[6] W. Kuyk, Generic construction of a class of non cyclic division algebras. J. Pure App. Algebra, 2 (1972), 121-130.

[7] F. M. J. Van Oystaeyen, Pseudo-places, algebras and the symmetric part of the Brauer group. Vrije Univ. Amsterdam, (Thesis), 1972.

[8] L. J. Risman, Group rings and series. Ring theory (Proceed, U. Antwerpen 1977), 113-128.

[9] L. J. Risman, Non cyclic division algebras. J. Pure App. Algebra 11 (1977-1978), 199-215.

[10] L. H. Rowen, Central simple algebras with involution ... . J. Alg. 63 (1980), 41-55.

[10a] L. H. Rowen, Central simple algebras with involution. Bull. A.M.S. 83 (1977), 1031-1032.

[11] L. H. Rowen, Central simple algebras. Israel J. Math. 29 (1978), 285-301.

[12] L. H. Rowen, Division algebra, counter-example of degree 8. Israel J. Math. To appear.

[13] L. H. Rowen, Cyclic division algebras. To appear.

[14] R. Elman, T. Y. Lam, J. P. Tignol, A. R. Wadsworth, Witt rings and Brauer groups under multiquadratic extensions I. Preprint.

[15] D. Shapiro, J. P. Tignol, A. R. Wadsworth, Witt rings and Brauer groups under multiquadratic extensions II. Preprint.

[16] J. P. Tignol, Products croises abeliens. Groupe de Brauer, Lecture notes in Math. 844 (1981), 1-34.

[17] J. P. Tignol, Sur les classes de similitude de corps de degrée 8. C. R. Acad. Sci. Paris A 286 (1978), 875-876.

[18] J. P. Tignol, Decomp, et descente de produit tensoriel d'algèbre de quaternions. Rapp. Sem. Math. Pure UCL 76 (1978), VII 1-26.

HEBREW UNIVERSITY
JERUSALEM, ISRAEL

Contemporary Mathematics
Volume 13, 1982

REPRESENTATION THEORY OF FINITE DIMENSIONAL ALGEBRAS

by

M. Auslander[1]

In honor of N. Jacobson

INTRODUCTION.

Until about ten years ago, little work was done in the area of the representation theory of finite dimensional associative nonsemisimple algebras except for the important work in the modular representations of finite groups whose main architect was R. Brauer. However, beginning around 1970 there has been growing interest in the representation theory of arbitrary finite dimensional algebras. It is my aim in this lecture to give a brief introduction to some of these developments.

No effort is made to give a complete survey of results. The main emphasis is on the least technically involved work concerning algebras of finite representation type. While no proofs are given, an effort has been made to make the lecture self-contained in the sense that the basic definitions and concepts needed to explain the results cited are given. Those readers desiring a more thorough acquaintanceship with the work being done in this area are referred to the Proceedings of the International Conferences on the Representation of Algebras I, II, III. ([17], [18], [19] and [20]), especially [18] which has an extensive bibliography.

We assume throughout this lecture that all algebras are finite dimensional associative algebras with identity over a fixed algebraically closed field k. While the hypothesis that the ground field is algebraically closed is not essential for many of our results, it simplifies the exposition to make this assumption. We also assume that all modules are unitary and unless stated to the contrary, are finite dimensional over the ground field k.

Suppose M is a module over the algebra Λ. Then M is said to be an <u>indecomposable</u> Λ-module if $M \neq (0)$ and if $M = M_1 \amalg M_2$ (direct sum) then either $M_1 = (0)$ or $M_2 = (0)$. Clearly a finite dimensional Λ-module is isomorphic to a finite sum (direct) of indecomposable Λ-modules. The

1) Written with the partial support of NSF.

0271-4132/82/0511/$04.25

Krull-Schmidt theorem says that this representation is unique in the sense explained below.

THEOREM (Krull-Schmidt). _Let_ $\{M_i\}_{i \in I}$ _be a complete set of non-isomorphic indecomposable_ Λ-_modules_. _Then for each_ Λ-_module_ M _there is a unique set of integers_ $n_i \geq 0$ _such that_ $M \approx \coprod_{i \in I} n_i M_i$, _where_ $n_i M_i$ _means the sum of_ n_i _copies of_ M_i.

Therefore to describe the Λ-modules it suffices to a) describe the indecomposable Λ-modules and b) give a method for decomposing a Λ-module into a sum of indecomposable Λ-modules. Acutally carrying out this program for individual or classes of algebras is one of the main problems of representation theory. To date most of the emphasis has been put on describing the indecomposable modules. The following are some well known examples of where this problem has been solved.

Examples: a) $\Lambda = M_n(k)$, the $n \times n$ - matrices over k. Then the n-dimensional vector space on which $M_n(k)$-operates naturally is the unique indecomposable (up to isomorphism) Λ-module.

b) $\Lambda = k[X]/f(x)\, k[X]$ where $f(X)$ is not constant. Then a complete set of nonisomorphic indecomposable Λ-modules is given as follows. Since k is algebraically closed, $f(X) = \prod_{j \in J} (X_j - \alpha_j)^{t_j}$ with the α_j uniquely determined distinct elements of k and $t_j \geq 1$. Then the set of Λ-modules $\bigcup_{j \in J} \{k[X]/(X_j - \alpha_j)^{u_j}\, k[X] \mid 1 \leq u_j \leq t_j\}$ is a complete set of nonisomorphic indecomposable Λ-modules.

c) $\Lambda = T_n(k)$, lower triangular $n \times n$ matrices over k, where $T_n(k)$ is the subalgebra of $M_n(k)$ consisting of all (a_{ij}) with $a_{ij} = 0$ for all $j > i$. Now it is easily checked that for each integer $1 \leq j \leq n$, the left ideal $\underline{a}_j(T_n(k))$ consisting of all matrices $(a_{k,\ell})$ in $T_n(k)$ with $a_{k,\ell} = 0$ if $\ell \neq j$ is an indecomposable $T_n(k)$ - module. Then it is well known that the set of indecomposable $T_n(k)$-modules $\bigcup_{i=1}^{n} \{\underline{a}_j(T_i(k))\}$, where the $\underline{a}_j(T_i(k))$ are viewed as $T_n(k)$ - modules by means of the obvious surjective algebra monomorphisms $T_n(k) \to T_i(k)$ for $1 \leq i \leq n$, is a complete set of nonisomorphic indecomposable $T_n(k)$ - modules.

It should be observed that in each of these examples there are only a finite number of nonisomorphic indecomposable modules. Such algebras are said to be of <u>finite representation type</u>, and these are the algebras of main concern to us in this lecture. However before settling down to our discussion of algebras of finite representation type, it seems appropriate to give some examples of algebras which are not necessarily of finite representation type.

<u>Examples</u>: d) Let $U_n(k)$ be the subalgebra of $M_n(k)$ consisting of all matrices with zero off the first column and main diagonal. Then $U_n(k)$ is of finite representation type if and only if $n \leq 4$. Also there is a complete description of the indecomposable modules for $n \leq 5$.

e) Let G be a finite group and p = characteristic of k. Then $k[G]$, the group ring of G over k, is semisimple, hence of finite representation type, if and only if $p \nmid$ order G. If $p \mid$ order G, then $k[G]$ is of finite finite representation type if and only if the p-Sylow subgroups of G are cyclic. In the case $k[G]$ is of finite representation type, the indecomposable modules have been described [15], [14].

The rest of this lecture is divided into two parts: the first part is devoted to pointing out strong similarities between the representation theory of semisimple algebras and the more general class of algebras of finite representation type; the second part is devoted to a discussion of some efforts at classifying classes of algebras of finite representation type.

§1. SEMISIMPLE AND FINITE REPRESENTATION TYPE ALGEBRAS

Our first connection between semisimple and finite representation type algebras is based on the following notion. We say that an algebra morphism $f: \Lambda \to \Sigma$ is a <u>semisimple approximation of</u> Λ if Σ is semisimple and given any algebra morphism $h: \Lambda \to \Sigma^1$ with Σ^1 semisimple there is an algebra morphism $g: \Sigma \to \Sigma^1$ which makes the diagram

$$\begin{array}{ccc} \Lambda & \xrightarrow{\ f\ } & \Sigma \\ & \searrow^{h} & \downarrow g \\ & & \Sigma^1 \end{array}$$

commute.

<u>Remark</u>: It should be observed that we are not assuming in this definition that there is a unique $g: \Sigma \to \Sigma^1$ such that $gf = h$, only that there is some such morphism.

With this terminology in mind we have the following characterization of algebras of finite representation type.

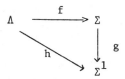

PROPOSITION 1.1. *An algebra* Λ *is of finite representation type if and only if it has a semisimple approximation* $f: \Lambda \to \Sigma$.

Suppose that Λ is of finite representation type. Then there are many semisimple approximations $\Lambda \to \Sigma$. However, it is not difficult to see that if $f_1: \Lambda \to \Sigma_1$ and $f_2: \Lambda \to \Sigma_2$ are two semisimple approximations of smallest dimension, then they are isomorphic in the sense that there is an algebra morphism $g: \Sigma_1 \to \Sigma_2$ such that the diagram

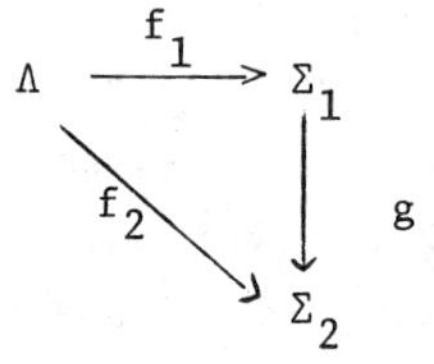

commutes and any such morphism is an isomorphism. In general there is more than one such morphism g. We now describe a way of constructing *minimal semisimple approximations of* Λ , i.e. semisimple approximations $\Lambda \to \Sigma$ with Σ of minimal dimension.

Let $M_1, \ldots, M_n$ be a complete set of nonisomorphic indecomposable Λ-modules. Associated with each M_i is the representation $f_i: \Lambda \to \mathrm{End}_k(M_i)$ given by $f_i(\lambda)(m) = m$ for all λ in Λ and m in M. It is easily seen that the induced algebra morphism $f: \Lambda \to \prod_{i=1}^{n} \mathrm{End}_k(M_i)$ is a minimal semisimple approximation of Λ . We denote this minimal semisimple approximation of Λ by $\Lambda \to \Sigma(\Lambda)$. We now give some properties of the minimal semisimple approximation $\Lambda \to \Sigma(\Lambda)$ which show that in some sense the representation theory of algebras of finite representation type can be reduced to the representation theory of semisimple algebras.

PROPOSITION 1.2. *Suppose* Λ *is of finite representation type and* $f: \Lambda \to \Sigma(\Lambda)$ *is a minimal semisimple approximation of* Λ . *Then*

a) f *is injective.*

b) *The restriction of the simple* $\Sigma(\Lambda)$ *modules to* Λ *gives a bijection between the isomorphism classes of simple* $\Sigma(\Lambda)$*-modules and the isomorphism classes of indecomposable* Λ*-modules.*

c) *Two* Λ*-modules* A *and* B *are isomorphic if and only if the induced modules* $\Sigma(\Lambda) \otimes_\Lambda A$ *and* $\Sigma(\Lambda) \otimes_\Lambda B$ *are isomorphic* $\Sigma(\Lambda)$*-modules.*

While the proofs of a) and b) of Proposition 1.2 are straightforward, the proof of c) is based on some results we will discuss later on (see Proposition 1.5).

Having established a close ring theoretic relationship between semisimple algebras and algebras of finite representation type, we now point out various similarities between the representation theory of these two types of algebras.

For semisimple algebras it is well known that every module, finite dimensional or not, is the (direct) sum of finite dimensional indecomposable modules, i.e. simple modules. As we now see the analogous property holds for an algebra if and only if it is of finite representation type.

PROPOSITION 1.3. *The following are equivalent for a k-algebra* Λ .

a) Λ *is of finite representation type.*

b) *Every* Λ*-module whether of finite dimension or not is a sum of finite dimensional indecomposable modules.*

c) Λ *has no infinite dimensional indecomposable modules.*

Proof: The proof is based on the fact that in the category of covariant additive functors from mod Λ , the category of finite dimensional Λ-modules, to abelian groups the simple functors are finitely presented, *i.e.*, if F is a nonzero functor with only (0) as a proper subfunctor, then there is an exact sequence $\mathrm{Hom}_\Lambda(A, \ ) \to \mathrm{Hom}_\Lambda(B, \ ) \to F \to 0$ of functors with A and B in mod Λ. (See [1]).

The following characterizations of algebras of finite representation type were first conjectured by Brauer - Thrall and served as motivation for some of the early work in the subject.

THEOREM 1.4. *The following are equivalent for an algebra* Λ .

a) Λ *is of finite representation type.*

b) *There is an integer* n *such that* $\dim_k M \leq n$ *for all indecomposable* Λ*-modules* M (finite dimensional of course).

c) *There is an integer* n *such that for each* $d \geq n$ *there is only a finite number of indecomposable* Λ*-modules* M *with* $\dim_k M = d$.

Proof. That b) implies a) was first shown by Roiter [23]. In view of the length and complexity of the paper by Nazarova and Roiter [16] on c) implies a), it is desirable to have a short proof of this implication.

Our next characterization of algebras of finite representation type is based on an analogue of the orthogonality relations for semisimple algebras which we state in the following form. Let Λ be a semisimple algebra and $\mathcal{I}$ a family of indecomposable Λ-modules. Then $\mathcal{I}$ has the property that each indecomposable Λ-module is isomorphic to some X in $\mathcal{I}$ if and only if two

Λ-modules A and B are isomorphic whenever $\dim_k \mathrm{Hom}_\Lambda(X, A) = \dim_k \mathrm{Hom}_\Lambda(X,B)$ for all X in $\mathcal{I}$. This property of semisimple algebras has the following direct generalization to all algebras.

PROPOSITION 1.5. Let Λ be an arbitrary algebra and $\mathcal{I}$ a family of indecomposable Λ-modules. Then $\mathcal{I}$ has the property that each indecomposable Λ-module is isomorphic to some X in $\mathcal{I}$ if and only if two Λ-modules A and B are isomorphic whenever $\dim_k \mathrm{Hom}_\Lambda(X,A) = \dim_k \mathrm{Hom}_\Lambda(X,B)$ for all X in $\mathcal{I}$.

Proof: Several proofs of this exist, all of which use the existence of almost split sequences at least in one of the implications. We recall that a nonsplit exact sequence $0 \to A \xrightarrow{f} B \xrightarrow{g} C \to 0$ is said to be almost aplit if A and C are indecomposable and satisfying either of the following equivalent conditions:

a) Given a nonisomorphism $h: X \to C$ with X indecomposable, there is a $t: X \to B$ such that $gt = h$.

b) Given any nonisomorphism $j: A \to Y$ with Y indecomposable, there is an $s: B \to Y$ such that $s = jg$.

The existence and uniqueness theorem says that given any nonprojective indecomposable module C there is a unique, up to isomorphism, almost split sequence $0 \to A \to B \to C \to 0$ and, dually, given any indecomposable noninjective module A there is a unique, up to isomorphism, almost split sequence $0 \to A \to B \to C \to 0$. See [4] for more details concerning almost split sequences.

Remark: The proof of part c) of Proposition 1.2 is based on Proposition 1.5.

Using Proposition 1.5 and 1.4 c) implies a) we obtain the following characterization of algebras of finite representation type.

PROPOSITION 1.6. *Λ is of finite representation type if and only if there is a finite family of indecomposable modules $\mathcal{I}$ with the property that two Λ-modules A and B are isomorphic if $\dim_k \mathrm{Hom}_\Lambda(X,A) = \dim_k \mathrm{Hom}_\Lambda(X,B)$ for all X in $\mathcal{I}$. Moreover if such a finite family $\mathcal{I}$ exists, then every indecomposable Λ-module is isomorphic to some module in $\mathcal{I}$.*

Next we point out the analogue for algebras of finite representation of the Wedderburn theorem for semisimple algebras. We recall that an algebra Λ is semisimple if and only if there is an algebra Γ which is a finite product $k \times \ldots \times k$ of copies of k and a Γ-module M such that $\Lambda \cong \mathrm{End}_\Gamma(M)$. Also we know that if Λ is semisimple $S_1, \ldots, S_n$ is a complete set of nonisomorphic indecomposable Λ-modules, then

$\Gamma = \mathrm{End}_\Lambda(\coprod_{i=1}^{n} S_i)^{op}$ and the functor mod $\Lambda \to$ mod Γ given by $X \to \mathrm{Hom}_\Lambda(S, X)$ for all X in mod Λ where $S = \coprod_{i=1}^{n} S_i$ is an equivalence of categories.

The analogue for algebras of finite representation type of the Wedderburn theorem for semisimple algebras is based on the following result.

THEOREM 1.7. *Let Λ be a nonsemisimple algebra of finite representation type and suppose $M = \coprod_{i=1}^{n} M_i$ where $M_1, \ldots, M_n$ is a complete set of non-isomorphic indecomposable Λ-modules. Then $\Gamma = \mathrm{End}_\Lambda(M)^{op}$ has the following properties*:

a) gl. dim.$\Gamma = 2$

b) dom. dim.$\Gamma = 2$, *i.e. if* $0 \to \Gamma \to I_0 \to I_1$ *is a minimal injective copresentation of* Γ , *then each* I_i *is projective as well as injective*.

c) *The functor* mod $\Lambda \to$ mod Γ *given by* $X \to \mathrm{Hom}\ (M,X)$ *is fully faithful and induces an equivalence* mod$\Lambda \to \underline{\underline{P}}(\Gamma)$, *where* $\underline{\underline{P}}(\Gamma)$ *is the category of all finitely generated projective* Γ-*modules*.

d) Λ *is Morita equivalent to* $\mathrm{End}_\Gamma(I)$ *where* I *is the sum of a complete set of nonisomorphic indecomposable* Γ-*modules which are both projective and injective*.

We can now state the Wedderburn theorem for algebras of finite representation type.

THEOREM 1.8. *The map which assigns to each nonsemisimple algebra of finite representation type* Λ *the algebra* $\Gamma = \mathrm{End}_\Lambda(\coprod_{L=1}^{n} M_i)^{op}$ *where* $M_1, \ldots, M_n$ *is a complete set of nonisomorphic indecomposable* Λ-*modules induces a bijection between the Morita equivalence classes of nonsemsimple algebras* Λ *of finite representation type and the Morita equivalence classes of algebras* Γ *with* gl. dim.$\Gamma = 2$ *and* dom. dim.$\Gamma = 2$.

Proof: The reader is referred to [2] for proofs of Proposition 1.7 and Theorem 1.8.

The Wedderburn theorem for nonsemisimple algebras of finite representation type, shows in a rather precise way that the problem of classifying algebras of finite representation type is the same thing as classifying algebras Γ with gl. dim.$\Gamma = 2$ and dom. dim.$\Gamma = 2$. The advantage this second problem has over the first, is that, in principle at least, determining whether or not an algebra Γ has gl. dim.$\Gamma = 2$ and dom. dim $\Gamma = 2$ is a finite problem while there is no finite procedure known for checking whether or not an algebra is of finite representation type. The fact that this advantage is not only theoretical but also practical was dramatically demonstrated by C. Riedtmann in her work on classifying selfinjective algebras, i.e. algebras where the projective and injective modules are the same. More generally, much of the recent work on algebras of finite type has involved studying the algebras of finite type Λ together with the algebras Γ of global dimension and dominant dimension two corresponding to Λ. The next section of this lecture is devoted to a brief introduction to some of the ideas being used in this work.

§2. QUIVERS OF ALGEBRA

Throughout this section we shall assume that all of our algebras are indecomposable and basic. We recall that a k-algebra Λ is basic if and only if $\Lambda/\underline{r} \cong k \times \ldots \times k$ where $\underline{r}$ is the radical of Λ or equivalently if we write Λ as a sum $\coprod_{i=1}^{n} P_i$ of indecomposable projective Λ-modules P_i, then $P_i \approx P_j$ implies $i = j$. Since every k-algebra is Morita equivalent to a finite product of indecomposable basic algebras, we lose no generality by restricting our attention to indecomposable, basic algebras.

The main new idea of this section is the notion of a <u>quiver of an algebra</u> whose importance in representation theory was first pointed out by P. Gabriel. We first recall the notion of a quiver. A <u>quiver</u> Q is nothing more than a directed graph, <u>i.e.</u>, Q consists of a set of points together with a finite number of arrows between the points. We will always assume that our quivers are finite which means that they have only a finite number of points. Some examples of quivers are the following:

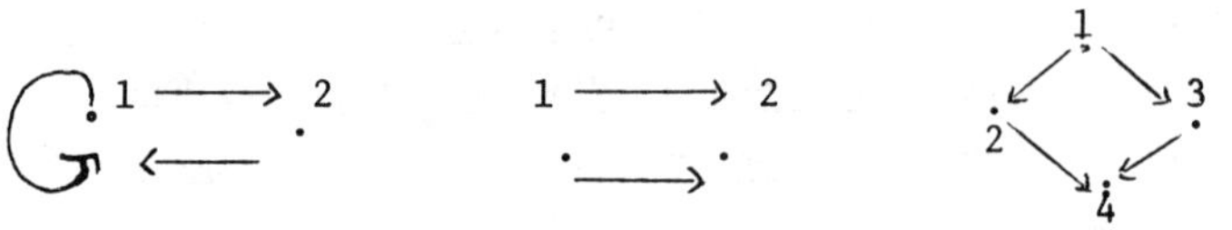

Suppose Λ is a k-algebra and let $e_1, \ldots, e_n$ be a complete set of orthogonal primitive idempotents in Λ.

Then $\Lambda = \coprod_{i=1}^{n} \Lambda e_i$ where each $P_i = \Lambda e_i$ is an indecomposable projective Λ-module. The quiver of Λ, which we denote by $Q(\Lambda)$, consists of n points, one for each P_i and the number of arrows from i to j is the number of times the simple module $S_j = P_j/\underline{r}\,P_j$ occurs as a summand in the semisimple Λ-module $\underline{r}P_i/\underline{r}^2P_i$. By way of illustration, we give the quivers of some of the algebras mentioned in the introduction to this lecture.

a) If $\Lambda = M_n(k)$, then $Q(\Lambda) = \cdot$, single point with no arrows

b) If $\Lambda = k[X]/(X^n)$ $(n \geq 2)$, then $Q(\Lambda) =$ 

c) If $\Lambda = T_n(k)$, then $Q(\Lambda) = \underset{1}{\cdot} \rightarrow \underset{2}{\cdot} \rightarrow \cdots \rightarrow \underset{n}{\cdot}$

d) If $\Lambda = U_n(k)$, then $Q(\Lambda) =$ 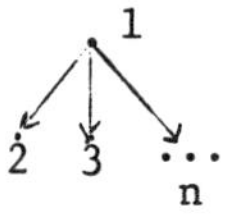

Using quivers with relations, a notion we are not going to elaborate on in this lecture, one can show that every quiver is the quiver of some algebra, and usually more than one. Obviously the quiver of Λ is the same as the quiver of $\Lambda/\underline{r}^2$. Nonetheless it tells quite a bit about algebras, at least in some special cases, even when the square of the radical is not zero. The following results illustrate this point.

Lemma 2.1. $Q(\Lambda)$ *has no oriented cycles if and only* Λ *is a factor of an hereditary algebra.*

THEOREM 2.2. *If* Λ *is hereditary, i.e.,* gl. dim.$\Lambda = 1$, *then* Λ *is of finite representation type if and only if the underlying graph of* $Q(\Lambda)$ *is one of the Dynkin diagrams* A_n, D_n, E_6, E_7, E_8.

Proof: This result is due to P. Gabriel [11] and is the starting point for much of the contemporary work in representation theory, especially in the form given by Bernstein, Gelfand and Ponomarev [7].

Gabriel's result leads one to believe that there should be fairly strong restrictions on the quiver of an algebra of finite representation type. As an illustration of this point we point out the following.

PROPOSITION 2.3. *Suppose* Λ *is of finite representation type. Then* $Q(\Lambda)$ *has the following easily verified local properties:*

a) *Between any two points distinct or not, there is at most one arrow.*

b) *There are at most three arrows coming into any point and at most three arrows coming out of any point.*

Now suppose Λ is of finite representation type. Let $\Gamma = \mathrm{End}_\Lambda(\coprod_{i=1}^{n} M_i)$ where $M_1, \ldots, M_n$ is a complete set of nonisomorphic indecomposable Λ-modules. Then Γ is also a basic indecomposable algebra and therefore in addition to the quiver $Q(\Lambda)$ we also have the quiver $Q(\Gamma)$ associated with the algebra Λ. Since the algebra Γ is described in terms of the morphisms between indecomposable Λ-modules, it would be expected that the quiver $Q(\Gamma)$ would have a description in similar terms. To explain what this is it is convenient to have the following alternative description of the quiver of an arbitrary algebra.

Suppose Λ is an arbitrary basic algebra and let $P_1, \ldots, P_n$ be a complete set of indecomposable nonisomorphic projective Λ-modules. Then $Q(\Lambda)$ can be described as follows. The points of $Q(\Lambda)$ are the indecomposable projective $P_1, \ldots, P_n$. The number of arrows from P_i to P_j is the number of times P_j appears as a summand of the projective cover of $\underline{r}P_i$.

We now return to the situation that Λ is an algebra of finite representation type and $\Gamma = \mathrm{End}_\Lambda(M_1 \amalg \cdots \amalg M_n)^{op}$ where $M_1, \ldots, M_n$ is a complete set of nonisomorphic indecomposable Λ-modules. The equivalence of categories $\mathrm{mod}\,\Lambda \to \underline{\underline{P}}(\Gamma)$, where $\underline{\underline{P}}(\Gamma)$ is the category of projective Γ-modules, given by $X \mapsto \mathrm{Hom}_\Lambda(\coprod_{i=1}^n M_i, X)$ for all X in $\mathrm{mod}\,\Lambda$, induces a bijection $M_j \mapsto \mathrm{Hom}(\coprod_{i=1}^n M_i, M_j)$ between the representatives $M_1, \ldots, M_n$ of the isomorphism classes of indecomposable Λ-modules and the representatives $\mathrm{Hom}_\Lambda(\amalg M_i, M_j), \ldots, \mathrm{Hom}_\Lambda(\amalg M_i, M_n)$ of the isomorphism classes of indecomposable projective Λ-modules. Thus the points of $Q(\Gamma)$ can be viewed as the indecomposable Γ-modules $M_1, \ldots, M_n$. It remains to interpret the number of arrows from M_i to M_j in terms of morphisms.

Suppose M_i is a projective Λ-module. Then the number of arrows from M_j to M_i is the multiplicity of M_j in the decomposition $\underline{r}M_i$ into indecomposables.

Suppose M_i is not projective. Then there is a unique, up to isomorphism, almost split sequence (see section 1) $0 \to A \to B \to M_i \to 0$. Then the multiplicity of M_j in the decomposition of B into indecomposables, is the number of arrows from M_i to M_j. So we see that the quiver $Q(\Gamma)$ is given by the structure of the almost split sequences and the projectives in $\mathrm{mod}\,\Lambda$. This construction is due to C. M. Ringel who did it for arbitrary artin algebras [22].

As we saw in section 1, the algebra Γ has rather restrictive, homological properties. This suggests that the quiver $Q(\Gamma)$ is not arbitrary which is indeed the case. We now list some of the special properties of $Q(\Gamma)$.

PROPOSITION 2.4. a) <u>No point is connected to itself by an arrow.</u>
b) <u>Between any two points there is at most one arrow.</u>
c) <u>For each nonprojective point C there is a picture of the form</u>

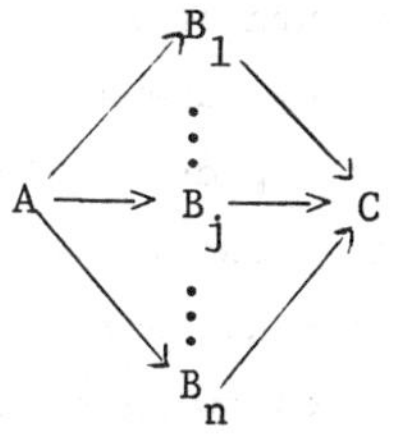

with A noninjective where the B_i are all the points in $Q(\Gamma)$ connected by an arrow going to C and the B_i are also all the points connected to A by arrows going out of A.

d) For each point P in $Q(\Gamma)$ there are at most four arrows going to P as well as at most four arrows going out of P. Furthermore, if there are four arrows going to P, then one of the points connected to P by these arrows must be projective and injective. Similarly, if there are four arrows coming out of P, then at least one of the points connected to P by these arrows must be projective and injective.

Proof: a) This is an elementary property of almost split sequences [4].

b) This is due to R. Bautista [5]. Another proof has been given by K. Bongartz [21].

c) This is also a consequence of the elementary properties of almost split sequences [4].

d) This result is due to R. Bautista and S. Brenner [6].

We end this discussion of the quivers $Q(\Lambda)$ and $Q(\Gamma)$ by giving a few examples of results obtained using these notions.

The first result is due to Happel and Ringel [13] and gives an application of these ideas to obtaining information about the structure of the indecomposable modules, at least in some interesting special cases. The proof involves the notion of tilting modules, whose genesis goes back to the paper of Bernstein, Gelfand and Ponomareva [7] (see [3], [10] and [13].)

PROPOSITION 2.5. *Suppose $Q(\Gamma)$ has no oriented cycles. Then*

a) *$Q(\Lambda)$ has no oriented cycles and*

b) *Every indecomposable Λ-module is completely determined, up to isomorphism, by its composition factors.*

Our last results concern classification theorems for selfinjective algebras of finite representation type and algebras of finite type with at most two simple modules. In both cases the theory of coverings of quivers is used to study the quivers of algebras which correspond to algebras of finite representation type under the Wedderburn correspondence. While the theory of coverings of quivers is essentially the same as the topological theory, the fact that it is an important tool in representation theory was first demonstrated by Gabriel-Riedtmann [12] and C. Riedtmann [21].

C. Riedtmann's work on classifying selfinjective algebras of finite representation type is based on the following result. Let Λ be a selfinjective algebra of finite representation type and Γ its Wedderburn correspondent. Then there is a uniquely determined Dynkin diagram of type A_n, D_n, E_6, E_7, or E_8 which occurs naturally as a subquiver of $Q(\Gamma)$ minus the projective points.

Finally using coverings of quivers, Bongartz-Gabriel have classified all algebras of finite representation type having at most two simple modules [9].

BIBLIOGRAPHY

[1] Auslander, M., Large modules over artin algebras, Algebra, Topology and Categories, Academic Press (1976), 1-17.

[2] Auslander, M., Representation Theory of Artin Algebras, II, Comm. Algebra 1 (1974), 269-310.

[3] Auslander, M., Platzeck, M. I. and Reiten, I., Coxeter functors without diagrams, Trans. Am. Math. Soc. Vol. 250, (1979), 1-46.

[4] Auslander, M. and Reiten, I., Representation theory of artin algebras III, Comm. Algebras 3 (1975), 239-294.

[5] Bautista, R., Irreducible maps and the radical of a category., Preprint.

[6] Bautista, R. and Brenner, S., On the number of terms in the middle of an almost split sequence, Proceedings of ICRA III, Springer Lecture Notes, to appear.

[7] Bernstein, I. N., Gelfand, I. M., Ponomarev, V. A., Coxeter functors and Gabriel's theorem, Uspechi, Mat. Nauk. 28 (1973). Translated in Russian Math. Surveys 28 (1973), 17-32.

[8] Bongartz, K., Tilted Algebras, Proceedings, ICRA III, Springer Lecture Notes, to appear.

[9] Bongartz, K. and Gabriel, P., Covering Spaces in representation-theory, Preprint.

[10] Brenner, S., and Butler, M.C.R., Generalizations of the Bernstein-Gelfand-Ponomarev reflection functors, Proceedings ICRA II, Ottawa, Lecture Notes, 832.

[11] Gabriel, P., Unzerlegbare Darstellungen I., Manuscripta Math. 6 (1972), 71-103.

[12] Gabriel, P., and Riedtmann, Ch., Group representations without groups, Comment. Math. Helv. 54 (1979), 240-287.

[13] Happel, D., and Ringel, C. M., Tilted algebras, To appear Trans. Amer. Math. Soc.

[14] Janusz, G., Indecomposable modules for finite groups, Ann. of Math. 89 (1969), 209-241.

[15] Kupisch, H., Unzerlegbare Moduln endlicher Gruppen mit zyklischet p-Sylow Gruppe, Math. Z. 108 (1969), 77-104.

[16] Nazarova, L. A., and Roiter, A. V., Categorical matrix problems and the Brauer-Thrall conjecture, Preprint, Inst. Math. Scad. Sci., Kiev 1973 translated in Mitt. Math. Sem. Giessen 115 (1975).

[17] Proceedings ICRA I, Springer Lecture Notes 488, 1975.

[18] Representation Theory I, Proceedings 1979, Edited by V. Dlab and P. Gabriel, Springer Lecture Notes, Vol. 831, 1980.

[19] Representation Theory II, Proceedings, 1979, Edited by V. Dlab and P. Gabriel, Springer Lecture Notes, Vol. 832, 1980.

[20] Proceedings ICRA III, 1980, Springer Lecture Notes, to appear.

[21] Riedtmann, Ch., Representation-finite self-injective algebras of Class A_n, Springer Lecture Notes 832, 449-520.

[22] Ringel, C. M., Finite-dimensional hereditary algebras of wild representation type, Math. Z., (1978), 235-255.

[23] Roiter, A. V., Unboundedness of the dimensions of the indecomposable representations of an algebra which has infinitely many indecomposable representations, Izv. Abad. SSSR. Math. 32 (1968), 1275-1282.

Contemporary Mathematics
Volume 13, 1982

THE POLYNOMIAL IDENTITIES OF MATRICES

Edward Formanek

To Nathan Jacobson

TABLE OF CONTENTS

0. Introduction
1. T-ideals
2. Central polynomials
3. Representation theory of the symmetric group and general linear group
4. Trace identities
5. Codimensions and cocharacters
6. The ring of generic matrices
7. Example: The ring of 2×2 generic matrices

0. INTRODUCTION

An excellent survey by Amitsur [4] shows that the theory of polynomial identities has its roots in the work of Dehn [12] on projective geometry in 1922. Amitsur also discovered that the first polynomial identities for matrices were given by Wagner [62] in 1937. Wagner's construction, which has a very modern flavor, was described by Amitsur in his survey. As Amitsur pointed out, many of Wagner's ideas were rediscovered later. However, his article was completely overlooked and had no influence on future developments.

The theory of polynomial identities as a well-defined field of study began with the well-known article of Kaplansky [27] which was based on work of M. Hall [20] and Jacobson [22]. Not long afterward came the Amitsur-Levitski Theorem [7] which is still a basic theorem on the polynomial identities of matrices. The next twenty years saw a steady but modest growth of this new theory, mainly from a ring theoretic point of view. As far as the polynomial identities of matrices are concerned, there were only two noteworthy developments: Identities based on the standard identity and the Cayley-Hamilton Theorem, due to Amitsur [2], and the introduction of the ring of generic matrices, which was shown to be a domain (but not explicitly defined) by Amitsur [1] and formally defined and studied further by Amitsur and Procesi [8,34].

0271-4132/82/0512/$10.75

The past decade has witnessed a great increase in popularity for PI-rings. One measure of this is the bibliography of the recent book of Rowen [56], which lists 104 research articles that appeared before 1970 and 239 that appeared in the period 1970-79. The impetus for this activity was the positive solution of a number of long-standing problems. These problems were not concerned solely with the identities of matrices but all were related to them.

The first of these major results was M. Artin's Theorem [9], which characterizes Azumaya algebras of constant rank n^2 as rings all of whose nonzero homomorphic images satisfy the identities of $M_n(\mathbb{Z})$ but not of $M_{n-1}(\mathbb{Z})$. Despite the hypothesis of this theorem, its applications are ring-theoretic rather than to the combinatorics of matrix identities. Nevertheless, Artin's article influenced subsequent research on matrix identities in two significant ways: First, the work of Procesi on trace identities was motivated by questions raised by Artin; second, later proofs of Artin's theorem were based on central polynomials and Capelli polynomials.

The next major result was the tensor product theorem of Regev [43]. He showed that the tensor product of two PI-algebras is a PI-algebra. The statement of his theorem says nothing about the identities satisfied by matrices, but its proof is based on a careful analysis of the T-ideal of identities of an arbitrary PI-ring. By applying his methods to the T-ideal of identities of $n \times n$ matrices Regev has recently obtained many remarkable quantitative results.

Shortly after came the discovery by Formanek [15] and Razmyslov [39] of central polynomials for $n \times n$ matrices. These proved a powerful tool in many aspects of PI-theory; from our point of view they are important because they increased our knowledge of identities.

The last major result was the discovery of trace identities for $n \times n$ matrices by Razmyslov [42] and Procesi [36]. These trace identities are a generalization of ordinary identities and they contain the ordinary identities of $n \times n$ matrices as an identifiable subset. Razmyslov and Procesi gave a complete description of trace identities in characteristic zero and thus a complete description of the ordinary identities of $n \times n$ matrices in characteristic zero, at least in principle. I have added the warning "in principle" since in practice their description has proved difficult to use effectively.

At the outset I quoted Amitsur to the effect that the idea of a polynomial identity first occurred in the work of Dehn in 1922. This is true, but as far as the polynomial identities of matrices are concerned, one can go back to the Cayley-Hamilton Theorem and classical invariant theory, for the polynomial identities of matrices are a part of invariant theory, as the approaches of both Razmyslov and Procesi to trace identities show. Razmyslov obtains trace identities by multilinearizing the characteristic polynomial while Procesi obtains them as a suitable interpretation of the multilinear invariants of tensor products of vector spaces.

In this survey I will adopt a combinatorial or quantitative viewpoint. A large part of the article will be concerned with recent research, much still in progress. To understand this research as well as trace identities, some knowledge of the representations of the symmetric and general linear groups is essential, so I will include a brief description of some of this theory.

Aside from the specific references in the text, the reader is directed to the following: The monographs [24,35,56] are general treatises on PI-rings, and the survey articles [4,5,6,23,25,37,38,46,53] overlap with this one.

1. T-IDEALS

We will work over a field K of characteristic zero and rings will be K-algebras with unit. We definitely require characteristic zero at many places but the particular field K will not matter, except at a few very explicit points. Let

$$K\langle X\rangle = K\langle x_1, x_2, \ldots\rangle$$

be a free associative algebra in countably many indeterminates. (Sometimes we will use other variables x, y, y_i for notational simplicity.)

<u>Definition</u>. A ring R satisfies a <u>polynomial identity</u> if there is a nonzero polynomial $f(x_1,\ldots,x_m) \in K\langle X\rangle$ which vanishes when evaluated on any $r_1,\ldots,r_m \in R$.

<u>Definition</u>. A T-ideal is an ideal of $K\langle X\rangle$ which is closed under K-endomorphisms of $K\langle X\rangle$.

<u>Definition</u>. If R is a ring,

$$T(R) = \{f(x_1,\ldots,x_m) \in K\langle X\rangle \mid f(r_1,\ldots,r_m) = 0 \quad \text{for all} \quad r_1,\ldots,r_m \in R\}.$$

It is easy to see that $T(R)$ is a T-ideal, called the T-ideal of identities of R. Conversely, if $\mathcal{J}$ is a T-ideal,

$$T(K\langle X\rangle/\mathcal{J}) = \mathcal{J},$$

so every T-ideal is of the form $T(R)$ for some R. We can already state a basic open problem

PROBLEM 1 (Specht [60]). <u>Does</u> $K\langle X\rangle$ <u>satisfy the ascending chain condition on T-ideals?</u>

Equivalently, are all the identities satisfied by a ring consequences of a finite number of them? The analogous problem for varieties of groups has a negative solution: There are varieties of groups which are not finitely based.

The free ring $K\langle X\rangle = K\langle x_1, x_2, \ldots\rangle$ can be graded by giving each x_i degree one and can be multigraded by assigning a separate degree function to each variable.

Definition. A polynomial $f(x_1,\ldots,x_m) \in K\langle X\rangle$ is homogeneous if each monomial in f has the same degree in each of $x_1,\ldots,x_m$. It is multilinear if it is homogeneous of degree one in each of $x_1,\ldots,x_m$.

Examples: $f(x_1,x_2) = x_1^2x_2 + x_1x_2x_1$ (homogeneous)

$$f(x_1,x_2,x_3) = x_1x_2x_3 - x_1x_3x_2 \quad \text{(multilinear)}$$

The next theorem effectively reduces the study of PI's to homogeneous or even multilinear ones.

THEOREM 2. Suppose $\mathcal{J}$ is a T-ideal of $K\langle X\rangle$. Then
(1) $\mathcal{J}$ is homogeneous. In other words, every homogeneous component of a polynomial in $\mathcal{J}$ lies in $\mathcal{J}$.
(2) If $\mathcal{J}$ contains a polynomial of degree d, it contains a multilinear polynomial of degree d. Moreover, $\mathcal{J}$ is generated by the multilinear polynomials it contains.

Remark: The first part of the theorem fails for finite fields: The finite field with q elements satisfies $x^q - x$ but not its homogeneous components. The second part fails for fields of characteristic $p > 0$: The T-ideal generated by x^p is not generated by the multilinear polynomials it contains.

Definition. The T-ideal of identities satisfied by $M_n(K)$ is $\mathfrak{M}_n = T(M_n(K))$.

Definition. The standard polynomial of degree m is the polynomial

$$\mathcal{S}_m(x_1,\ldots,x_m) = \sum_{\pi \in S_m} (\operatorname{sign} \pi) x_{\pi(1)} \cdots x_{\pi(m)}$$

(S_m denotes the symmetric group of degree m.)

The Amitsur-Levitski Theorem shows that

$$\mathfrak{M}_1 > \mathfrak{M}_2 > \mathfrak{M}_3 > \cdots$$

is a properly descending chain of T-ideals whose intersection is zero and also gives the least degree of a polynomial satisfied by $M_n(K)$.

THEOREM 3. (Amitsur-Levitski [7]).
(1) $M_n(K)$ satisfies the standard polynomial of degree $2n$.
(2) $M_n(K)$ satisfies no polynomial of degree $< 2n$.
(3) If $f(x_1,\ldots,x_{2n})$ is a polynomial of degree $2n$ satisfied by $M_n(K)$, then it is a scalar multiple of the standard identity.

The original proof of Theorem 3 was a clever induction using matrix units but subsequently there have been proofs by Kostant [30] based on the cohomology of $M_n[\mathbb{C}]$ as a Lie algebra and by Razmyslov [42] and Rosset [55] based on the Cayley-Hamilton Theorem. We will outline Razmyslov's proof in section 4 on trace identities. The trace identities of section 4 lead to a description of all the identities in $\mathfrak{M}_n$ in section 5. In spite of this description Problem 1 is still unresolved for the ideals $\mathfrak{M}_n$.

PROBLEM 4. Is $\mathfrak{M}_n$ finitely generated as a T-ideal?

It is trivial that $\mathfrak{M}_1$ is generated as a T-ideal by

$$f(x_1,x_2) = x_1x_2 - x_2x_1 = [x_1,x_2].$$

Razmyslov [40] has shown that $\mathfrak{M}_2$ is also finitely generated as a T-ideal, and Drensky [13] has shown that the two polynomials

$$\mathcal{S}_4(x_1,x_2,x_3,x_4) \quad \text{and} \quad [[x_1,x_2]^2,x_1]$$

suffice to generate it.

From a ring theoretic viewpoint, the ideals $\mathfrak{M}_n$ are quite special.

THEOREM 5. (Amitsur [1]).
(1) The only nonzero prime T-ideals are the ideals $\mathfrak{M}_n$.
(2) For all n, $K\langle X\rangle/\mathfrak{M}_n$ has no zero divisors.

Part of our study of $\mathfrak{M}_n$ will be devoted to the relatively free algebra $K\langle X\rangle/\mathfrak{M}_n$. It turns out that this algebra has a useful concrete representation. Let

$$\{u_{ij}(r) \mid 1 \le i,j \le n,\ r = 1,2,\ldots\}$$

be independent commuting indeterminates over K, and let

$$U(r) = (u_{ij}(r)) \in M_n(K[u_{ij}(r)])$$

be $n \times n$ matrices whose entries are these indeterminates.

Definition. The matrices $U(r)$ are called $n \times n$ generic matrices over K. The K-algebra they generate is called a ring of generic matrices. It is denoted $K[U(r)]$.

We can define a surjective homomorphism $K\langle X\rangle \to K[U(r)]$ by $x_r \to U(r)$. It is not difficult to see that the kernel of this homomorphism is $\mathfrak{M}_n$, so we have an exact sequence

$$0 \to \mathfrak{M}_n \to K\langle X\rangle \to K[U(r)] \to 0.$$

Finally, although we will stick to characteristic zero, there is one question about polynomial identities for matrices in positive characteristics which deserves mention. For any commutative ring A (not necessarily a K-algebra), let $M_n(A)$ denote the ideal in $A\langle X\rangle$ of identities satisfied by $M_n(A)$. Then $\mathfrak{M}_n(\mathbb{Z})$ generates $\mathfrak{M}_n(L)$ as an L-vector space for any field L of characteristic zero but we do not know what happens in characteristic $p > 0$.

PROBLEM 6. *If L is an infinite field of characteristic $p > 0$, are all identities of $M_n(L)$ consequences of identities of $M_n(\mathbb{Z})$? More precisely, does the image of the natural map $\mathfrak{M}_n(\mathbb{Z}) \to \mathfrak{M}_n(L)$ generate $\mathfrak{M}_n(L)$ as a vector space over* L?

This is trivially true for $n = 1$ and also true for 2-variable identities for 2×2 matrices as a consequence of Formanek, Halpin and Li [18] or Li [32].

2. CENTRAL POLYNOMIALS

Definition. A polynomial $f(x_1,\ldots,x_m) \in K\langle X\rangle$ is a *central polynomial* for a ring R if

(1) All evaluations $f(r_1,\ldots,r_m)$ $(r_i \in R)$ lie in the center of R.

(2) When evaluated on R, f assumes at least two distinct values. (This insures that f is not a scalar plus a PI.)

For 2×2 matrices, the Cayley-Hamilton Theorem shows that

$$f(x,y) = (xy - yx)^2$$

is a central polynomial, since $xy - yx$ has trace zero. This was already noted by Wagner [62] in 1937. Later, Kaplansky [28] asked if central polynomials exist for $n \times n$ matrices, $n \geq 3$. This was answered affirmatively.

THEOREM 7. (Formanek [15], Razmyslov [39]). *For all* n, *$M_n(K)$ has a central polynomial*.

The central polynomial $f(x,y_1,\ldots,y_n)$ discovered by Formanek is homogeneous of degree $n^2 - n$ in x and degree 1 in each of $y_1,\ldots,y_n$. Its total degree n^2 is the least known degree of a central polynomial for $M_n(K)$. The two-variable polynomial

$$g(x,y) = f(x,y,\ldots,y)$$

is also a (nonvanishing) central polynomial for $M_n(K)$.

PROBLEM 8. <u>What is the minimal degree of a central polynomial for</u> $M_n(K)$?

PROBLEM 9. <u>What is the minimal degree of a two-variable central polynomial for</u> $M_n(K)$?

It is known only that n^2 is the minimal degree for $n = 1,2$. It can be shown that if $G(x,y)$ is a two-variable central polynomial for $M_n(K)$, then it has degree $\geq n$ in each variable. Moreover, if its degree in y is exactly n, then the minimal degree in x is $n^2 - n$, and the only central polynomial $G(x,y)$ of degree $n^2 - n$ in x and degree n in y is essentially the polynomial $g(x,y)$ above. Incidentally, although the Amitsur-Levitzki Theorem shows that $2n$ is the least degree of a PI for $M_n(K)$, the minimal degree of a two-variable identity is unknown.

PROBLEM 10. <u>What is the minimal degree of a two-variable polynomial identity satisfied by</u> $M_n(K)$?

Again, one can show that if $f(x,y)$ is a two-variable identity for $M_n(K)$ then it has degree $\geq n$ in each variable. Moreover, there are two-variable identities of degree n in y, and among them there is a unique one (up to a scalar multiple) of minimal degree in x, namely

$$\mathcal{S}_n([x^n,y],\ [x^{n-1},y],\ldots,[x,y])$$

which has total degree $\frac{1}{2}(n^2+ 3n)$. This is the minimal total degree if $n \leq 3$. But for $n \geq 7$ suitable substitutions of two-variable monomials in $\mathcal{S}_{2n}$ give a PI of total degree $< \frac{1}{2}(n^2+ 3n)$ for $M_n(K)$ (see [56], p. 104). So there is not yet a reasonable conjecture for the answer to Problem 10.

Returning to central polynomials, Razmyslov's construction gave a multilinear polynomial of degree $3n^2- 1$. His proof was based on the notion of a weak identity.

<u>Definition</u>. A polynomial $f(x_1,\ldots,x_m)$ is a <u>weak identity</u> for $M_n(K)$ if $f(X_1,\ldots,X_m) = 0$ whenever $X_1,\ldots,X_m$ are $n \times n$ matrices of trace zero. An <u>essential weak identity</u> is a weak identity which is not a polynomial identity.

For example, $[x^2,y]$ is an essential weak identity for $M_2(K)$. Razmyslov showed that an essential weak identity of degree m gives rise to a central polynomial of degree less than $2m$. Thus the minimal degree of a central polynomial is related to the minimal degree of an essential weak identity.

PROBLEM 11. What is the minimal degree of an essential weak identity for $M_n(K)$?

A reasonable candidate for an essential weak identity of minimal degree has been found by Halpin [21]. Let X be an $n \times n$ matrix with trace zero. Then X satisfies its characteristic polynomial

$$X^n + a_2X^{n-2}+\cdots+ a_{n-1}X + a_n = 0,$$

so for any matrix Y,

$$[X^n,Y] + a_2[X^{n-2},Y] +\cdots+ a_{n-1}[X,Y] = 0.$$

This is a linear dependence relation over K, so the fact that the standard identity vanishes under linearly dependent substitutions shows that

$$S_{n-1}([x^n,y],[x^{n-2},y],\ldots,[x,y])$$

is a weak identity for $M_n(K)$ of degree $1/2(n^2+ n)$ and one can verify that it is an essential weak identity. For $n \geq 3$, it is an essential weak identity of minimal degree. Halpin shows that Razmyslov's method applied to this polynomial yields a central polynomial of degree n^2.

Finally, we mention the important Capelli polynomial which was introduced by Razmyslov [41].

Definition. The nth Capelli polynomial is

$$C_n(x_1,\ldots,x_n,y_1,\ldots,y_{n-1}) = \sum_{\pi \in S_n} x_{\pi(1)}y_1x_{\pi(n-1)}y_{n-1}x_{\pi(n)}.$$

We will give only a few properties of C_n here. A survey by Amitsur [5] gives a much more complete discussion, and Rowen [56, pp. 23-31] has based his development of central polynomials on the Capelli polynomial. The combinatorial properties of the Capelli polynomial make it a very useful tool in application to structure theory. For example, the shortest proof of Artin's Theorem, due to Schelter [57], [5, p. 10], uses the Capelli polynomial. Work of Regev and Kemer to be mentioned later will also show the significance of this polynomial.

Since $C(x_1,\ldots,x_n,y_1,\ldots,y_{n-1})$ is multilinear and alternating as a function of $x_1,\ldots,x_n$, it vanishes whenever $x_1,\ldots,x_n$ are linearly dependent. Hence C_{n^2+1} is a PI for $M_n(K)$ and C_{n^2} vanishes on any proper subspace of $M_n(K)$;

an explicit substitution shows that C_{n^2} is not a PI for $M_n(K)$. Using these determinant-like properties Amitsur constructed a central polynomial using C_{n^2}.

THEOREM 12. (Amitsur [5, Theorem 7]). <u>Let</u> $X_1,\ldots,X_{n^2}, Y_1,\ldots,Y_{n^2}$ <u>be</u> $n \times n$ <u>matrices over</u> K. <u>Then</u>

$$\mathrm{Tr}(C_{n^2}(X_1,\ldots,X_{n^2},Y_1,\ldots,Y_{n^2-1}))\cdot\mathrm{Tr}(Y_{n^2})$$

$$= \sum_{j=1}^{n^2} (-1)^{j(n^2-1)} Y_j C_{n^2}(X_1,\ldots,X_{n^2},Y_{j+1}\cdots Y_{n^2-1},1,Y_1,\ldots,Y_{j-2},(Y_{j-1}Y_{n^2})).$$

(Here and later $\mathrm{Tr}(X)$ denotes the trace of a matrix X; moreover an equation like

$$\mathrm{Tr}(f(X_i)) = g(X_i),$$

where f and g are polynomials in $K\langle X\rangle$ evaluated on matrices $X_i \in M_n(K)$, means that the left hand side is identified with the $n \times n$ scalar matrix $\mathrm{Tr}(f(X_i))\cdot I$, where I is the $n \times n$ identity matrix.)

3. REPRESENTATION THEORY OF THE SYMMETRIC GROUP AND GENERAL LINEAR GROUP

In the two sections after this one the main results are stated in terms of the representations of S_m, the symmetric group on m letters, and $GL(r,K)$, the group of invertible $r \times r$ matrices over K, so we will collect some standard facts here. These can be found in [26].

Let $K[S_m]$ be the group algebra of the symmetric group on m letters. It is isomorphic to the direct product

$$M_{m_1}(K) \times\cdots\times M_{m_k}(K)$$

of full matrix rings over K. The number k of factors is the number of distinct irreducible representations of S_m, and the numbers $m_1,\ldots,m_k$ are the dimensions of the corresponding simple modules. The number of simple factors is equal to the number of partitions of m. Moreover, there are combinatorial algorithms which associate representations with partitions in a natural way. Suppose

$$\lambda = (\lambda_1,\ldots,\lambda_s) \quad (\lambda_1 \geq\cdots\geq \lambda_s > 0,\ \lambda_1 +\cdots+ \lambda_s = m)$$

is a partition of m. Associated with λ is a <u>Young diagram</u> $D(\lambda)$, which is an array of m boxes, arranged in s rows, with λ_i boxes in the i-th row.

Example: $\lambda = (4,2,2,1)$ $D(\lambda)$ = 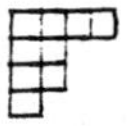

For our purposes it is unnecessary to know the algorithm for constructing a representation from a Young diagram. Note, however, that $K[S_m]$ $(m \geq 2)$ has exactly two one-dimensional representations, the trivial representation and the sign representation, and that these correspond to the two extreme partitions of m.

$\lambda = (m)$ $D(\lambda)$ =  Trivial representation.

$\lambda = (1,1,\ldots,1)$ $D(\lambda)$ = 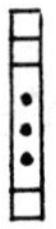 Sign representation.

The set of Young diagrams of all sizes is partially ordered by defining $D_1 \leq D_2$ if D_2 can be obtained from D_1 by adding boxes.

Lattice of Young Diagrams

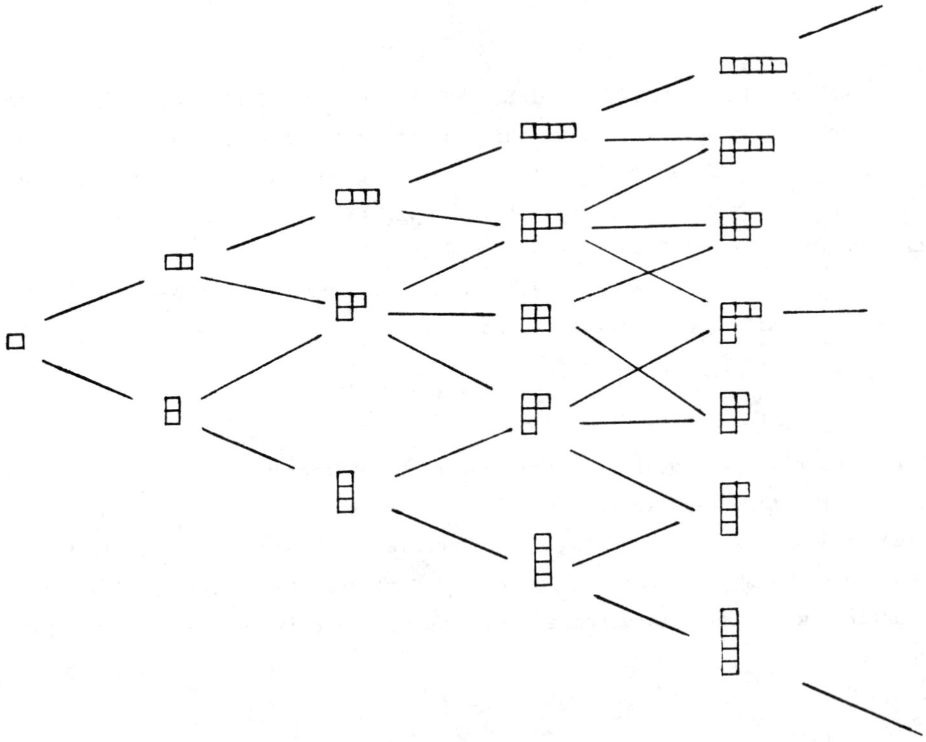

For any Young diagram $D = D(\lambda)$ of size m, let $M(D)$ denote the corresponding simple S_m-module. Note that

$$S_1 < S_2 < S_3 < \cdots .$$

If M is an S_m-module and $\ell < m$, then by restriction M is an S_ℓ-module. This module is denoted $M|S_\ell$. If $\ell > m$ one associates to M an induced module over S_ℓ, which is defined by

$$M^{S_\ell} = K[S_\ell] \otimes_{K[S_m]} M.$$

The important branching theorem describes restriction and induction in terms of the lattice of Young diagrams.

THEOREM 13. (BRANCHING THEOREM). Suppose $M(D)$ is an irreducible S_m-module. Let $C_1,\ldots,C_s$ be all the Young diagrams of size $m-1$ which precede D in the lattice of Young diagrams, and let $E_1,\ldots,E_t$ be all the Young diagrams of size $m+1$ which follow it. Then

(1) $M(D)|S_{m-1} \cong M(C_1) \oplus\cdots\oplus M(C_s)$

(2) $M(D)^{S_{m+1}} \cong M(E_1) \oplus\cdots\oplus M(E_t)$.

Remark: The two conclusions of the theorem are equivalent, by Frobenius reciprocity.

Any two-sided ideal in $K[S_n]$ is the direct product of some of the matrix algebras in the direct product decomposition of $K[S_m]$, and each such matrix algebra corresponds to a Young diagram. If we consider

$$K[S_1] < K[S_2] < K[S_3] < \cdots$$

and let $I(D)$ denote the minimal two-sided ideal of $K[S_m]$ corresponding to a diagram D, then the branching theorem implies that the ideal of $K[S] = \cup K[S_r]$ generated by $I(D)$ is generated as a vector space over K by the ideals $I(D')$ for which $D' \geq D$. Later the ideal of $K[S]$ generated by $D(\lambda)$, where $\lambda = (1,\ldots,1)$ corresponds to the sign representation of S_m will have great importance. In this case $D' \geq D(\lambda)$ if and only if D' has $\geq n$ rows.

We now turn to a description of the finite-dimensional representations of $GL(r,K)$. This also involves Young diagrams and is intimately connected to the representation theory of the symmetric group.

Let V be a vector space of dimension r -- that is, V is the standard $GL(r,K)$-module. Let $GL(r,K)$ act diagonally on $V^{\otimes m}$:

$$g(v_1 \otimes\cdots\otimes v_m) = gv_1 \otimes\cdots\otimes gv_m.$$

Let S_m act on $V^{\otimes m}$ by permuting positions:

$$\pi(v_1 \otimes\cdots\otimes v_m) = v_{\pi^{-1}(1)} \otimes\cdots\otimes v_{\pi^{-1}(m)}.$$

(The inverse is needed so that $\pi(\rho(v)) = (\pi\rho)(v)$.) It is clear that these two actions centralize each other, but more is true.

THEOREM 14. <u>Suppose</u> $GL(r,K)$ <u>and</u> S_m <u>act on</u> $V^{\otimes m}$ <u>as above. Let</u>
A = <u>Subalgebra of</u> $End(V^{\otimes m})$ <u>generated by</u> $GL(r,K)$.
B = <u>Subalgebra of</u> $End(V^{\otimes m})$ <u>generated by</u> S_m.

<u>Then</u> A = centralizer B, <u>and</u> B = centralizer A.

(Note that since A and B are semisimple, the double centralizer theorem implies that A = centralizer B if and only if B = centralizer A.)

Theorem 14 makes it possible to describe the structure of $V^{\otimes m}$ as a $GL(r,K)$-module. As noted above, the irreducible S_m-modules correspond to Young diagrams with m boxes. Let

$$M(D_1),\ldots,M(D_t)$$

be a full set of irreducible S_m-modules, and let

$$m_i = \dim_K(M(D_i))$$
$$n_i = \text{multiplicity of } M(D_i) \text{ in } V^{\otimes m}.$$

Then

$$V^{\otimes m} = U_1 \oplus\cdots\oplus U_t \cong n_1M(D_1) \oplus\cdots\oplus n_tM(D_t),$$

where U_i is the sum of all irreducible S_m-submodules of $V^{\otimes m}$ isomorphic to $M(D_i)$.

It follows from the fact that the subalgebras of $End(V^{\otimes m})$ generated by the actions of $GL(r,K)$ and S_m are the centralizers of one another that

(1) Each U_i is invariant under $GL(r,K)$.

(2) $V^{\otimes m} = U_1 \oplus\cdots\oplus U_t \cong m_1N(D_1) \oplus\cdots\oplus m_tN(D_t)$,

where $N(D_1),\ldots,N(D_t)$ are non-isomorphic irreducible $GL(r,K)$-modules (or zero) and U_i (if nonzero) is the sum of all irreducible $GL(r,K)$-submodules of $V^{\otimes m}$ isomorphic to $N(D_i)$.

(3) $\dim_K(N(D_i)) = n_i$.

The integers m_i and n_i can both be computed from the Young diagrams D_i by ingenious algorithms. The integers m_i are all nonzero because they are the dimensions of irreducible S_m-modules. But their multiplicities n_i in $V^{\otimes m}$ may be zero. In fact n_i is nonzero if and only if the Young diagram D_i has $\le r$ rows, where $r = \dim_K(V)$. Thus irreducible $GL(r,K)$-submodules of $V^{\otimes m}$ arise only from Young diagrams with $\le r$ rows.

THEOREM 15. *There is a 1-1 correspondence between Young diagrams with* m *boxes and* $\le r$ *rows and irreducible* $GL(r,K)$-*submodules of* $V^{\otimes m}$. *More precisely, as a* $GL(r,K)$-*module*

$$V^{\otimes m} \cong m_1 N(D_1) \oplus \cdots \oplus m_s N(D_s),$$

where $D_1,\ldots,D_s$ *are all Young diagrams with* m *boxes and* $\le r$ *rows,* $N(D_1),\ldots,N(D_s)$ *are non-isomorphic irreducible* $GL(r,K)$-*modules, and the multiplicity* m_i *of* $N(D_i)$ *in* $V^{\otimes m}$ *is the dimension of the irreducible* S_m-*module* $M(D_i)$.

It is easy to show that for distinct m, the irreducible $GL(r,K)$-modules which occur are non-isomorphic, because the action of $GL(r,K)$ on $V^{\otimes m}$ is given by a "polynomial map of degree m". More precisely, $V^{\otimes m}$ is a vector space of dimension mr, so the action of $GL(r,K)$ on V^{m} gives rise to a homomorphism

$$GL(r,K) \to GL(mr,K)$$

$$(a_{ij}) \to (f_{pq}(a_{ij})),$$

where the "coordinate functions" f_{pq} are polynomials of degree m. A basic result of the theory is that all finite-dimensional $GL(r,K)$-modules (essentially) arise from this construction. The only $GL(r,K)$ modules we will be interested in will be explicitly given as submodules or quotient modules of $V^{\otimes m}$. They will come from the action of $GL(r,K)$ on the free algebra $K\langle x_1,\ldots,x_r\rangle$ which can be identified with the tensor algebra

$$K \oplus V \oplus (V\otimes V) \oplus (V\otimes V\otimes V) \oplus \cdots \quad .$$

We conclude with an illustration of Theorem 15.

Example: $r = \dim_K V$ $\quad m = 4$

Young diagram	S_4-module	$\dim_K(M_i)$	$GL(r,K)$-module	$\dim_K(N_i)$
D_i	M_i	m_i	N_i	n_i
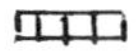	M_1	1	N_1	$\frac{1}{24}r(r+1)(r+2)(r+3)$

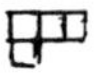	M_2	3	N_2	$\frac{1}{8}(r-1)r(r+1)(r+2)$
	M_3	2	N_3	$\frac{1}{12}(r-1)r^2(r+1)$
	M_4	3	N_4	$\frac{1}{8}(r-2)(r-1)r(r+1)$
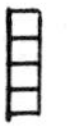	M_5	1	N_5	$\frac{1}{24}(r-3)(r-2)(r-1)r$

$$m_1n_1 + m_2n_2 + m_3n_3 + m_4n_4 + m_5n_5 = r^4 = \dim_K V^{\otimes 4}$$

The formula for n_i shows that $n_i = 0$ if the number of rows in D_i is larger than r.

4. TRACE IDENTITIES

One of the most important developments of the past decade in PI-theory was the discovery of trace identities for $n \times n$ matrices by Razmyslov [42] and Procesi [36]. These trace identities include ordinary polynomial identities as an identifiable proper subset and thus give an explicit description of the set of all polynomial identities of $n \times n$ matrices. Our treatment will be brief, and the reader who wishes to learn more is directed to the articles cited above and to a more recent survey by Procesi [37].

We will begin with Razmyslov's approach via the Cayley-Hamilton Theorem but state the main theorem in Procesi's terminology. Along the way we will give Razmyslov's proof of the Amitsur-Levitzki Theorem. It turns out that the case 2×2 matrices honestly represents the general case of $n \times n$ matrices so we will consider 2×2 matrices for simplicity.

Let $\mathrm{Tr}(X)$ and $\mathrm{Det}(X)$ denote the trace and determinant of a matrix X. Any $X \in M_2(K)$ satisfies its characteristic polynomial

$$\varphi(X) = X^2 - \mathrm{Tr}(X)X + \mathrm{Det}(X)I$$

This is almost what we will call a trace polynomial. We need to express the determinant of X in terms of traces, which can be done by using Newton's formulas to express the elementary symmetric functions in terms of the symmetric power functions. More precisely, let α and β be the characteristic roots of X. Then

$$\mathrm{Tr}(X) = \alpha + \beta, \quad \mathrm{Det}(X) = \alpha\beta, \quad \mathrm{Tr}(X^2) = \alpha^2 + \beta^2,$$

and

$$\mathrm{Tr}(X^2) = \mathrm{Tr}(X)^2 - 2\,\mathrm{Det}(X)$$
$$\mathrm{Det}(X) = \tfrac{1}{2}(\mathrm{Tr}(X)^2 - \mathrm{Tr}(X^2)).$$

We can therefore rewrite the characteristic polynomial of X:

$$\varphi(X) = X^2 - \mathrm{Tr}(X)X + \tfrac{1}{2}(\mathrm{Tr}(X)^2 - \mathrm{Tr}(X^2))I.$$

We will call expressions like the above, which involve both variables and traces of variables, _mixed trace polynomials_. Since we are more interested in multilinear trace polynomials we substitute $X = X_1 + X_2$ and form

$$\begin{aligned}\Phi(X_1,X_2) &= \varphi(X_1+X_2) - \varphi(X_1) - \varphi(X_2)\\ &= X_1X_2 + X_2X_1 - \mathrm{Tr}(X_1)X_2 - \mathrm{Tr}(X_2)X_1 + \mathrm{Tr}(X_1)\mathrm{Tr}(X_2) - \mathrm{Tr}(X_1X_2).\end{aligned}$$

This polynomial is multilinear as a function of X_1 and X_2 and vanishes if $X_1,X_2 \in M_2(K)$. It is what is called a _multilinear mixed trace identity_ for $M_2(K)$. Note that the "2" in the denominator in $\varphi(X)$ has disappeared from $\Phi(X_1,X_2)$. This is no accident.

At this point we follow Razmyslov [42] and use $\Phi(X_1,X_2)$ to show that $M_2(K)$ satisfies the standard identity of degree four. Substituting $X_1 = Y_1Y_2$, $X_2 = Y_3Y_4$ gives

$$\Phi(Y_1Y_2,Y_3Y_4) =$$

$$Y_1Y_2Y_3Y_4 + Y_3Y_4Y_1Y_2 - \mathrm{Tr}(Y_1Y_2)Y_3Y_4 - \mathrm{Tr}(Y_3Y_4)Y_1Y_2 + \mathrm{Tr}(Y_1Y_2)\mathrm{Tr}(Y_3Y_4) - \mathrm{Tr}(Y_1Y_2Y_3Y_4)$$

Now we make two crucial observations: First, considered as products of four letters, $Y_1Y_2Y_3Y_4$ and $Y_3Y_4Y_1Y_2$ differ by an _even_ permutation. Second, for any _even_ integer $2k$ and $n \times n$ matrices $Z_1,\dots,Z_{2k}$ of any size n, $\mathcal{S}_{2k}(Z_1,\dots,Z_{2k})$ is a matrix of trace zero. Next form the sum

$$\theta(Y_1,Y_2,Y_3,Y_4) = \sum_{\pi\in S_4} (\mathrm{sign}\,\pi)\Phi(Y_{\pi(1)}Y_{\pi(2)}Y_{\pi(3)}Y_{\pi(4)})$$

Our first observation shows that the monomials of Φ which do not involve traces will sum up to give $2\mathcal{S}_4(Y_1,Y_2,Y_3,Y_4)$. The second observation shows that the remaining terms will contribute zero when evaluated on matrices of any size. Since Φ vanishes on $M_2(K)$, we have proved

THEOREM 16. _If_ $Y_1,Y_2,Y_3,Y_4 \in M_2(K)$, _then_

$$2\mathcal{S}_4(Y_1,Y_2,Y_3,Y_4) = \theta(Y_1,Y_2,Y_3,Y_4) = 0.$$

This is the Amitsur-Levitzki Theorem for 2×2 matrices, and the same proof, starting with the characteristic polynomial of an $n \times n$ matrix, works for any n.

We now return to the multilinear characteristic polynomial $\Phi(X_1,X_2)$. Let us multiply $\Phi(X_1,X_2)$ by a new variable X_3 and take the trace. Reordering the monomials for a reason which will immediately become clear, we obtain a <u>multilinear pure trace polynomial</u> $\tau(X_1,X_2,X_3)$ which is satisfied by $M_2(K)$:

$$\tau(X_1,X_2,X_3) = \underset{(1)\ (2)\ (3)}{\mathrm{Tr}(X_1)\mathrm{Tr}(X_2)\mathrm{Tr}(X_3)} + \underset{(123)}{\mathrm{Tr}(X_1X_2X_3)} + \underset{(213)}{\mathrm{Tr}(X_2X_1X_3)}$$

$$- \underset{(1)\quad (23)}{\mathrm{Tr}(X_1)\mathrm{Tr}(X_2X_3)} - \underset{(2)\quad (13)}{\mathrm{Tr}(X_2)\mathrm{Tr}(X_1X_3)} - \underset{(3)\quad (12)}{\mathrm{Tr}(X_3)\mathrm{Tr}(X_1X_2)}$$

The form of τ motivates a definition.

<u>Definition</u>. Suppose $\pi \in S_m$, the symmetric group of permutations of $\{1,\ldots,m\}$, and let

$$\pi = (a_1,\ldots,a_i)(b_1,\ldots,b_j)(c_1,\ldots,c_k)\ldots$$

express π as a product of disjoint cycles (including 1-cycles). Then the <u>associated trace monomial</u> of π is T_π, where

$$T_\pi(X_1,\ldots,X_m) = \mathrm{Tr}(X_{a_1},\ldots,X_{a_i})\mathrm{Tr}(X_{b_1},\ldots,X_{b_j})\mathrm{Tr}(X_{c_1},\ldots,X_{c_k})\ldots$$

Note that although the expression for π as a product of disjoint cycles is not unique inasmuch as disjoint cycles may be permuted and any individual cycle can be replaced by a cyclic permutation, $T_\pi(X_1,\ldots,X_m)$ is uniquely defined as a function on matrices since traces commute and cyclic permutations of a monomial have the same trace.

In terms of our new definition the trace polynomial $\tau(X_1,X_2,X_3)$ becomes

$$\tau(X_1,X_2,X_3) = \sum_{\pi \in S_3} (\mathrm{sign}\ \pi)T_\pi(X_1,X_2,X_3)$$

It is clear that we can formally define pure trace polynomials so that we have a vector space isomorphism

$$K[S_m] \to \text{multilinear pure trace polynomials in } X_1,\ldots,X_m$$

$$\Sigma a_\pi \pi \to \Sigma a_\pi T_\pi(X_1,\ldots,X_m).$$

The trace polynomial $\tau(X_1,X_2,X_3)$ is satisfied by $M_2(K)$. It corresponds to the sign representation of S_3. More generally, $M_n(K)$ satisfies the pure trace polynomial in $n + 1$ variables which corresponds to the sign representation of S_{n+1}.

It is natural to ask which elements of $K[S_m]$ correspond to pure trace polynomials satisfied by $M_n(K)$. It is clear that they form a **vector space left ideal subspace in $K[S_m]$, invariant under conjugation, since**

$$T_{\sigma\pi\sigma^{-1}}(X_1,\ldots,X_n) = T_\pi(X_{\sigma(1)},\ldots,X_{\sigma(n)}).$$

In fact they form a two-sided ideal which is completely described by the following fundamental theorem.

THEOREM 17. (Razmyslov [42, Proposition 1], Procesi [36, Theorem 4.3]). Let $\Sigma a_\pi \pi \in K[S_m]$. Then $\Sigma a_\pi T_\pi(X_1,\ldots,X_m)$ is a pure trace polynomial satisfied by $M_n(K)$ if and only if $m \geq n + 1$ and $\Sigma a_\pi \pi$ belongs to the two-sided ideal of $K[S_m]$ generated by $\sum_{\pi \in S_{n+1}} (\text{sign } \pi)\pi$.

Equivalently, the ideal of $K[S_m]$ is the one associated with all Young diagrams which have at least $n + 1$ rows. In particular, there are no pure trace polynomials of degree $\leq n$ satisfied by $M_n(K)$ and a unique one (up to scalar multiplication) of degree $n + 1$ which $M_n(K)$ satisfies. This is precisely the pure trace polynomial obtained by multilinearizing the characteristic polynomial of an $n \times n$ matrix, multiplying by a new variable X_{n+1}, and taking the trace. It can be shown that all pure trace polynomials can be obtained by suitable operations on this basic trace polynomial and hence that all pure trace polynomials satisfied by $M_n(K)$ are consequences (in an appropriate sense) of the Cayley-Hamilton Theorem.

5. CODIMENSIONS AND COCHARACTERS

The object of this section is to describe how quantitative information about T-ideals, particularly about $\mathfrak{M}_n$, the T-ideal of identities of $M_n(K)$, can be obtained by studying the actions of S_m and $GL(r,K)$ on certain subspaces of $K\langle X\rangle$.

Definition. V_m = K-linear span of multilinear polynomials in $x_1,\ldots,x_m$. $W(r)$ = K-linear span of $x_1,\ldots,x_r$. (We will usually assume r is fixed and write W for $W(r)$).

The $m!$ monomials $\{x_{\pi(1)},\ldots,x_{\pi(m)} | \pi \in S_m\}$ form a basis for V_m, and

$$\Sigma a_\pi \pi \to \Sigma a_\pi x_{\pi(1)}\cdots x_{\pi(m)}$$

defines a vector space isomorphism $K[S_m] \to V_m$. Via this isomorphism, V_m is both a left and a right S_m-module. As with trace identities, left multiplication by elements of S_m corresponds to a substitution of variables. More precisely, if

$$\Sigma a_\pi \pi \to \Sigma a_\pi x_{\pi(1)} \cdots x_{\pi(m)} = f(x_1,\ldots,x_m),$$

then for any $\sigma \in S_m$

$$\Sigma a_\pi \sigma\pi \to \Sigma a_\pi x_{\sigma\pi(1)} \cdots x_{\sigma\pi(m)} = f(x_{\sigma(1)},\ldots,x_{\sigma(m)}).$$

This implies that for any T-ideal J, $J \cap V_m$ is a left S_m-module. On the other hand, right multiplication by elements of S_m corresponds to a place permutation on monomials which does not in general leave $J \cap V_m$ invariant. Hence we will only make use of the left action of S_m on V_m.

Since $W = W(r)$ is a vector space of dimension r, it is a module over $GL(r,K)$, and so is $W^{\otimes m}$ for any m, where $GL(r,K)$ acts diagonally. We will identify $K\langle x_1,\ldots,x_r\rangle$ with the graded tensor algebra

$$K \oplus W \oplus (W\otimes W) \oplus (W\otimes W\otimes W) \oplus \cdots ,$$

so that $GL(r,K)$ acts as a group of automorphisms of $K\langle x_1,\ldots,x_r\rangle$. It is clear that $J \cap W^{\otimes m}$ is invariant under the action of $GL(r,K)$.

PROBLEM 18: <u>For any T-ideal J, determine the structure of $J \cap V_m$ as an S_m-module, and the structure of $J \cap W^{\otimes m}$ as a $GL(r,K)$-module. In particular, do this for $J = \mathfrak{M}_n$, the T-ideal of identities of $M_n(K)$.</u>

Problem 18 is probably impossible to solve in complete generality and even looks out of reach for $\mathfrak{M}_n$ at present. The reader should interpret this problem as: Do what you can! We will pose more precise and less ambitious questions as we proceed.

In practice, it has proved more useful to study the factor modules

$$V_m/(J \cap V_m) \quad \text{and} \quad W^{\otimes m}/(J \cap W^{\otimes m}).$$

Furthermore, any finite-dimensional module over S_m or $GL(r,K)$ is determined up to isomorphism by its character (denoted χ_{S_m} for S_m and χ_{GL} for $GL(r,K)$), and the usual practice in the literature (which we will follow) is to use characters rather than modules.

<u>Definition</u>. Suppose J is a T-ideal and let $W = W(r)$, $GL = GL(r,K)$.

(1) The m-th <u>multilinear codimension</u> of J is $\dim_K[V_m/(J \cap V_m)]$.

(2) The m-th <u>multilinear cocharacter</u> (or <u>S_m-cocharacter</u>) of J is $\chi_{S_m}[V_m/(J \cap V_m)]$.

(3) The m-th homogeneous codimension of $\mathcal{J}$ in r variables is $\dim_K[W^{\otimes m}/(\mathcal{J} \cap W^{\otimes m})]$.

(4) The m-th homogeneous cocharacter (or GL-cocharacter) of $\mathcal{J}$ in r variables in $\chi_{GL}[W^{\otimes m}/(\mathcal{J} \cap W^{\otimes m})]$.

Any T-ideal J is generated as a T-ideal by the multilinear polynomials it contains (Theorem 2), so it is completely determined by the subspaces $J \cap V_m$. It therefore comes as no surprise that the S_m-cocharacters of $\mathcal{J}$ determine the GL(r,K)-cocharacters of J. What is surprising is the extremely simple and elegant relationship between them recently discovered independently by Berele and Drensky, which we now describe.

Recall (Section 3) that irreducible S_m-modules are indexed by Young diagrams with m boxes (or partitions of m) and irreducible GL(r,K)-submodules of $W^{\otimes m}$ are indexed by Young diagrams with m boxes and not more than r rows (or partitions of m into not more than r parts). For any Young diagram D with m boxes, let

$$\chi_{S_m}(D), \quad \chi_{GL}(D)$$

be the characters of the corresponding irreducible modules, with the convention that $\chi_{GL}(D) = 0$ if D has more than r rows.

THEOREM 19. (Berele [10, Theorem 2.7], Drensky [14, Lemma 1.1]). Let J be a T-ideal. Suppose the m-th multilinear cocharacter of J is

$$\chi_{S_m}[V_m/(J \cap V_m)] = \sum_D a(D)\chi_{S_m}(D),$$

where D ranges over all Young diagrams with m boxes and the a(D) are non-negative integers. Then for any r, the m-th homogeneous cocharacter of J in r variables is

$$\chi_{GL}[W^{\otimes m}/(J \cap W^{\otimes m})] = \sum_D a(D)\chi_{GL}(D).$$

Of course the theorem also implies that the GL(r,K)-cocharacter of J determines its S_m-cocharacter, provided $m \leq r$. They have also shown that the homogeneous codimensions have polynomially bounded growth.

THEOREM 20. (Berele [10, Corollary 4.12], Drensky [14, Proposition 1.2]). Suppose that J is a nonzero T-ideal and let W = W(r), where r is fixed. Then there is a constant c and an integer d such that for all m

$$\dim_K[W^{\otimes m}/(J \cap W^{\otimes m})] < c\, m^d.$$

Berele and Drensky have obtained other interesting quantitative results. Drensky [14, Theorem 2.1] has computed the homogeneous codimensions for $\mathfrak{J} = \mathfrak{M}_2$, the T-ideal of identities of $M_2(K)$, and all r. It is possible to define a more refined homogeneous codimension for a T-ideal which reflects the multigrading of $K\langle x_1,\ldots,x_r\rangle$, not just total degree. Berele [10, Corollary 3.4] has shown that the multigraded codimensions determine the GL(r,K)-cocharacter of J. These multigraded codimensions were computed for $\mathfrak{M}_2$ and $r = 2$ by Formanek, Halpin and Li [18, Theorem 9] and by Drensky [14, Theorem 2.2], and using them Berele [10, Theorem 3.5] computed the GL(2,K)-cocharacters of $\mathfrak{M}_2$. We will give the precise codimensions and cocharacters in Section 7, where we carefully analyze the ring of 2×2 generic matrices.

The above results on homogeneous codimensions and GL-cocharacters are very recent and I believe they are the first steps in a promising new direction. The rest of this section will be devoted to multilinear codimensions and S_m-cocharacters, which have been studied much more extensively, mainly by Regev. Indeed, the formalism of codimensions and cocharacters is due to him. Codimensions originated in his seminal article [43] where he showed that the tensor product of two PI-rings is a PI-ring. The following result is crucial to the proof.

THEOREM 21. (Regev [43], Latyshev [31]). _Let_ $\mathcal{J}$ _be a T-ideal. If_ $\mathcal{J}$ _contains a polynomial of degree_ d, _then_

$$\dim_K[V_m/(\mathcal{J} \cap V_m)] \leq (d-1)^{2m}.$$

The original bound of Regev was $(3\cdot 4^{d-3})^m$ and Latyshev later obtained the much better bound $(d-1)^{2m}$. For the purpose of proving the tensor product theorem it is sufficient to bound $\dim_K[V_m/(\mathcal{J} \cap V_m)]$ by c^m for some constant c.

THEOREM 22. (Regev [43]. See [56, p. 239-242]). _If_ A _and_ B _are PI-algebras, so is_ $A \otimes_K B$.

Outline of Proof: It clearly suffices to prove the theorem in case

$$A = K\langle X\rangle/\mathcal{I}, \quad B = K\langle X\rangle/\mathcal{J},$$

where $\mathcal{I}$ and $\mathcal{J}$ are nonzero T-ideals. It can be shown (although it is not obvious) that $A \otimes_K B$ will satisfy a PI provided that for some m

$$\dim_K[V_m/(\mathcal{I} \cap V_m)]\cdot\dim_K[V_m/(\mathcal{J} \cap V_m)] < n!$$

Theorem 21 guarantees that this happens.

In the preceding section we gave the Razmyslov-Procesi description of all trace identities satisfied by $M_n(K)$ which also involved an identification with the group algebra of the symmetric group. We can relate ordinary polynomial identities of $M_n(K)$ to trace identities via an observation first made by Kostant in his proof of the Amitsur-Levitzki Theorem [30]. Recall that trace is a non-degenerate bilinear form on $M_n(K)$; that is, if $U \in M_n(K)$, then $U = 0$ if and only if $Tr(UZ) = 0$ for all $Z \in M_n(K)$. This implies the following relation between ordinary identities and pure trace identities for $M_n(K)$.

THEOREM 23. <u>Suppose</u> $f(x_1,\ldots,x_m) \in K\langle X\rangle$. <u>Then</u> $f(x_1,\ldots,x_m)$ <u>is a</u> PI <u>for</u> $M_n(K)$ <u>if and only if</u> $Tr(f(x_1,\ldots,x_m)x_{m+1})$ <u>is a pure trace identity for</u> $M_n(K)$.

Now consider the following diagram where A and C are vector space isomorphisms and B is one-to-one:

$$\begin{array}{ccl} K[S_m] & \xrightarrow{A} & V_m = \text{multilinear polynomials in } x_1,\ldots,x_m \\ & & \qquad\qquad \downarrow B \\ K[S_{m+1}] & \xrightarrow{C} & \text{multilinear pure trace polynomials in } x_1,\ldots,x_{m+1} \end{array}$$

The maps A,B,C are defined by

$$A(\Sigma a_\pi \pi) = \Sigma a_\pi x_{\pi(1)}\cdots x_{\pi(m)}$$

$$B(f(x_1,\ldots,x_m)) = Tr(f(x_1,\ldots,x_m)x_{m+1})$$

$$C(\Sigma a_\pi \pi) = \Sigma a_\pi T_\pi(x_1,\ldots,x_{m+1})$$

A little effort shows that the composition $C^{-1}BA:K[S_m] \to K[S_{m+1}]$ is given by

$$\Sigma a_\pi \pi \to \Sigma a_\pi \pi(12\ldots m+1)\pi^{-1}.$$

In other words, the image of $C^{-1}BA$ is the linear subspace of $K[S_{m+1}]$ spanned by the (m+1)-cycles. The Razmyslov-Procesi theorem tells us that if $Z \in K[S_{m+1}]$ then $C(Z)$ is a trace identity for $M_n(K)$ if and only if $m \geq n$ and Z lie in the ideal of $K[S_{m+1}]$ generated by $\Sigma\{\pm\pi \mid \pi \in S_{n+1}\}$. This makes the calculation of multilinear codimensions and cocharacters for $\mathfrak{M}_n$ a technical problem but so far it has not been possible to find closed formulas for all the codimensions and cocharacters of $\mathfrak{M}_n$. There seem to be two major obstacles.

The first difficulty is that we do not have sufficiently good exact formulas for the dimension of certain ideals in $K[S_m]$. To be precise, let $D(m,n)$ denote the two-sided ideal of $K[S_m]$ corresponding to Young diagrams with $\leq n$ rows. Then $\dim_K[D(m,n)]$ deserves to be called the m-th multilinear codimension for pure trace identities for $M_n(K)$.

PROBLEM 24. <u>Find an explicit formula for</u> $\dim_K[D(m,n)]$ <u>as a function of</u> m <u>with</u> n <u>fixed</u>.

It is known that

$$\dim_K[D(m,1)] = 1 \quad \text{and} \quad \dim_K[D(m,2)] = \frac{1}{m+1}\binom{2m}{m},$$

but for $n \geq 3$ no such closed formula is known. Incidentally, the generating functions for $n = 1,2$ represent quite reasonable analytic functions:

n = 1.
$$1 + \Sigma \dim_K[D(m,1)]t^m =$$
$$1 + t + t^2 + \cdots = \frac{1}{1-t}$$

n = 2.
$$1 + \Sigma \dim_K[D(m,2)]t^m =$$
$$1 + t + 2t^2 + 5t^3 + \cdots = \frac{1-\sqrt{1-4t}}{2t} .$$

The second difficulty in applying the Razmyslov-Procesi Theorem to ordinary PI's for $M_n(K)$ is that the ordinary PI's correspond (via the map $C^{-1}B$ in the diagram after Theorem 23) to a subspace of the linear span of the (m+1)-cycles in $K[S_{m+1}]$ and this set does not sit very nicely in $K[S_{m+1}]$.

Instead of looking for exact formulas for the codimensions of $\mathfrak{M}_n$ or for $\dim_K[D(m,n)]$ one can study their asymptotic behaviour. For gaining information about the identities of matrices or the structure of the ring of generic matrices, the asymptotic behaviour of the codimensions may be more useful than exact formulas.

<u>Definition.</u> Let ϕ and θ be real-valued functions on the positive integers. Then ϕ and θ are <u>asymptotically equal</u> (denoted $\phi \sim \theta$) if

$$\lim_{m\to\infty}[\phi(m)/\theta(m)] = 1.$$

Regev [48, Theorem 2.10] has determined the asymptotic behaviour of $\dim_K[D(m,n)]$ as a function of m with n fixed. It seems reasonable that the multilinear codimensions of $\mathfrak{M}_n$ will exhibit the same asymptotic behaviour.

PROBLEM 25. *Show that for fixed* n,

$$\dim_K[V_m/(M_n \cap V_m)] \sim \dim_K[D(m+1,n)].$$

If $n = 1$, both dimensions are equal to one for all m. The only non-trivial case for which Problem 25 has been solved is $n = 2$.

THEOREM 26. (Regev [53, Theorem 2.2]).

$$\dim_K[V_m/(M_2 \cap V_m)] \sim \dim_K[D(m+1,2)] \sim \frac{4^{m+1}}{m\sqrt{m}} .$$

In a series of articles [43-53], Regev has obtained purely algebraic results from information about codimensions and cocharacters, and has developed techniques for making asymptotic estimates for codimensions and cocharacters of $\mathfrak{M}_n$. His methods involve algebra, combinatorics and analysis-including the evaluation of integrals. For the reader who wants to learn more, [6], [46], [53], and the first section of [47] are expository. Among the results Regev has obtained are the following. (In Theorem 27, $K[S_m] \overset{A}{\to} V_m$ is the usual isomorphism.)

THEOREM 27. (1) [44, Theorem 4.2]. *Let* $\mathcal{J}$ *be any nonzero T-ideal. Then for sufficiently large* m, $A^{-1}(\mathcal{J} \cap V_m)$ *contains a nonzero two-sided ideal of* $K[S_m]$.

(2) [45, Theorem 3]. $A^{-1}(\mathfrak{M}_n \cap V_m)$ *contains the two-sided ideal of* $K[S_m]$ *corresponding to Young diagrams with more than* n^2 *rows. Equivalently, the* S_m*-cocharacter of* M_n *only involves characters corresponding to Young diagrams with* $\le n^2$ *rows.*

(3) [45, Theorem 2] (with Amitsur). *Let* $\mathcal{J}$ *be a T-ideal. Then the Capelli polynomial* C_{r+1} *(defined in Section 2) lies in* $\mathcal{J}$ *if and only if for all* m, *the* S_m*-cocharacter of* $\mathcal{J}$ *only involves characters corresponding to Young diagrams with* $\le r$ *rows.*

(4) [49, Theorem (6)]. *Let* A *and* B *be K-algebras. If* A *satisfies* C_{r+1} *and* B *satisfies* C_{s+1}, *then* $A \otimes_K B$ *satisfies* C_{rs+1}.

Parts (3) and (4) have added significance since A. R. Kemer [29] has recently proved the important result that every finitely generated PI-algebra satisfies some Capelli polynomial.

Regev has also obtained asymptotic lower and upper bounds for the multilinear codimensions of $\mathfrak{M}_n$ [47], although only for $\mathfrak{M}_1$ and $\mathfrak{M}_2$ are there sharp estimates (Theorem 26).

Finally, we mention a result which may be of use in trying to show that T-ideals are finitely generated.

THEOREM 28. (Razmyslov [41, Proposition 2]). Let J be a T-ideal and suppose that the Capelli polynomial C_{r+1} lies in J. Then J is generated as a T-ideal by C_{r+1} and $J \cap K\langle x_1,\ldots,x_r\rangle$.

6. THE RING OF GENERIC MATRICES

Recall from the introduction that the ring of $n \times n$ generic matrices is a concrete realization of the relatively free algebra $K\langle X\rangle/\mathfrak{M}_n$, where $\mathfrak{M}_n$ is the ideal of polynomial identities satisfied by $M_n(K)$. It is the K-subalgebra $K[U(1), U(2),\ldots]$ of $M_n(K[u_{ij}(r)])$ generated by the $n \times n$ generic matrices $U(r) = (u_{ij}(r))$, where

$$\{u_{ij}(r) \mid 1 \leq i,j \leq n,\ r = 1,2,\ldots\}$$

are independent commuting indeterminates over K. We will sometimes consider subalgebras generated by a finite number $r \geq 2$ of generic matrices. It turns out that they have all the essential properties of the generic matrix ring generated by countably many generic matrices.

We have already mentioned (Theorem 5) one of the fundamental properties of the ring of generic matrices: It has no zero divisors. This implies that it is a prime PI-ring, so Posner's Theorem, a basic structure theorem of PI-theory, applies.

THEOREM 29. (Posner [33], see [56, p. 53]). Suppose that R is a prime PI-ring with center C. Let $S = C - \{0\}$ be the monoid of nonzero elements of C. Then $S^{-1}R$ is a finite dimensional central simple algebra with center $S^{-1}C$, the field of quotients of C. Moreover, R and $S^{-1}R$ satisfy exactly the same polynomial identities, namely those in $\mathfrak{M}_n$ for some n, and the dimension of $S^{-1}R$ over $S^{-1}C$ is n^2.

Since the ring of $n \times n$ generic matrices has no zero divisors, the quotient ring obtained from it via Posner's Theorem is a division ring.

Definition. The generic division ring is the quotient ring of the ring of generic matrices.

For the rest of this section we will use the following notation:

$R = K[U(1),U(2),\ldots]$ = ring of generic matrices

C = center of R

$Q(R) = K(U(1),U(2),\ldots)$ = generic division ring

$Q(C)$ = field of quotients of C = center of $Q(R)$

Note that there are inclusions

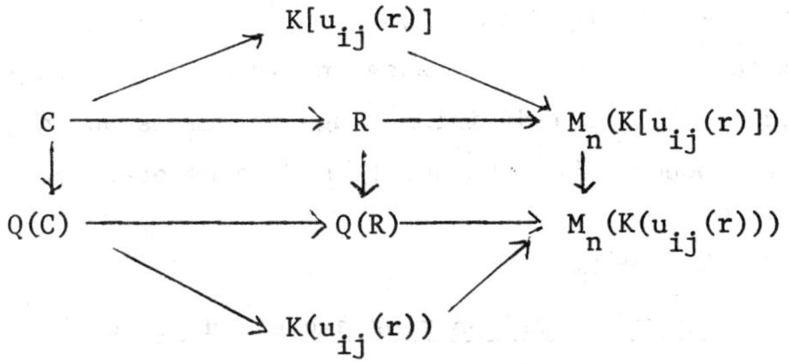

and we consider R and $Q(R)$ to be concretely realized as subrings of $M_n(K(u_{ij}(r))$. Elements of C and $Q(C)$ are then scalar matrices zI, where $z \in K(u_{ij}(r))$, and often z and zI will be identified.

The most important application of the ring of generic matrices is Amitsur's construction of division algebras which are not crossed products [3]. Recall that a division ring D of dimension n^2 over its center F is said to be a <u>crossed product with Galois group</u> G if D has a maximal subfield L such that L is Galois over F and G is the Galois group of L over F. Such a group G is necessarily of order n, since all maximal subfields of D are n-dimensional over F. The group G may not be unique in the sense that D may have a Galois maximal subfield L' such that the Galois group of L' over F is not isomorphic to G.

Amitsur's method relies on the following two steps:

I. If the $n \times n$ generic division ring $Q(R)$ is a crossed product relative to a certain group G, then all division algebras of dimension n^2 over their centers are crossed products relative to the same group G.

II. Suppose $p^3 | n$ if $p = 2$ or $p^2 | n$ for some odd prime p. Then there exist division algebras D_1, D_2 of dimension n^2 over their centers which satisfy:

(A) D_1 is a crossed product, but only with respect to a group G_1 whose Sylow p-subgroup is cyclic.

(B) D_2 is a crossed product, but only with respect to a group G_2 whose Sylow p-subgroup is elementary abelian.

It is easy to see that (1) and (2) together imply

THEOREM 30. (Amitsur [3]). *Let* $K = Q$, *the field of national numbers. Suppose* $8|n$ *or* $p^2|n$ *for an odd prime* p. *Then the* $n \times n$ *generic division ring* $Q(R) = K(U(1),U(2),\ldots)$ *is not a crossed product.*

The restriction $K = Q$ is used in proving (II). Without this assumption Theorem 30 has been proved with $p^3|n$.

Amitsur's method does not apply when n is a prime. It is classical that for $n = 1,2,3,4,6,12$, all division rings of dimension n^2 over their centers are crossed products, but the question remains open for primes $p \geq 5$.

PROBLEM 31. *Let* p *be a prime,* $p \geq 5$.

(1) *Is the* $p \times p$ *generic division ring a crossed product?*

(2) *Does the* $p \times p$ *generic division ring contain an element* z *such that* z *is not in its center but* z^p *is?*

(3) *Does there exist a* p-th *power central polynomial for* $M_p(K)$? *More precisely, does there exist* $f \in K\langle X\rangle$ *such that* f *is not a central polynomial for* $M_p(K)$ *but* f^p *is?*

The three parts of the problem are equivalent. The first two are equivalent by virtue of the Skolem-Noether Theorem. If L is a maximal subfield of $Q(R)$ which is Galois over its center $Q(C)$, then the Galois group of L over $Q(C)$ can only be cyclic of order p. The Skolem-Noether Theorem asserts that any automorphism of L over $Q(C)$ is realized by conjugation by some $z \in Q(R)$. This gives rise to a non-central element z in $Q(R)$ such that z^p is central.

By Posner's Theorem, the element z may be chosen from R, the ring of $n \times n$ generic matrices. Then we get the equivalence of (2) and (3) by observing that the isomorphism

$$R = K[U(1),U(2),\ldots] \cong K\langle X\rangle/\mathfrak{M}_n$$

identifies the center C of R with

$$\{\text{central polynomials}\}/\mathfrak{M}_n .$$

We will now give several descriptions of $Q(C)$, which is both the field of quotients of C and the center of $Q(R)$. It is easier to describe than C itself. Later we will see that important questions about the Brauer group of a field are related to the structure of $Q(C)$.

One description is in terms of the notion of invariants of $n \times n$ matrices. Consider $K[u_{ij}(r)]$, the commutative polynomial ring generated by the entries of all the generic matrices $U(1), U(2), \ldots$. There is a group action of $GL(n,K)$ - really of $PGL(n,K)$ - on $K[u_{ij}(r)]$ as follows: If $P \in GL(n,K)$ and $U(r) = (u_{ij}(r))$ is an $n \times n$ generic matrix, let

$$PU(r)P^{-1} = (\bar{u}_{ij}(r)),$$

where the $\bar{u}_{ij}(r)$ are K-linear combinations of the $u_{ij}(r)$. Then $u_{ij}(r) \to \bar{u}_{ij}(r)$ induces automorphisms of $K[u_{ij}(r)]$ and its field of quotients $K(u_{ij}(r))$ which we denote by $f \to f^P$.

<u>Definition</u>. A polynomial $f(u_{ij}(r)) \in K[u_{ij}(r)]$ is an <u>invariant</u> of $n \times n$ matrices if $f^P = f$ for all $P \in GL(n,K)$. A rational function $\varphi \in K(u_{ij}(r))$ is called a <u>rational invariant</u> of $n \times n$ matrices if $\varphi^P = \varphi$ for all $P \in GL(n,K)$. The set of all invariants (rational invariants) is called the <u>ring (field) of invariants of</u> $n \times n$ <u>matrices</u>. They are denoted, respectively, $K[u_{ij}(r)]^{GL}$ and $K(u_{ij}(r))^{GL}$.

It is not hard to show that the field of invariants is the field of quotients of the ring of invariants.

If we only have a single $n \times n$ generic matrix $U(1)$ - in other words, if we restrict the action of $GL(n,K)$ to the polynomial ring $K[u_{ij}(1)]$ in n^2 commuting variables, then it is a classical result that the ring of invariants $K[u_{ij}(1)]^{GL}$ is a polynomial ring in n variables, the coefficients of the characteristic polynomial of $U(1)$. In characteristic zero Newton's formulas show that all coefficients of the characteristic polynomial of $U(1)$ can be expressed in terms of the n traces

$$Tr(U(1)), Tr(U(1)^2), \ldots, Tr(U(1)^n).$$

Hence $K[u_{ij}(1)]^{GL}$ is also the polynomial ring generated by these traces.

More generally, suppose $f = f(U(1), \ldots, U(m))$ is any element of R, the ring of generic matrices. Then $Tr(f) \in K[u_{ij}(r)]$, and

$$\begin{aligned}(Tr[f(U(1),\ldots,U(m))])^P &= Tr[f(PU(1)P^{-1},\ldots,PU(m)P^{-1})] \\ &= Tr[f(U(1),\ldots,U(m))],\end{aligned}$$

so $Tr(f) \in K[u_{ij}(r)]^{GL}$. Conversely, all invariants can be expressed in terms of traces.

THEOREM 32. (Gurevich [19, Theorem 16.2], Siberskii [60, Theorem 1], Procesi [36, Theorem 1.3]). _The ring of invariants of_ $n \times n$ _matrices_, $K[u_{ij}(r)]^{GL}$, _is generated by all traces_ $Tr[U(i_1)\ldots U(i_n)]$ _of products of generic matrices_.

The attribution of the theorem deserves some explanation. The so-called "First Fundamental Theorem" of vector invariants [19, Theorem 16.2] gives a generating set for the invariants of m vectors and m covectors - i.e. invariants of $GL(n,K)$ acting on the symmetric algebra of

$$(V^{\otimes m}) \otimes (V^*)^{\otimes m},$$

where V is an n-dimensional vector space, the standard $GL(n,K)$-module. This theorem is quite old but the first complete proof seems to be that of Gurevich. Theorem 32 as stated is a translation of the First Fundamental Theorem originaly due to Siberskii and rediscovered by Procesi, using the following dictionary:

$$V \otimes V^* \cong M_n(K)$$

$$\text{Vector invariant} = \text{trace}.$$

For the details, see [36, pp. 310-313].

Now suppose that $f = f(U(1),\ldots,U(m))$ is a central polynomial for $M_n(K)$ or, more precisely, the evaluation of a central polynomial on $n \times n$ generic matrices $U(1),\ldots,U(m)$. Then $f = zI$ for some $z \in K[u_{ij}(r)]$. If $P \in GL(n,K)$, then

$$\begin{aligned} z^P I &= f(PU(1)P^{-1},\ldots,PU(m)P^{-1}) \\ &= Pf(U(1),\ldots,U(m))P^{-1} \\ &= P(zI)P^{-1} = zI, \end{aligned}$$

so $z \in K[u_{ij}(r)]^{GL}$. In other words, the center C of the ring of $n \times n$ generic matrices is a subring of the ring of invariants of $n \times n$ matrices.

Next consider $Q(R)$, the generic division ring. It is central simple of dimension n^2 over its center $Q(C)$ and the theory of central simple algebras tells us that there is a natural map $Q(R) \to Q(C)$, called the _reduced trace_. In fact it is just our trace Tr, so every invariant is a quotient of elements of C. Combining the above, we have

THEOREM 33. _Let_ $R = K[U(r)] \le M_n(K[u_{ij}(r)])$ _be the ring of_ $n \times n$ _generic matrices and let_ C _be its center_. _Then_

$$C \subseteq \text{Ring generated by traces of elements of } R$$
$$= \text{Ring of invariants} = K[u_{ij}(r)]^{GL}.$$

Moreover, both rings have the same field of quotients, namely $Q(C)$, *the center of* $Q(R)$.

For 1×1 matrices, $R = C$ = ring of invariants. But for $n \times n$ matrices, $n \geq 2$, $Tr(U(1))$ is an invariant but not an element of C, so C is then a proper subring of the ring of invariants.

PROBLEM 34. *Give presentations (generators and relations) for* C *and/or the ring of invariants.*

Procesi [36, Theorem 3.3] has shown that the ring of invariants is generated by the traces of monomials of degree $\leq 2^n - 1$, but not all of them are needed. The Razmyslov-Procesi Theorem on trace identities gives all relations among traces in principle, but Problem 34 asks for something more explicit. On the other hand, very little is known about identifying C as a subring of the ring of invariants. Even finding the lowest degree nonconstant polynomials in C amounts to solving Problem 8 on central polynomials of minimal degree. Only for the ring generated by two 2×2 generic matrices is our knowledge really complete - see Section 7.

Finally we give one last description of $Q(C)$ which seems the most useful for making explicit calculations. It is obtained by an argument of Procesi [34] as elaborated by Formanek [16]. For simplicity we reduce to two $n \times n$ generic matrices $X = (x_{ij})$ and $Y = (y_{ij})$. If we pass to an algebraic closure of $K(x_{ij}, y_{ij})$, then the matrix X is conjugate to a diagonal one. The conjugation is an isomorphism of K-algebras, so the K-algebra generated by

$$X = \begin{pmatrix} x_1 & & & & 0 \\ & \cdot & & & \\ & & \cdot & & \\ & & & \cdot & \\ 0 & & & & \cdot \\ & & & & & x_n \end{pmatrix}, \quad Y = (y_{ij})$$

(where the nonzero entries are independent commuting indeterminates over K) is isomorphic to the original generic matrix ring and now its center can be described as the fixed field of S_n rather than $GL(n,K)$.

THEOREM 35. (Formanek [16, Theorem 3]). *Suppose* R *is a generic matrix ring generated by two* $n \times n$ *generic matrices, and* C, $Q(C)$ *and* $Q(R)$

are as above. Let $\{x_i, y_{ij} | 1 \le i,j \le n\}$ *be independent commuting indeterminates over* K *and let* L *be the subfield of* $K(x_i, y_{ij})$ *generated by*

$$\{x_i, y_{ii}, y_{ij}y_{ji}, y_{ij}y_{jk}y_{ki} | 1 \le i,j,k \le n\}.$$

Then:

(1) L *is a rational function field (pure transcendental extension) over* K *of transcendence degree* $n^2 + 1$.

(2) *The action of* S_n *on* $\{x_i, y_{ij}\}$ *defined by*

$$\pi(x_i) = x_{\pi(i)}, \ \pi(y_{ij}) = y_{\pi(i)\pi(j)}$$

induces an action of S_n *as a group of* K-*automorphisms of* L.

(3) *The center of the generic division ring,* $Q(C)$, *is isomorphic to the fixed field of* S_n *acting on* L.

PROBLEM 36. *Is* $Q(C)$, *the center of the* $n \times n$ *generic division ring, a rational function field over* K?

Problem 36 has been solved affirmatively only for $n \le 4$. For $n = 1$ it is trivial and for $n = 2$ it is due to Procesi [34, Theorem 2.2] although an equivalent statement can be found in an article of Sylvester [61] written in 1883. For $n = 3,4$, it was proved by Formanek [16,17] using the description of $Q(C)$ in Theorem 35.

A subfield (containing K) of a rational function field over K is said to be *unirational* over K. Only in the early 1970's were unirational fields which are not rational first shown to exist over $\mathbb{C}$, giving negative solution to the so-called Lüroth problem. The identifications

$$Q(C) = K(u_{ij}(r))^{GL} \cong L^{S_n} \qquad \text{(Theorem 36)},$$

show that the center of the generic division ring is unirational over K. If it is not rational, taking $K = \mathbb{C}$ provides a naturally occurring counter-example to the Lüroth problem. Rosset [54] attempted to show that the rationality of $Q(C)$ has implications for Brauer groups which are known to be false. Although his argument was faulty, Snider [59] showed that his idea has merit, in that the rationality of $Q(C)$ does have implications - albeit weaker - for Brauer groups. Consequently we have neither a theorem nor a counter-example, but a new way of looking at a major open problem on the generation of the Brauer group of a field.

We will give Snider's result as modified by Procesi [38] so as to apply to the generic division ring rather than the "generic crossed product" defined by Snider. The expository article [38] gives a more detailed account of what we are discussing here and includes some proofs.

The starting point is a theorem of Bloch [11, Theorem 1.1] on the m-th-power norm residue symbol, which is a natural map from K_2 of a field K to its Brauer group, which is defined when K contains a primitive m-th root of unity. His theorem has the following corollary. (Recall that a central simple algebra is _cyclic_ if it is a crossed product with a cyclic Galois group.)

THEOREM 37. [38, Teorema 6]. _Suppose that K is a field which contains all roots of unity and L is a rational function field over K. If the Brauer group of K is generated by cyclic algebras, so is the Brauer group of L._

The next step is a suitable universal property for the generic division ring.

THEOREM 38. ([38, Corollario 8]. Compare [59, Theorem 2]). _Suppose that the $n \times n$ generic division ring Q(R) is equivalent to a product of cyclic algebras in the Brauer group of its center Q(C). Then every central simple algebra of dimension n^2 over its center L, where L contains K, is equivalent to a product of cyclic algebras in the Brauer group of L._

It is clear that these two theorems imply that if Q(C) is a rational function field over K (Problem 36) then the following basic problem has a positive solution.

PROBLEM 39. _Suppose K is a field which contains all roots of unity. Is the Brauer group of K generated by cyclic algebras?_

7. EXAMPLE: THE RING OF 2×2 GENERIC MATRICES

In this final section we will analyze in detail the generic matrix ring generated by two 2×2 generic matrices, in the hope of illuminating what has come before. We follow Formanek, Halpin, and Li [18]. A similar treatment of the corresponding generic division ring was made by Procesi [34]. In fact, the main ideas are implicit in Sylvester [61].

The ring generated by two 2×2 generic matrices is the smallest non-commutative generic matrix ring and also the only one for which a detailed analysis has been made. A similar analysis for three or more 2×2 generic matrices appears substantially more difficult and 3×3 matrices present an even higher order of difficulty.

Notation will be a slight modification of previous usage. The two generic matrices will be

$$X = \begin{pmatrix} x_{11} & x_{12} \\ x_{21} & x_{22} \end{pmatrix} \qquad Y = \begin{pmatrix} y_{11} & y_{12} \\ y_{21} & y_{22} \end{pmatrix}$$

The ring of generic matrices $K[X,Y]$, its center, and their quotient rings will be denoted, respectively, R, C, $Q(R)$, $Q(C)$, as in Section 6. The appropriate free ring will now be $K\langle x,y\rangle$ and there is now an isomorphism

$$R = K[X,Y] \cong K\langle x,y\rangle/\bar{\mathfrak{m}}_2,$$

where $\bar{\mathfrak{m}}_2$ denotes the ideal of two-variable polynomial identities for $M_2(K)$. As earlier, we regard R and $Q(R)$ as subrings of $M_2(K(x_{ij},y_{ij}))$.

Our first goal is to find the ring of invariants of R. We know that $Tr(X)$, $Tr(Y)$, $Det(X)$, $Det(Y)$, and $Tr(XY)$ are all invariants, and it is not hard to show that they are algebraically independent over K. We want to prove that

$$B = K[Tr(X),Tr(Y),Det(X),Det(Y),Tr(XY)]$$

is the full ring of invariants of R, a fact already known to Sylvester [61].

By Posner's Theorem, $Q(R)$ is a four-dimensional division algebra over $Q(C)$. The fact that a certain 4×4 determinant is nonzero shows that I, X, Y, XY form a basis for $Q(R)$ over $Q(C)$. We give this determinant explicitly since it is a basic invariant. Let $XY = (z_{ij})$. Then

$$\mathrm{Det}\begin{pmatrix} 1 & 0 & 0 & 1 \\ x_{11} & x_{21} & x_{12} & x_{22} \\ y_{11} & y_{21} & y_{12} & y_{22} \\ z_{11} & z_{21} & z_{12} & z_{22} \end{pmatrix} = \mathrm{Det}(XY-YX) = (XY-YX)^2$$

$$= Tr(XY)^2 - Tr(X)Tr(Y)Tr(XY) + Tr(X)^2Det(Y) + Tr(Y)^2Det(X) - 4\,Det(X)Det(Y).$$

Set

(*) $\quad S = BI + BX + BY + BXY,$

the free left B-module on the indicated four generators. We claim that S is a ring. To see this it suffices to show that products of pairs of the basis elements I, X, Y, XY lie in S. The ordinary characteristic polynomial in one variable takes care of all such products except YX (e.g. $X^2 = \mathrm{Tr}(X)X - \mathrm{Det}(X)$). To show that $YX \in S$, rewrite the equation $\Phi(X,Y) = 0$, where Φ is the multilinear characteristic polynomial derived in Section 4, as

$$YX = \mathrm{Tr}(XY) - \mathrm{Tr}(X)\mathrm{Tr}(Y) + \mathrm{Tr}(Y)X + \mathrm{Tr}(X)Y - XY.$$

Thus $YX \in S$, so S is a ring, as claimed. Since S is a ring, it contains R, so (*) shows that the trace of every element of R lies in B, which implies that B is the ring of invariants of R. Moreover, S = BR is the ring obtained by adjoining the invariants of R to R.

We would like to identify R as a subring of S, and its center C as a subring of B. A series of identities again derived from the Cayley-Hamilton Theorem

$$\text{(e.g.} \qquad (XY-YX)Y = \mathrm{Tr}(Y)(XY-YX) - Y(XY-YX))$$

show that the commutator ideal of R has a simple description in terms of S, namely

$$[R,R] = R(XY-YX)R = S(XY-YX) = [S,S]$$

Summing up the above and analyzing the center of R by similar means, we have

THEOREM 40. (see [18]). *Let* R = K[X,Y] *be the ring generated by two* 2 × 2 *generic matrices. Then*

(1) *The ring of invariants of* R *is a polynomial ring over* K *in five variables:*

$$B = K[\mathrm{Tr}(X),\mathrm{Tr}(Y),\mathrm{Det}(X),\mathrm{Det}(Y),\mathrm{Tr}(XY)].$$

(2) *The ring* S = BR *obtained by adjoining the invariants of* R *to* R *is a free* B-*module of rank* 4,

$$S = BI + BX + BY + BXY.$$

(3) *The commutator ideal of* R *is equal to* S(XY-YX). *Hence, as a vector space over* K,

$$R = \bigoplus_{i,j\geq 0} X^iY^j \oplus [R,R] = \bigoplus_{i,j\geq 0} X^iY^j \oplus S(XY-YX).$$

(4) The center of R is the direct sum of K and a principal ideal in B:

$$C = K \oplus B(XY-YX)^2.$$

(5) Q(C) = Q(B) is a rational function field in five variables.

One feature of Theorem 40 is that it describes R in terms of the much simpler ring B. This makes it possible to calculate the homogeneous codimensions and cocharacters of $\mathfrak{M}_2$ in two variables (see Section 5 for the definition).

To be precise, let W be the K-vector space spanned by x and y, with the standard action of GL(2,K). It should be emphasized that we have an action of GL(2,K) because we have two variables x, y, not because we are studying 2×2 generic matrices. As in Section 5, the free algebra $K\langle x,y\rangle$ will be identified with the graded tensor algebra

$$K \oplus W \oplus (W \otimes W) \oplus (W \otimes W \otimes W) \oplus \cdots$$

Since $\bar{\mathfrak{M}}_2$ is a homogeneous ideal (Theorem 2), the isomorphism $R \cong K\langle x,y\rangle/\bar{\mathfrak{M}}_2$ gives rise to a decomposition of R as a graded ring.

$$R = K \oplus W/(\bar{\mathfrak{M}}_2 \cap W) \oplus W^{\otimes 2}/(\bar{\mathfrak{M}}_2 \cap W^{\otimes 2}) \oplus \cdots .$$

Moreover, the dimensions and GL(2,K)-characters of its homogeneous parts are precisely what we called the homogeneous codimensions and cocharacters of $\mathfrak{M}_2$ in two variables.

The above grading on R = K[X,Y] amounts to giving both X and Y degree one. We need the more precise information obtained by bigrading R: Let X have degree (1,0) and Y have degree (0,1). This bigrading is induced by the corresponding bigrading on $K\langle x,y\rangle$. It can be induced on R as a subring of $M_2(K[x_{ij},y_{ij}])$ by bigrading the latter by giving all the x_{ij} degree (1,0) and all the y_{ij} degree (0,1). The advantage of this is that B and S inherit the bigrading. For example, Tr(X), Tr(Y), Det(X), Det(Y), Tr(XY), have degrees (1,0), (0,1), (2,0), (0,2), (1,1), respectively.

A standard tool used in connection with graded rings is the Poincaré series.

Definition. Let A be a graded (bigraded) K-module with homogeneous components A_p (A_{pq}). Then the Poincare series of A relative to the grading (bigrading) is the formal power series

$$P(A) = \Sigma \dim_K(A_p)t^p$$
$$(P(A) = \Sigma \dim_K(A_{pq})s^p t^q).$$

Since B is a commutative polynomial ring generated by five independent variables of degrees (1,0), (0,1), (2,0), (0,2), (1,1), its Poincaré series can be written down by inspection.

$$P(B) = [(1-s)(1-t)(1-s^2)(1-t^2)(1-st)]^{-1}.$$

Given the Poincaré series of B and Theorem 40 which describes R in terms of B, it is easy to find the Poincaré series of R.

THEOREM 41. (Formanek, Halpin, Li [18, Theorem 9], Drensky [14, Theorem 2.2]). *Let* $R = K[X,Y]$ *be the ring generated by two* 2×2 *generic matrices, bigraded by giving* X *degree* (1,0) *and* Y *degree* (0,1). *Then the Poincare series of* R *is*

$$P(R) = (1-s)^{-1}(1-t)^{-1} + st(1-s)^{-2}(1-t)^{-2}(1-st)^{-1}.$$

PROBLEM 42. *Determine the Poincare series of a generic matrix ring* $K[U(1),\ldots,U(r)]$, *where* $r \geq 2$ *and* $U(1),\ldots,U(r)$ *are* $n \times n$ *generic matrices. Is it a rational function?*

Of course, finding this Poincare series is the same as finding the homogeneous codimensions of $\mathfrak{M}_n$ in r variables. The only result beyond Theorem 41 is due to Drensky [14, Theorem 2.1], who found the Poincare series in one variable for any number of 2×2 matrices. His general formula is rather complicated. For three 2×2 matrices it is

$$(1-t)^{-3}[(1-t^2)^{-3}(1-t)^{-3} - (1-t)^{-3} + 1 - t^3].$$

We mentioned in Section 5 that Berele proved a Theorem [10, Corollary 3.4] which makes it possible to compute the homogeneous cocharacter of a T-ideal from its multigraded codimensions. Theorem 41 gives the multigraded codimensions for $\mathfrak{M}_2$ in two variables and Berele applied that information and his result to the action of GL(2,K) on $R \cong k\langle x,y\rangle/\bar{\mathfrak{M}}_2$ (i.e. the homogeneous cocharacter of $\mathfrak{M}_2$ in two variables). Note that by Theorem 15, irreducible GL(2,K)-modules correspond to Young diagrams with ≤ 2 rows. Let D(p,q) denote the Young diagram with p boxes in the first row and q boxes in the second row.

THEOREM 43. (Berele [10, Theorem 3.5]). *The multiplicity of the character of* D(p,q) *in* R *as a* GL(2,K)*-module is* 1 *if* $q = 0$ *and* $(p-q+1)q$ *if* $q > 0$.

So far we have looked at $R \cong k\langle x,y\rangle/\bar{\mathfrak{M}}_2$ and we have ignored $\bar{\mathfrak{M}}_2$, the ideal of two-variable identities of $M_2(K)$. We have already noted that the

T-ideal of identities of $M_2(K)$ in countably many variables is finitely generated <u>as a</u> T-<u>ideal</u>. Here we are interested in generators of $\bar{\mathfrak{M}}_2$ as a left ideal. Li [32, Theorem 2] has shown that $\bar{\mathfrak{M}}_2$ is generated as a left ideal in $K\langle x,y\rangle$ by all

$$[x^i,u][x^j,v] - [x^j,u][x^i,v]$$
$$[y^i,u][y^j,v] - [y^j,u][y^i,v],$$

where i and j are positive integers and u,v vary over $K\langle x,y\rangle$. Moreover, she found a K-basis for the bigraded homogeneous components of $\bar{\mathfrak{M}}_2$. By counting these basis elements she obtained the Poincaré series of $\bar{\mathfrak{M}}_2$ and thus an alternate proof of Theorem 41.

REFERENCES

1. S. A. Amitsur, The T-ideals of the free ring, J. Lond. Math. Soc. 30 (1955), 470-475.

2. S. A. Amitsur, Identities and generators of matrix rings, Bull. Res. Council, Israel 5a (1955), 1-10.

3. S. A. Amitsur, On central division algebras, Israel J. Math. 12 (1972), 408-420.

4. S. A. Amitsur, Polynomial identities, Israel J. Math. 19 (1974), 183-199.

5. S. A. Amitsur, Alternating identities, in "Ring Theory", Proceedings of the Ohio University Conference, S. K. Jain, editor, pp. 1-14, Lecture notes in mathematics No. 25, Dekker, New York, 1977.

6. S. A. Amitsur, The polynomial identities of associative rings, in "Noetherian Rings and Rings with Polynomial Identities", pp. 1-38, Proceedings of L.M.S. Symposium, Durham, 1979.

7. S. A. Amitsur and J. Levitski, Minimal identities for algebras, Proc. Amer. Math. Soc. 1 (1950), 449-463.

8. S. A. Amitsur and C. Procesi, Jacobson rings and Hilbert algebras with polynomial identities, Ann. Mat. Pura. Appl. (4) 71 (1966), 61-72.

9. M. Artin, On Azumaya algebras and finite-dimensional representations of rings, J. Alg. 11 (1969), 532-563.

10. A. Berele, Homogeneous polynomial identities, Israel J. Math. (to appear).

11. S. Bloch, Torsion algebraic cycles, K_2, and the Brauer group of function fields, Bull. Amer. Math. Soc. 80 (1974), 941-945.

12. M. Dehn, Über die Grundlagen der projectiven Geometrie und allgemeine Zahlsysteme, Math. Ann. 85 (1922), 184-193.

13. V. S. Drensky, A minimal basis for the identities of 2 × 2 matrices over a field of characteristic zero, Alg. i. Logika (to appear) (Russian).

14. V. S. Drensky, Codimensions of T-ideals and Hilbert series of relatively free algebras, Comptes Rendus de l'Academic Bulgare des Sciences (to appear).

15. E. Formanek, Central polynomials for matrix rings, J. Alg. 23 (1972), 129-132.

16. E. Formanek, The center of the ring of 3×3 generic matrices, Lin. Mult. Alg. 7 (1979), 203-212.

17. E. Formanek, The center of the ring of 4×4 generic matrices, J. Alg. 62 (1980), 304-319.

18. E. Formanek, P. Halpin, and W.-C. W. Li, The Poincaré series of the ring of 2×2 generic matrices, J. Alg. 69 (1981), 105-112.

19. G. B. Gurevich, Foundations of the Theory of Algebraic Invariants, P. Noordhoff Ltd., Groningen, 1964.

20. M. Hall, Projective planes, Trans. Amer. Math. Soc. 54 (1943), 229-277.

21. P. Halpin, Polynomial identities and weak identities of matrices, Thesis, Pennsylvania State University, 1981.

22. N. Jacobson, Structure theory for algebraic algebras of bounded degree, Ann. of Math. 46 (1945), 695-707.

23. N. Jacobson, PI-algebras, in "Ring Theory", Proceedings of the Oklahoma Conference, B. R. McDonald, A. R. Magid, and K. C. Smith, editors, pp. 1-30, Lecture notes in mathematics No. 7, Dekker, New York, 1974.

24. N. Jacobson, PI-algebras, An Introduction, Lecture notes in mathematics No. 441, Springer-Verlag, Berlin, New York, 1975.

25. N. Jacobson, Some recent developments in the theory of algebras with polynomial identity, in "Topics in Algebra", Proceedings, 18th Summer Research Institute of the Australian Mathematical Society, M. F. Newman, editor, pp. 8-46, Lecture notes in mathematics No. 697, Springer-Verlag, Berlin, Heidelberg, New York, 1978).

26. G. D. James, The Representation Theory of the Symmetric Groups, Lecture notes in mathematics No. 682, Springer-Verlag, Berlin, Heidelberg, New York, 1978.

27. I. Kaplansky, Rings with a polynomial identity, Bull. Amer. Math. Soc. 54 (1948), 575-580.

28. I. Kaplansky, "Problems in the theory of rings", revisited, Amer. Math. Monthly 77 (1970), 445-454.

29. A. R. Kemer, Capelli identities and nilpotency of the radical of a finitely generated PI-algebra, Dokl. Akad. Nauk. SSSR 255 (1980), 793-797 (Russian).

30. B. Kostant, A theorem of Frobenius, a theorem of Amitsur-Levitski and cohomology theory, J. of Math. and Mech. 7 (1958), 237-264.

31. V. N. Latyshev, On Regev's theorem on identities in a tensor product of PI-algebras, Uspekhi Mat. Nauk. 27 (4) (1972), 213-214 (Russian).

32. W.-C. W. Li, Generators of the ideal of polynomial identities satisfied by 2×2 matrices. J. Alg. (to appear).

33. E. C. Posner, Prime rings satisfying a polynomial identity, Proc. Amer. Math. Soc. 11 (1960), 180-184.

34. C. Procesi, Non-commutative affine rings, Atti Acc. Naz. Lincei, S. VIII, v. VIII, fo. 6 (1967), 239-255.

35. C. Procesi, Rings with Polynomial Identities, Dekker, New York, 1973.

36. C. Procesi, The invariant theory of $n \times n$ matrices, Adv. in Math. 19 (1976), 306-381.

37. C. Procesi, Trace identities and standard diagrams, in "Ring Theory", Proceedings of the 1978 Antwerp conference, F. van Oystaeyen, editor, pp. 191-218, Lecture notes in mathematics No. 51, Dekker, New York, 1979.

38. C. Procesi, Relazioni tra geometrica algebrica ed algebra non commutativa. Algebre cicliche e problema di Luroth, Boll. Unione Mat. Ital. (5) 18-A (1981), 1-10.

39. Y. P. Razmyslov, On a problem of Kaplansky, Izv. Akad. Nauk. SSSR Ser. Mat. 37 (1973), 483-501 (Russian). Translation: Math. USSR Izv. 7 (1973), 479-496.

40. Y. P. Razmyslov, Finite basing for the identities of a matrix algebra of second order over a field of characteristic zero, Alg. i. Logika 12 (1973), 83-113 (Russian). Translation: Alg. and Logic 12 (1973),47-63.

41. Y. P. Razmyslov, The Jacobson radical in PI-algebras, Alg. i. Logika 13 (1974), 337-360 (Russian). Translation: Alg. and Logic 13 (1974), 192-204.

42. Y. P. Razmyslov, Trace identities of full matrix algebras over a field of characteristic zero, Izv. Akad. Nauk. SSSR Ser. Mat. 38 (1974), 723-756 (Russian). Translation: Math. USSR Izv. 8 (1974), 727-760.

43. A. Regev, Existence of identities in $A \otimes B$, Israel J. Math. 11 (1972), 131-152.

44. A. Regev, The representations of S_n and explicit identities for PI-algebras, J. Alg. 51 (1978), 25-40.

45. A. Regev, Algebras satisfying a Capelli identity, Israel J. Math. 33 (1979), 149-154.

46. A. Regev, The identities of matrices, described by characters of the symmetric groups, in "Ring Theory", Proceedings of the 1978 Antwerp conference, F. van Oystaeyen, editor, pp. 233-241, Lecture notes in mathematics No. 51, Dekker, New York, 1979.

47. A. Regev, The polynomial identities of matrices in characteristic zero, Comm. in Alg. 8 (1980), 1417-1467.

48. A. Regev, Asymptotic values for degrees associated with strips of Young diagrams, Adv. in Math. (to appear).

49. A. Regev, The Kronecker product of S_n-characters and an $A \otimes B$ theorem for Capelli identities, J. Alg. 66 (1980), 505-510.

50. A. Regev, On the height of the Kronecker product of S_n-characters (preprint).

51. A. Regev, A polynomial rate of growth for the multiplicities in cocharacters of matrices (preprint).

52. A. Regev, Combinatorial sums, identities and trace identities of 2×2 matrices (preprint).

53. A. Regev, Young tableaux and PI-algebras (preprint).

54. S. Rosset, Generic matrices, K_2, and unirational fields, Bull. Amer. Math Soc. 81 (1975), 707-708.

55. S. Rosset, A short proof of the Amitsur-Levitski Theorem, Israel J. Math 23 (1976), 187-188.

56. L. H. Rowen, Polynomial Identities in Ring Theory, Academic Press, New York, 1980.

57. W. Schelter, Azumaya algebras and Artin's theorem, J. Alg. 46 (1977), 303-304.

58. K. S. Siberskii, Algebraic invariants for a set of matrices, Sib. Mat. Zhurnal 9 (1) (1968), 152-164 (Russian). Translation: Sib. Math. Jour. 9 (1968), 115-124.

59. R. L. Snider, Is the Brauer group generated by cyclic algebras? In "Ring Theory, Waterloo, 1978", pp. 279-301. Lecture notes in mathematics N_o.734, Springer-Verlag, Berlin, Heidelberg, New York, 1979.

60. W. Specht, Gesetze in Ringen I, Math. Z. 52 (1950), 557-589.

61. J. J. Sylvester, On the involution of two matrices of the second order, British Association Report, Southport (1883), 430-432. Reprinted in: Collected Mathematical Papers, Vol. IV, Chelsea, New York, 1973, pp. 115-117.

62. W. Wagner, Uber die Grundlagen der projectiven Geometrie und allgemeine Zahlsysteme, Math. Z. 113 (1937), 528-567.

The Pennsylvania State University
University Park,
Pennsylvania 16802

(Note: reference [48] should appear shortly).

Contemporary Mathematics
Volume 13, 1982

THE ORDER OF A FINITE GROUP OF LIE TYPE

T. A. Springer
Department of Mathematics
University of Utrecht

1. Let G be a connected reductive linear algebraic group defined over the algebraic closure of a finite field of characteristic p. Let σ be an endomorphism of G in the sense of algebraic groups which is surjective and has the property that its set of fixed points G_σ is finite. The finite groups G_σ so obtained are the "finite groups of Lie type", including the twisted forms. A thorough discussion of the situation is given in [11]. We quote some results which we shall need. If A is a σ-stable subgroup of G, then A_σ denotes the subgroup of fixed points for σ.

1.1. (i) σ fixes a Borel subgroup B and a maximal torus $T \subset B$. Any two such couples (B,T) are conjugate by an element of G_σ;

(ii) T being as in (i), let N be its normalizer in G. Then (B_σ, N_σ) is a Tits system in G_σ.

For (i) see [2, p. 175], (ii) is proved in [11,p.72-73].

For the notion of a Tits system see [3].

Let T and B be as in 1.1. Let X be the character group of T (in the sense of algebraic groups). Put $V = X \otimes_{\mathbb{Z}} \mathbb{C}$, $W = N/T$. We identify W with a Weyl group in V and the root system R of G, with respect to T, with the root system of that Weyl group. Put $\dim V = n$. Let σ^* be the extension to V of the endomorphism of X induced by the restriction of σ to T.

Let S be the algebra of complex polynomial functions on V (the symmetric algebra on the dual of V), and S^W the algebra of W-invariant elements of S. It is known that there are n algebraically independent homogeneous elements $f_1,\dots,f_n$ of S^W which generate that algebra [3, Ch. V, §5].

Denote by I the ideal in S generated by the non-constant homogeneous elements of S^W, and put $J = I/I^2$. This is a graded vector space of dimension n, on which σ^* operates naturally. Let σ_J^* be the induced map.

One also knows (see [loc. cit., p. 113]) that the differential form of degree n, $\omega = df_1 \wedge \cdots \wedge df_n$ is unique up to a nonzero scalar. Since σ^* operates on the differential forms, there is a constant c, independent of the

0271-4132/82/0513/$03.25

choice of the f_i, such that

$$\sigma^* \omega = c\omega.$$

We can now state the general formula for the order $|G_\sigma|$ of the finite group G_σ.

1.2. <u>Theorem</u>. $|G_\sigma| = c \det(\sigma_J^* - \mathrm{id}_J)$.

This formula is stated and proved in [11], see [loc. cit., p. 80]. Perhaps the best proof nowadays available is the one using Grothendieck's formula for the number of rational points of an algebraic variety defined over a finite field. For this one has to know the ℓ-adic cohomology of the algebraic variety G, and the action of the appropriate Frobenius automorphism on that cohomology. A proof along these lines is sketched in [6, pp. 230-231].

It is easy to see that it suffices to prove the theorem in the case that G is quasi-simple (i.e. its proper closed normal subgroups are finite and central). This we shall assume from now on.

In that case one can unravel the formula of the theorem as follows (see [11]).

There is a real number $q > 1$ and a linear transformation ρ of finite order of V such that $\sigma^* = q\rho$. We have $\rho W \rho^{-1} = W$. Let $\widetilde{W} \subset GL(V)$ be the finite group generated by W and ρ.

ρ acts on the algebra S. We may assume the generators f_i of S^G to be chosen such that, for $1 \le i \le n$,

$$\rho . f_i = \varepsilon_i f_i,$$

where the ε_i are roots of unity (see [11, p. 17]).

The formula for $|G_\sigma|$ is now equivalent to

(1) $$|G_\sigma| = q^N \prod_{i=1}^{n} (q^{d_i} - \varepsilon_i).$$

If G is defined over a finite field k and if σ is the corresponding Frobenius endomorphism, then $q = |k|$. If we are not in that situation the characteristic p is either 2 or 3 and q^2 is an odd power of p. The corresponding G_σ is then a Suzuki or Ree group; see [11, p. 76] for details. For Chevalley groups ($\rho = \mathrm{id}$) (1) was proved by Chevalley in [5], using tools from topology. An algebraic proof in that case was given by Solomon [9]. It was extended to a proof of (1) in [11]. Other proofs are given in [4] and [8] (and in [6], as was already mentioned). See also [7]. The proof to be given below uses results about the orders of Sylow subgroups of G_σ. It is based on the results of [10] about eigenvalues of elements of W.

2. There is a polynomial P of degree $\dim G$ with integral coefficients such that the right-hand side of (1) equals $P(q)$. Moreover $P(q)$ is an integer. These facts are readily checked, using the facts stated in [11].

If ℓ is a prime number and n an integer, we denote by $o_\ell(n)$ the order of n for ℓ. If S is a finite set we put $o_\ell(S) = o_\ell(|S|)$.

We claim that (1) is a consequence of the following result.

2.1. <u>Proposition</u>. (i) We have $o_\ell(G_\sigma) = o_\ell(P(q))$ in the following cases: (a) $\ell = p$ (the characteristic of the underlying field), (b) ℓ does not divide the order $|\widetilde{W}|$;

(ii) If ℓ divides $|\widetilde{W}|$, there exists a constant c_ℓ, independent of q, such that

$$|o_\ell(G_\sigma) - o_\ell(P(q))| \leq c .$$

We now prove our claim that 2.1 implies (1). It follows from 1.1(ii) that $|G_\sigma|$ is a polynomial function of q, of degree dim G (see [11, pp. 76-77]). Then 2.1 implies that there are only finitely many possibilities for the rational number $|G_\sigma|\cdot|P(q)|^{-1}$. One knows that, G and σ being given, arbitrarily large q can occur. It then follows that (1) holds if q is sufficiently large. But, both sides of (1) being polynomial functions, (1) must hold for all q.

So it remains to establish 2.1. Assertion (a) is known to be true [loc. cit.]. So we may assume from now on, in proving 2.1, that $\ell \neq p$. To establish the remaining assertions of 2.1, we need some results about an ℓ-Sylow group S_ℓ of G_σ. These are established in [2].

2.2. <u>Lemma</u>. S_ℓ normalizes a maximal torus of G which is fixed by σ. This is proved in [loc. cit., p. 212].

2.3. <u>Lemma</u>. (i) There is a bijection of the set of G-conjugacy classes of maximal tori of G fixed by σ onto the set of orbits of W in $W\rho$ (W acting by inner automorphisms).

Let A be a σ-stable maximal torus of G, whose conjugacy class corresponds to $w\rho$ by the bijection of (i); let $N(A)$ be its normalizer in G.

(ii) $N(A)_\sigma/A_\sigma$ is isomorphic to the stabilizer of $w\rho$ in W;

(iii) If $f_{w\rho}$ is the characteristic polynomial of $w\rho$, then the order of A_σ equals $|f_{w\rho}(q)|$.

For the proof see [loc. cit., pp. 186-188].

2.4. We next collect some elementary facts about cyclotomic polynomials.

Let Φ_m be the cyclotomic polynomial whose roots are the primitive m-th roots of unity. Recall that

$$\Phi_m(T) = \prod_{d|m} (T^d - 1)^{\mu(m/d)}, \tag{2}$$

where μ is Möbius' function.

Let ℓ be a prime, let t be an integer prime to ℓ. Let d be the smallest integer > 0 such that

$$t^d \equiv 1 \pmod{\ell},$$

then $0 < d < \ell$ and d divides $\ell-1$. Put

$$\alpha = o_\ell(t^d-1).$$

If $\ell = 2$ and $\alpha = 1$, we put

$$\beta = o_\ell(t+1).$$

2.5. Lemma. (i) If $\ell \neq 2$ or if $\ell = 2$, $\alpha > 1$, we have $o_\ell(t^x-1) = 0$ if $d \nmid x$ and $o_\ell(t^{dx}-1) = \alpha + o_\ell(x)$;

(ii) If $\ell = 2$, $\alpha = 1$, we have $o_\ell(t^x-1) = 1$ if x is odd and $o_\ell(t^{2x}-1) = \beta + 1 + o_\ell(x)$;

(iii) If ℓ and t are as in (i) then $o_\ell(\Phi_m(t)) = 0$ unless $m = \ell^h d$, for some $h \geq 0$, moreover $o_\ell(\Phi_{\ell^h d}(t))$ equals α if $h = 0$ and 1 if $h > 0$;

(iv) If ℓ and t are as in (ii) then $o_\ell(\Phi_m(t)) = 0$ unless $m = 2^h$ for some $h \geq 0$, moreover $o_2(\Phi_{2^h}(t))$ equals β if $h = 1$ and 1 if $h \neq 1$.

The proof of 2.5 is left to the reader. For similar results see [1, p. 388].

We can now deal with $o_\ell(P(q))$ for $\ell \neq p$. Let d be the smallest integer > 0 such that q^d is an integer congruent to 1 modulo ℓ. If $q \in \mathbb{Z}$ then d is a divisor of $\ell-1$. If $q \notin \mathbb{Z}$ then d is even.

If $\ell \neq 2$ or $o_\ell(q^d-1) > 1$ we put $\gamma = o_\ell(q^d-1)$. If $\ell = 2$ and $o_2(q-1) = 1$ put $\gamma = o_2(q^2-1) - 1$.

Fix a primitive d-th root of unity ζ_d and let $A(d,\rho)$ be the set of i with $1 \leq i \leq n$ and $\varepsilon_i \zeta_d^{d_i} = 1$. The number of elements of $A(d,\rho)$ is denoted by $a(d,\rho)$. Let $b(d,\ell)$ be the number of i with $1 \leq i \leq n$ and $d|d_i, \varepsilon_i = (e^{2\pi i \ell})^{-1}$ and c the number of i with $-\varepsilon_i = (-1)^{d_i}$. Notice that $c = n - a(2,\rho)$ if $\rho^2 = 1$.

2.6. Lemma. Let $\ell \neq p$ be a prime.

(i) If ℓ is odd or if $\ell = 2$, $o_2(q^d-1) > 1$ then

$$o_\ell(P(q)) = \gamma a(d,\rho) + b(d,\ell) + \sum_{i \in A(d,\rho)} o_\ell(d_i);$$

(ii) If $\ell = 2$, $q \in \mathbb{Z}$ and $o_2(q-1) = 1$ then

$$o_2(P(q)) = \gamma a(2,\rho) + c + \sum_{i \in A(1,\rho)} o_2(d_i);$$

(iii) If $\ell \nmid |\widetilde{W}|$ we have $o_\ell(P(q)) = \gamma a(d,\rho)$;

(iv) If $\ell \mid |\widetilde{W}|$ there is a constant c_ℓ', independent of q, such that

$$|o_\ell(P(q)) - \gamma a(d,\rho)| \leq c_\ell' \quad \text{or}$$
$$|o_\ell(P(q)) - \gamma a(2,\rho)| \leq c_\ell',$$

in the cases of (i) and (ii), respectively.

It is known that $\varepsilon_1^{-1},\ldots,\varepsilon_n^{-1}$ are the eigenvalues of ρ (see [10, 6.5]).

If $q \in \mathbb{Z}$ then ρ is defined over $\mathbb{Q}$, for some $\mathbb{Q}$-structure on V. It follows that in that case, if a factor $q^{d_i} - \varepsilon_i$ occurs in the right-hand side of (1), all $q^{d_i} - \varepsilon_i'$, where ε_i' is a conjugate of ε_i over $\mathbb{Q}$, will also occur. As a matter of fact, the classification shows that ρ has order 2 or 3, the latter case occurring only if G has type D_4.

If $q \notin \mathbb{Z}$ then $q^2 \in \mathbb{Z}$ and the classification shows that all d_i are even and ρ has order 2.

Hence we see that we can write $q^{-N}P(q)$ as a product of factors $\Phi_e(q^f)$ where f is even if $q \notin \mathbb{Z}$, $q^2 \in \mathbb{Z}$. It follows from 2.5 that unless $\ell = 2$, and f is odd,

$$(3) \qquad o_\ell(\Phi_e(q^f)) = \begin{cases} \gamma + o_\ell(f) & \text{if } e = (d,f)^{-1}d, \\ 1 & \text{if } e \equiv (d,f)^{-1}\ell^h d,\quad h > 0, \\ 0 & \text{otherwise} \end{cases}$$

If $\ell = 2$ and f is odd we have

$$(3)' \qquad o_\ell(\Phi_e(q^f)) = \begin{cases} \gamma & e = 2, \\ 1 & e = 2^h,\ h \neq 1, \\ 0 & \text{otherwise.} \end{cases}$$

The statements of 2.6 are consequences of (3) and (3)'. We omit the verification. Notice that if $\ell \nmid |\widetilde{W}|$ only the factors of the form $q^{d_i}-1$ in the product for $P(q)$ give a contribution to $o_\ell(P(q))$ (because $e = 1,2,3$). Assertion (iv) follows from (i) and (ii) except in the case that G is a Ree group of type $^2G_2$, $\ell = 2$. In that case an easy direct check is required, using that then $q^2 = 3^{2m+1}$, $\varepsilon_1 = 1$, $\varepsilon_2 = -1$ (see [11, p. 75, p. 82]).

The next lemma deals with the ℓ-order of the left-hand side of (1). We denote by $V(w\rho,\zeta)$ the eigenspace of $w\rho$ for the eigenvalue ζ and by ζ_e a primitive e-th root of unity. If $w \in W$, $Z(w\rho)$ is the centralizer in W of $w\rho$.

2.7. Lemma. Let $\ell \neq p$ be a prime.

(i) If ℓ is odd or if $\ell = 2$, $o_2(q^d-1) > 1$ then

$$o_\ell(G_\sigma) = \max_{w\in W}\{\gamma \dim V(w\rho,\zeta_d) + \sum_{h>0} \dim V(w\rho,\zeta_{\ell^h d}) + o_\ell(Z(w\rho))\};$$

(ii) If $\ell = 2$, $q \in \mathbb{Z}$ and $0_2(q-1) = 1$ then

$$o_2(G_\sigma) = \max_{w\in W}\{\gamma \dim V(w\rho,\zeta_2) + \sum_{h\neq 1} \dim V(w\rho,\zeta_{2^h}) + o_2(Z(w\rho))\};$$

(iii) If $\ell \nmid |\widetilde{W}|$ we have $o_\ell(G_\sigma) = \gamma \max_{w\in W} \dim V(w\rho,\zeta_d)$;

(iv) If $\ell \,|\, |\widetilde{W}|$ there is a constant c_ℓ'' independent of q such that

$$|o_\ell(G_\sigma) - \gamma \max \dim V(w\rho,\zeta_d)| \leq c''_\ell$$
$$|o_\ell(G_\sigma) - \gamma \max \dim V(w\rho,\zeta_2)| \leq c''_\ell,$$

in the cases of (i) and (ii), respectively.

Let S_ℓ be an ℓ-Sylow subgroup of G_σ, so $o_\ell(G_\sigma) = o_\ell(S_\ell)$. By 2.2, S_ℓ normalizes a σ-stable maximal torus A of G. We use the notations of 2.3.

First assume that q is an integer. Then $w\rho$ is defined over the field of rationals $\mathbb{Q}$ (for some $\mathbb{Q}$-structure on V). Hence, if ζ_e is a primitive e-th root of unity, $\dim V(w\rho,\zeta_e)$ is independent of the choice of ζ_e, and

(4) $$o_\ell(f_{w\rho}(q)) = \sum_e \dim V(w\rho,\zeta_e) o_\ell(\Phi_e(q)).$$

The statements of 2.7 now follow from 2.4 and 2.5.

There remains the case that $q \notin \mathbb{Z}$. Then G_σ is a Suzuki or Ree group, and G is of type B_2, F_4, G_2, respectively. The possibilities for $w\rho$ are discussed in [11, p. 75, p. 82]. In the first two cases $q\rho$ is defined over $K = \mathbb{Q}(\sqrt{2})$, in the last one over $K = \mathbb{Q}(\sqrt{3})$. To deal with these cases we use the following auxiliary result.

2.8. <u>Lemma</u>. Assume that $q \notin \mathbb{Z}$, let K be as above. Let the primitive root of unity ζ be an eigenvalue of some $w\rho \in W\rho$. Then either Φ_e is irreducible as a polynomial with coefficients in K, or Φ_e is a product of two irreducible polynomials Φ'_e, Φ''_e with coefficients in K, such that $\Phi'_e(q)$ and $\Phi''_e(q)$ are relatively prime integers.

This can be proved by checking the possible cases. The possibilities for e can be found from what was established in [10, p. 183]. In type B_2 they are 2,8; in type G_2: 2,12 and in type F_4: 2,3,8,12,24. For example, in the latter case, with e = 24, the factorization of Φ_{24} is the one indicated in [2, p. 213]:

$$T^8 - T^4 + 1 = (T^4 - \sqrt{2}T^3 + T^2 - \sqrt{2}T+1)(T^4 + \sqrt{2}T^3 + T^2 + \sqrt{2}T+1),$$

and in this situation the assertion of the lemma is easily checked. 2.8 implies that if $q \notin \mathbb{Z}$, formula (4) remains true and 2.7 is proved as before.

2.9. To conclude the proof of 2.1, and hence of (1), it suffices to invoke results proved in [10]. It is shown there that, with the notations of 2.6 and 2.7, we have

$$\max_{w \in W} \dim V(w\rho,\zeta_d) = a(d,\rho)$$

(see [loc.cit., 6.2(i)]).

2.1 now readily follows from 2.6 and 2.7. This concludes the proof of (1).

3. For the proof of (1) one doesn't really need the precise statements (i), (ii) of 2.6 and 2.7, but only the cruder statements (iii), (iv). But the precise statements have some other consequences, which we shall briefly discuss now.

3.1. It follows from (1), 2.6(i) and 2.7(i) that if ℓ is an odd prime we have

$$(5)\quad \begin{cases} \max_{w\in W}\{\gamma \dim V(w\rho,\zeta_d) + \sum_{h>0} \dim V(w\rho,\zeta_{\ell^h d}) + o_\ell(Z(w\rho))\} = \\ = \gamma a(d,\rho) + b(d,\ell) + \sum_{i\in A(d,\rho)} o_\ell(d_i), \end{cases}$$

the notations being as before. If $\ell \nmid |\tilde{W}|$ then (5) just asserts that

$$\max_{w\in W} V(w\rho,\zeta_d) = a(d,\rho),$$

which is a result proved in [10] and used in no. 2 to prove (1). We shall use the general form of (5) to obtain a result not covered by [10].

Let W be a Weyl group, with root system R. Assume that ρ is a linear transformation of the underlying vector space which permutes the elements of R and stabilizes a set of positive roots R^+.

3.2. Proposition. Let ℓ be an odd prime and d a divisor of $\ell-1$.

(i) For any $w \in W$ we have $\dim V(w\rho,\zeta_d) \le a(d,\rho)$;

(ii) There exists $w \in W$ such that

$$(6)\quad \begin{cases} \dim V(w\rho,\zeta_d) = a(d,\rho), \\ \sum_{h>0} \dim V(w\rho,\zeta_{\ell^h d}) + o_\ell(Z(w\rho)) = b(d,\ell) + \sum_{i\in A(d,\rho)} o_\ell(d_i). \end{cases}$$

(i) is part of the result of [10] alluded to above. To prove (ii) we proceed as follows. Choose $a \in \mathbb{Z}$ such that $a + \ell\mathbb{Z}$ has order d in the multiplicative group of the finite field F_ℓ^* and that $o_\ell(a^d-1) = 1$. By Dirichlet's theorem there is a prime number p which has the form $a + x\ell$. There exists a connected reductive group G over F_p such that its Weyl group W and the transformation ρ of no. 1 are the given ones. Applying (5) with this G and the ground fields F_{p^m} $(m \ge 1)$ we obtain for each $m \ge 1$ an element $w_m \in W$ such that

$$m \dim V(w_m\rho,\zeta_d) + \sum_{h>0} \dim V(w_m\rho,\zeta_{\ell^h d}) + o_\ell(Z(w_m\rho)) =$$
$$= ma(d,\rho) + b(d,\ell) + \sum_{i\in A(d,\rho)} o_\ell(d_i).$$

Choose m,m' such that $m \ne m'$, $w_m = w_{m'} = w$, say. This element w is then as required.

3.3. Let w be as in 3.2. If $w\rho$ has a regular eigenvector [10, p. 170] with eigenvalue ζ_d, the first formula (6) holds by [loc.cit., 6.4(ii)] and the second one can also be deduced from [loc.cit., 6.4]. But there are cases in which the element w has no regular eigenvector. As an example, take W of type A_n, with $\rho = 1$. Let ℓ be a prime $\leq n$ and d a divisor of ℓ which does not divide n or $n+1$. The discussion of [loc.cit., p. 175] shows that w cannot have a regular eigenvector. In cases as this one 3.2 gives a result which is not covered by [10].

The question arises whether 3.2 can be proved by the algebro-geometric methods of [10].

3.4. If $\ell = 2$ a statement like 3.2 is still true. But it can best be given another form.

Recall (see 2.3) that the G_σ-conjugacy classes of σ-stable maximal tori of G are parametrized by the orbits of W in $W\rho$, under inner automorphisms. We shall say that a σ-stable maximal torus A in G has <u>type</u> $w\rho$ if its conjugacy class under G_σ corresponds to the class of $w\rho$.

The linear transformation ρ stabilizes a Weyl chamber (because of [11, 11.2, 11.14]); let w_0 be the corresponding element of W of maximal length.

For $\ell = 2$, the statements of 2.6(i), (ii) and 2.7(i),(ii) can be used to locate the 2-Sylow subgroups of G_σ. The precise result is as follows.

3.5. <u>Proposition</u>. Let $p \neq 2$. A 2-Sylow subgroup of G_σ normalizes a maximal torus of type ρ if $o_2(q^d-1) > 1$ and a maximal torus of type $w_0\rho$ if $o_2(q^d-1) = 1$.

Assume $o_2(q^d-1) > 1$. Using 2.6(i), 2.7(i) we see that in this case the proposition will follow if we show that

$$\dim V(\rho,1) = a(1,\rho),$$
$$\sum_{h>0} \dim V(\rho,\zeta_2 h) + o_2(Z(\rho)) = b(1,2) + \sum_{i \in A(1,\rho)} o_2(d_i).$$

These formulas readily follow from [10,6.5]. If $o_2(q^d-1) = 1$, the proof is similar. We omit it.

REFERENCES

[1] E. Artin, The orders of the linear groups. Coll. Papers, pp. 387-397, Addison-Wesley, 1965.

[2] A. Borel et. al., Seminar on algebraic groups and related finite groups, Lect. Notes in Math. no. 131, Springer-Verlag, 1970.

[3] N. Bourbaki, Groupes et algébres de Lie, Ch. IV, V, VI, Hermann, Paris, 1968.

[4] R. W. Carter, Weyl groups and finite Chevalley groups, Proc. Cambridge Philos. Soc. 67 (1970), 269-276.

[5] C. Chevalley, Sur certains groupes simples, Tôhoku Math. J. 7 (1955), 14-66.

[6] P. Deligne, Applications de la formules des traces aux sommes trigonométriques, in: SGA 4 1/2, pp.168-232, Lect. Notes in Math. no. 569, Springer-Verlag, 1977.

[7] M. Demazure, Invariants symétriques entiers des groupes de Weyl et torsion, Inv. Math. 21 (1973), 287-301.

[8] I. G. Mac Donald, The Poincaré series of a Coxeter group, Math. Ann. 199 (1972), 161-174.

[9] L. Solomon, The orders of the finite Chevalley groups, J. Alg. 3 (1966), 376-393.

[10] T. A. Springer, Regular elements of finite reflection groups, Inv. Math. 25 (1974), 159-193.

[11] R. Steinberg, Endomorphisms of linear algebraic groups, Mem. Amer. Math. Soc. no. 80 (1968).

Mathematisch Instituut
der Rijksuniversiteit
Utrecht

Contemporary Mathematics
Volume 13, 1982

RESTRICTED LIE ALGEBRAS (AND BEYOND)

J. E. Humphreys*

In a paper [14] published 40 years ago, Jacobson introduced the notion of "restricted Lie algebra" over a field of prime characteristic. This notion of course had antecedents in the study of certain linear Lie algebras (such as derivation algebras), cf. [13]. The present survey is an attempt to summarize briefly some of the areas in which progress has been made and to formulate problems (some quite familiar) which deserve further study. I have made no attempt to be comprehensive, limiting myself to a few areas with which I am familiar and quoting only a sample of the extensive literature.

1. RESTRICTED LIE ALGEBRAS [8, 13, 14, 15, 28, 29]

For convenience, we discuss only finite dimensional Lie algebras over an algebraically closed field k of prime characteristic p. By analyzing carefully the behavior of a Lie subalgebra of an associative algebra which is closed under the (associative) p^{th} power, Jacobson was able to formulate conditions on an abstract Lie algebra L which guarantee similar behavior. Besides the Lie bracket, L should have an operation $x \mapsto x^{[p]}$, subject to the following axioms:

(1) $(\mathrm{ad}\, x)^p = \mathrm{ad}\, x^{[p]}$ for all $x \in L$;

(2) $(\lambda x)^{[p]} = \lambda^p x^{[p]}$ for all $x \in L$, $\lambda \in k$;

(3) $(x+y)^{[p]} = x^{[p]} + y^{[p]} + \sum_{i=1}^{p-1} s_i(x,y)$, where $is_i(x,y)$ is the coefficient of λ^{i-1} in $x(\mathrm{ad}(\lambda x+y))^{p-1}$, for all $x, y \in L$.

A number of natural examples come to mind: an associative algebra viewed as Lie algebra, with $x^{[p]}$ being the associative p^{th} power; the derivation algebra of a nonassociative algebra, with the usual p^{th} power; the Lie algebra $\mathrm{Lie}(G)$ of an affine algebraic group. In particular, various restricted simple Lie algebras arise: those of "classical type" which are

---

*Research partially supported by NSF grant MCS 79-02738.

closely associated with the Lie algebras of simple algebraic groups, those of "Cartan type" (such as the Jacobson-Witt algebras W_n) which occur as derivation algebras.

Let us note a few features of restricted Lie algebras which help to justify the abstract definition.

(a) If we factor out of the universal enveloping algebra U(L) of L the two-sided ideal generated by the (central!) elements $x^{[p]} - x^p$ $(x \in L)$, we obtain a finite dimensional associative algebra $\underline{u}(L)$, the $\underline{u}$-algebra or restricted universal enveloping algebra. This includes L as a Lie subalgebra, so that in particular L has a faithful finite dimensional representation. More precisely, if $(x_1, \ldots, x_n)$ is an ordered basis of L, the images in $\underline{u}(L)$ of the monomials $x_1^{i_1} \ldots x_n^{i_n}$ $(0 \le i_j < p)$ form a basis of $\underline{u}(L)$. This algebra plays for restricted representations of L the role played by U(L) for arbitrary representations.

(b) For an arbitrary Lie algebra L to have the structure of restricted Lie algebra, it is necessary and sufficient that L have a basis $x_1, \ldots, x_n$ such that each derivation $(\mathrm{ad}\ x_i)^p$ is inner.

(c) If L is an arbitrary Lie algebra with center 0, so that it can be embedded in its derivation algebra Der L, the smallest restricted subalgebra $\overline{L}$ of Der L containing L is a convenient tool for the study of L: L is an ideal in $\overline{L}$, and the quotient is abelian. For example, an arbitrary simple Lie algebra over k can be compared in this way with a restricted Lie algebra, which sometimes facilitates the study of Cartan subalgebras of L.

Problem 1: Find useful sufficient or necessary conditions on a restricted Lie algebra L which imply that L is (or is not) isomorphic to the Lie algebra of an affine algebraic group.

A number of necessary conditions can be deduced from a study of the Lie algebras Lie(G), cf. [8], [29]. For example, in an algebraic Lie algebra L all Cartan subalgebras must be conjugate under Aut L. For this (or other) reason, the simple algebras of Cartan type cannot be algebraic.

2. STRUCTURE AND CLASSIFICATION [2,4,5,26,28,37].

Problem 2: Determine all restricted simple Lie algebras.

This is of course a problem of long standing, which in recent years has been intensively studied by Block, Wilson, Kac, Weisfeiler, Schue and others (usually in the wider context of classifying all simple Lie algebras). At least when $p > 7$, the new work of Block and Wilson (cf. [2]) lends appreciable support to the conjecture of Kostrikin-Šafarevič that the only restricted simple

Lie algebras will be those of classical or Cartan type. (The primes 2,3 need to be treated separately, while 5 and 7 should behave much like the larger primes.)

The work of Block and Wilson also suggests that one may be able to get good information about semisimple algebras along the way, even though these are not always direct sums of simple ones.

Problem 3: Find good structural invariants of an arbitrary restricted Lie algebra.

This problem, though admittedly not well formulated, arises inevitably in the classification problem discussed above. The analogy with characteristic 0 is not always helpful; for example, in characteristic p Cartan subalgebras need not have the same dimension or be conjugate. But there are special features of a restricted Lie algebra L which allow one to imitate the study of algebraic groups, notably the Jordan decomposition $x = x_s + x_n$ for elements x of L. Here x_s is "semisimple" in the sense of being contained in the k-span of its various p^{th} powers, x_n is "nilpotent" in the sense that $x_n^{[p^e]} = 0$ for some e, and $[x_s x_n] = 0$. (Cf. [26], [28].)

The notion of semisimple element allows one to define a "torus" or "toral subalgebra" of L to be a subalgebra (necessarily abelian) consisting of semisimple elements. Then a Cartan subalgebra of L is the same thing as the centralizer of a maximal torus [28]. In case L is solvable, Winter [37] showed that all maximal tori have the same dimension (even though they need not all be conjugate). For the simple Lie algebras of classical type, this is also true, and for those of Cartan type, the same conclusion follows from the work of Demuškin [4], [5]. It seems possible that the dimension of a maximal torus is an invariant of L for any restricted Lie algebra L. (On the other hand, maximal nil subalgebras seem to vary considerably in dimension.)

3. REPRESENTATIONS [9,20,21,24,27,30,31,35,36,38]

In studying the representations of a Lie algebra L over k, it is perhaps most natural to look first for the irreducible ones. Whether L is restricted or not, Zassenhaus [38] showed that the irreducible L-modules are finite dimensional and of dimension $\le p^m$, the upper bound p^m being achieved for "most" modules. Here m is described somewhat indirectly by the formula: p^{2m} = degree of the quotient division algebra of $U(L)$ over the quotient field of the center of $U(L)$. When L is simple of classical type, excluding a few small primes, Rudakov [24] was able to show that m equals the number of positive roots; cf. also [36]. Here a restricted representation of the maximum degree exists: the Steinberg representation.

Further investigations by Veisfeiler-Kac [35] and Mil'ner [20], [21] have led to a general characterization of the integer m for a restricted Lie algebra L, in terms of the internal structure of L: $2m = n-r$, where $n = \dim L$ and $r = \min \dim L_\lambda$ (λ in the dual space L^*), with L_λ defined to be the stabilizer of λ in L under the co-adjoint representation, i.e. $\{z \in L \mid \lambda([z,L]) = 0\}$. When L is the simple Witt algebra W_1 of dimension p, $r = 1$; but on the other hand, the maximum dimension of a restricted irreducible L-module is just p (cf. [30]). This suggests another problem:

Problem 4: For an arbitrary restricted Lie algebra L, determine the maximum dimension of an irreducible restricted L-module.

It would be reasonable to expect this dimension to be a power of p. In any case, further study of the methods of Mil'ner (only sketched briefly in [21]) should be profitable.

Unless L is a torus, it has restricted representations which fail to be completely reducible, so it is also worthwhile to consider the nature of indecomposable L-modules other than irreducible ones. For example, what can be said about the indecomposable injective (= projective) $\underline{u}(L)$-modules? For L of classical type, these have been explored to some extent in [9] and have been found to have dimensions divisible by p^m (m as above); the minimum dimension p^m is attained by the Steinberg module.

Problem 5: If L is a restricted Lie algebra, describe the indecomposable injective modules for $\underline{u}(L)$. In particular, determine their minimum dimension and the highest power of p dividing all their dimensions.

In case L is solvable, the results of [27] and [31] may yield some insight into this problem.

4. RESTRICTED ENVELOPING ALGEBRAS [1,11,15,18,23,25]

The algebras $\underline{u}(L)$ provide interesting examples for the theory of finite dimensional associative (or Hopf) algebras, somewhat comparable to the group algebras of finite groups. For example, it was shown by Berkson [1] that $\underline{u}(L)$ is always a Frobenius algebra, i.e., has a nondegenerate associative bilinear form. Such a form can be obtained by choosing an ordered basis $(x_1, \ldots, x_n)$ of L and defining (u,v) to be $f(uv)$, where f is the linear function on $\underline{u}(L)$ taking value 1 at $x_1^{p-1} \ldots x_n^{p-1}$ and value 0 at other basis monomials. Schue [25] showed that the form is even symmetric (so $\underline{u}(L)$ is a symmetric algebra) in case $\mathrm{tr}(\mathrm{ad}\, x) = 0$ for all $x \in L$. In fact, results of Larson-Sweedler [19] (cf. [11]) show more generally that any finite dimensional Hopf algebra is Frobenius, and allow one to deduce that $\underline{u}(L)$ is symmetric if and only if Schue's condition holds. There is an interesting connection here with the center of $\underline{u}(L)$, via the notion of "integral" in a Hopf algebra.

Problem 6: When is $\underline{u}(L)$ of finite representation type, i.e., when are its indecomposable modules of bounded dimension?

This problem was studied by Pollack [23], who found in many cases that $\underline{u}(L)$ has indecomposable modules of arbitrarily high dimension. It seems difficult to find a plausible condition on L insuring that $\underline{u}(L)$ is of finite representation type, in view of the examples in [23], so it may be misleading to think of this as a problem about Lie algebras.

5. COHOMOLOGY [3,7,22,33,34]

Given a Lie algebra L and an L-module M, there are cohomology groups $H^i(L,M)$. When both L and M are restricted, it is natural to seek an adaptation of the definition taking the additional structure into account. Hochschild [7] introduced restricted Lie algebra cohomology groups $H^i_*(L,M)$ and showed how to interpret them in low dimensions in terms of extensions. In view of the frequent failure of complete reducibility, H^1_* can be expected to be nonzero in many cases.

Problem 7: Find good criteria for the vanishing or nonvanishing of restricted Lie algebra cohomology.

It is natural to attempt some comparisons between the restricted and ordinary Lie algebra cohomology, to get better insight into the former. There is a natural map from H^i_* to H^i, which Hochschild showed to be injective when $i = 1$; but for larger i things get more complicated.

When L is the Lie algebra of an affine algebraic group G, it is also natural to attempt a comparison between restricted Lie algebra cohomology and the (Hochschild) cohomology of G, cf. [3], [32]-[34], and recent work of J. O'Halloran. But this comparison is rather unsatisfactory until one widens the context in the way discussed below.

6. HYPERALGEBRAS [3,6,9,10,11,12,16,17,18,32]

It has long been clear that the connection between an algebraic group G and its Lie algebra L is somewhat weak in prime characteristic. Let us take G to be simply connected and semisimple, so L is obtained by reduction mod p from the $\mathbb{Z}$-span of a Chevalley basis in a complex semisimple Lie algebra. The reduction process suggests a way to go beyond $\underline{u}_1 = \underline{u}(L)$, obtaining a sequence of finite dimensional Hopf algebras $\underline{u}_r$ so that $\underline{u}_1 \subset \underline{u}_2 \subset \dots$. Here $\dim \underline{u}_r = (p^r)^{\dim L}$, with a basis analogous to that of $\underline{u}_1$, cf. [10]. The union of the $\underline{u}_r$ is just the reduction mod p of the $\mathbb{Z}$-form defined by Kostant for the universal enveloping algebra in characteristic 0, generated by elements of the form $x_\alpha{}^t/t!$ for root vectors x_α in the Chevalley basis. More intrinsically, $\underline{u}_r$ is an algebra of "distributions" associated with the rth Frobenius kernel G_r of G (an infinitesimal group scheme).

The algebras $\underline{u}_r$ have proved to be useful in the study of G and its representations; for example, the injective G-modules are (usually) direct limits of injective $\underline{u}_r$-modules. The $\underline{u}_r$ also provide a link between the Lie algebra cohomology discussed above and the cohomology of G. But many problems remain to be studied.

Problem 8: Determine the center of $\underline{u}_r$ (and of the Kostant $\mathbb{Z}$-form after reduction mod p).

This is closely related to the question of determining "blocks" of $\underline{u}_r$, which has been settled already [12], [16]. Haboush [6] found some explicit elements in the center of $\underline{u}_r$, but the full description seems to be elusive.

One concluding remark: There seems to be at least a rough analogy between the process just described for passing beyond the restricted enveloping algebra of a classical Lie algebra, and the process involved in generalizing the restricted simple Lie algebras of Cartan type to those of "generalized Cartan type".

REFERENCES

1. A. Berkson, The u-algebra of a restricted Lie algbera is Frobenius, Proc. Amer. Math. Soc. 15 (1964), 14-15.

2. R. E. Block and R. L. Wilson, The simple Lie p-algebras of rank two, Ann.of Math., 115 (1982), 93-168.

3. E. Cline, B. Parshall, L. Scott, Cohomology, hyperalgebras and representations, J. Algebra 63 (1980), 98-123.

4. S. P. Demuškin, Cartan subalgebras of the simple Lie p-algebras W_n and S_n, Sibirsk. Mat. Ž. 11 (1970), 310-325 = Siberian Math. J. 11 (1970), 233-245.

5. ______________, Cartan subalgebras of simple nonclassical Lie p-algebras, Izv. Akad. Nauk SSSR Ser. Mat. 36 (1972), 915-932 = Math. USSR-Izv. 6 (1972), 905-924.

6. W. J. Haboush, Central differential operators on split semisimple groups over fields of positive characteristic, in: Lect. Notes in Math. 795, Springer, 1980, pp. 35-85.

7. G. Hochschild, Cohomology of restricted Lie algebras, Amer. J. Math. 76 (1954), 555-580.

8. J. E. Humphreys, Algebraic groups and modular Lie algebras, Mem. Amer. Math. Soc. 71 (1967).

9. ______________, Modular representations of classical Lie algebras and semisimple groups, J. Algebra 19 (1971), 51-79.

10. ______________, On the hyperalgebra of a semisimple algebraic group, in: Contributions to Algebra: A Collection of Papers Dedicated to Ellis Kolchin, Academic Press, New York, 1977, pp. 203-210.

11. ______________, Symmetry for finite dimensional Hopf algebras, Proc. Amer. Math. Soc. 68 (1978), 143-146.

12. J. E. Humphreys, J. C. Jantzen, Blocks and indecomposable modules for semisimple algebraic groups. J. Algebra 54 (1978), 494-503.

13. N. Jacobson, Abstract derivation and Lie algebras, Trans. Amer. Math. Soc. 42 (1937), 206-224.

14. __________, Restricted Lie algebras of characteristic p, Trans. Amer. Math. Soc. 50 (1941), 15-25.

15. __________, Lie Algebras, Interscience, New York, 1962.

16. J. C. Jantzen, Über Darstellungen höherer Frobenius-Kerne halbeinfacher algebraischer Gruppen, Math. Z. 164 (1979), 271-292.

17. ____________, Darstellungen halbeinfacher Gruppen und ihrer Frobenius-Kerne, J. Reine Angew. Math. 317 (1980), 157-199.

18. V. Kac, B. Weisfeiler, Coadjoint action of a semi-simple algebraic group and the center of the enveloping algebra in characteristic p, Indag. Math. 38 (1976), 136-151.

19. R. G. Larson, M. E. Sweedler, An associative orthogonal bilinear form for Hopf algebras, Amer. J. Math. 91 (1969), 75-94.

20. A. A. Mil'ner, Irreducible representations of modular Lie algebras, Izv. Math. Akad. Nauk SSSR Ser. Mat. 39 (1975), 1240-1259 = Math. USSR-Izv. 9 (1975), 1169-1187.

21. A. A. Mil'ner, Maximal degree of irreducible Lie algebra representations over a field of positive characteristic, Funkcional. Anal. i Priložen. 14 (1980), no. 2, 67-68 = Functional Anal. Appl. 14 (1980), 136-137.

22. B. Pareigis, Kohomologie von p-Lie-Algebren, Math. Z. 104, (1968), 281-336.

23. R. D. Pollack, Restricted Lie algebras of bounded type, Bull. Amer. Math. Soc. 74 (1968), 326-331.

24. A. N. Rudakov, On representations of classical semisimple Lie algebras of characteristic p, Izv. Akad. Nauk SSSR Ser. Mat. 34 (1970), 735-743 = Math. USSR-Izv. 4 (1970), 741-750.

25. J. R. Schue, Symmetry for the enveloping algebra of a restricted Lie algebra, Proc. Amer. Math. Soc. 16 (1965), 1123-1124.

26. ________, Cartan decompositions for Lie algebras of prime characteristic, J. Algebra 11 (1969), 25-52; 13 (1969), 558.

27. ________, Representations of solvable Lie p-algebras, J. Algebra 38 (1976), 253-267.

28. G. B. Seligman, Modular Lie Algebras, Springer, Berlin, 1967.

29. ________, Algebraic Lie algebras, Bull. Amer. Math. Soc. 74 (1968), 1051-1065.

30. H. Strade, Representations of the Witt algebra, J. Algebra 49 (1977), 595-605.

31. ________, Darstellungen auflösbarer Lie-p-Algebren, Math. Ann. 232 (1978), 15-32.

32. J. B. Sullivan, Relations between the cohomology of an algebraic group and its infinitesimal subgroups, Amer. J. Math. 100 (1978), 995-1014.

33. ________, Lie algebra cohomology at irreducible modules, Ill. J. Math. 23 (1979), 363-373.

34. ________, The second Lie algebra cohomology group and Weyl modules, Pacific J. Math. 86 (1980), 321-326.

35. B. Yu. Veisfeiler, V. G. Kac, The irreducible representations of Lie p-algebras, Funkcional Anal. i Priložen. 5 (1971), no. 2, 28-36 = Functional Anal. Appl. 5 (1970), 111-117.

36. F. D. Veldkamp, The center of the universal enveloping algebra of a Lie algebra in characteristic p, Ann. Sci. École Norm. Sup.(4) 5 (1972), 217-240.

37. D. J. Winter, On the toral structure of Lie p-algebras, Acta Math. 123 (1969), 69-81.

38. H. Zassenhaus, The representations of Lie algebras of prime characteristic, Proc. Glasgow Math. Assn. 2 (1954), 1-36.

University of Massachusetts
Amherst, Massachusetts 01003

Contemporary Mathematics
Volume 13, 1982

DETERMINANTS ON FREE FIELDS

by P. M. Cohn

My object today will be the determination of K_1 of a free field, and I am grateful to the Mathematics Department of Yale University for inviting me to take part in these celebrations to mark the retirement of Professor N. Jacobson. Here, as in so many other parts of algebra, Jake's influence has been very noticeable; my own introduction to his work was in his 1943 book "Theory of Rings" [12], where in a little over 25 pages he gives a concise but essentially complete treatment of principal ideal domains. I want to describe a generalization of the theory presented there, and it was the concision and generality of that account which stimulated much of the later development.

1. Let R be a commutative ring and $M_n(R)$ the set of all $n \times n$ matrices over R, then the determinant can be described as a monoid homomorphism which is universal for homomorphisms into commutative monoids Σ :

$$\begin{array}{ccc} M_n(R) & \xrightarrow{\det} & R \\ & \searrow & \vdots \\ & & \Sigma \end{array}$$

For simplicity we ignore the additivity property, though this will play a role later. We remark that it is unusual to have a universal construction which is <u>not</u> a left adjoint in any natural way.

There have been many attempts to extend the definition of determinant to the non-commutative case, usually under special assumptions. E.g., if R is a finite-dimensional algebra over a field, we can use the norm to define a determinant. The general case was first treated by Dieudonne [8] in 1943; he observed that we can really form determinants over any skew field, as long as the target group is abelian. This is of course necessary because we shall want the matrices $\begin{pmatrix} a & 0 \\ 0 & b \end{pmatrix}$ and $\begin{pmatrix} b & 0 \\ 0 & a \end{pmatrix}$ (which differ only by elementary transforma-

0271-4132/82/0515/$03.50

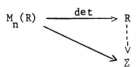

tions) to have the same determinant.

Thus let K be any skew field and consider $GL_n(K)$, the group of all invertible $n \times n$ matrices over K. Any matrix A in $GL_n(K)$ can be reduced to diagonal form by elementary transformations; more precisely, we can write

$$A = DU, \quad \text{where} \quad D = \begin{pmatrix} I & 0 \\ 0 & \mu \end{pmatrix}, \quad U \in E_n(K),$$

where $E_n(K)$ is the group generated by all elementary matrices and $\mu \in K^*(= K \setminus \{0\})$. The expression for A is not in general unique: if $A = D_1U_1$ is another, where $D_1 = \begin{pmatrix} I & 0 \\ 0 & \mu_1 \end{pmatrix}$, then $\mu_1\mu^{-1} \in K^{*\prime}$, the derived group of K^*. Hence the coset of μ in $K^{*ab} = K^* / K^{*\prime}$ is an invariant of the matrix A, and this is just the <u>Dieudonné determinant</u>. So we now have the following commutative diagram, where Det is universal for homomorphisms into abelian groups Γ :

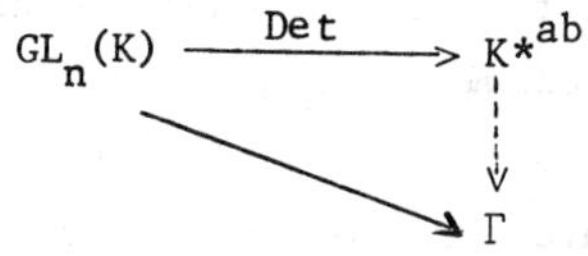

We thus have an isomorphism $GL_n(K)^{ab} \simeq K^{*ab}$, valid for any skew field K and any $n \geq 1$ (except $n = 2$ and $K = F_2$). This makes it of interest to determine K^{*ab} for various skew fields K.

To give an example, if K is existentially closed over k (i.e. every consistent system of equations with coefficients in K, centralizing k, has a solution in K), then any $c \in K^*$ is a commutator:

$$c = a^{-1}b^{-1}ab \quad \text{for some} \quad a,b \in K.$$

For we can find $a \in K$ such that $ac = ca$ and a is transcendental over $K(c)$. Hence a, ac are both transcendental over k and so are conjugate (cf. [2]): $ac = b^{-1}ab$, so $c = a^{-1}b^{-1}ab$. It follows that Det is trivial: $K^{*ab} = 1$.

2.1 want in particular to consider the free field over k on a set X: $k(X)$. This is defined as follows: Let us first take the case $|X| = 1$, say $X = \{x\}$. Then $k[x]$ is just the usual polynomial ring, a principal ideal domain, and $k(x)$, the rational function field, is the field we want. When X has more than one element, $k\langle X\rangle$ is the <u>free algebra</u> on X (non-commutative polynomial ring), thus if $X = \{x_1, \ldots, x_d\}$, then the elements of $k\langle X\rangle$ have the form

$$\Sigma c_{i_1 \ldots i_r} x_{i_1} \ldots x_{i_r} .$$

This ring $k\langle X\rangle$ can be embedded in a field in many different ways, i.e. there are several non-isomorphic fields of fractions (cf. [9], [4], p. 15), but there is a universal field of fractions, U, from which all others are obtained by specialization ([2], Ch. 7). This universal field of fractions is written k(X) and is called the free field on X over k. Formally it is obtained from $k\langle X\rangle$ by inverting all full matrices, where A is full if it cannot be written as A = PQ, where P is $n \times n-1$ and Q is $n-1 \times n$. The problem I want to discuss is the structure of $k(X)^{*ab}$. A priori one does not even know that the natural map $X \to k(X)^{*ab}$ is injective. It could hardly be otherwise, but this is not quite trivial to prove.

3. To solve our problem we need to describe the process by which $U = k(X)$ is obtained from $R = k\langle X\rangle$. We remark that R is a fir (= free ideal ring), i.e. every left ideal and every right ideal of R is free, as R-module, of unique rank ([2], Ch. 1). Every fir R has a universal field of fractions U ([2], Ch. 7), and all I have to say will apply to any fir and its universal field of fractions.

Let us first take the commutative case: a commutative fir is just a principal ideal domain, and for principal ideal domains (even non-commutative ones) we have the following basic results. First some definitions: An atom in a ring R is a non-unit which cannot be written as a product of two non-units. Two non-zero elements a and b are called similar if $R / aR \cong R / bR$. In an integral domain this condition is left-right symmetric.

a) Any principal ideal domain is a unique factorization domain: any non-zero non-unit c is a product of atoms, and two such complete factorizations of c have the same number of terms and the terms are unique up to order and similarity ([12], Chp. 3, Th. 5, p. 34).

This is essentially the Jördan-Holder theorem for groups with operators.

b) Every matrix over a principal ideal domain R is associated to a diagonal matrix:

$$A = PDQ, \quad A \in {}^mR^n, \; P \in GL_m(R), \; Q \in GL_n(R), \; D \text{ diagonal}.$$

With a suitable condition on the diagonal elements the matrix D becomes unique, but this is not of importance for us here. The result was first established for Z and k[x], then for Euclidean domains, principal ideal domains, and finally for non-commutative principal ideal domains (Frobenius-Stickelberger [10], Wedderburn [16], Jacobson [11], Teichmüller [15], Nakayama [13]).

There is a more explicit description of similarity. If we generalize to matrices at the same time, we have the

Definition. Two matrices A, B over a ring R are stably associated if there exist invertible matrices P, Q such that

$$\begin{pmatrix} A & 0 \\ 0 & I \end{pmatrix} = P \begin{pmatrix} B & 0 \\ 0 & I \end{pmatrix} Q.$$

Here the unit matrices on the two sides need not have the same size.

Now matrix atoms can be defined as before, as atoms in $M_n(R)$. We restrict attention to rings with invariant basis number (where any invertible matrix is square), then any matrix atom is necessarily square. Then a)-b) generalize as follows ([2], Th. 5. 6. 4):

THEOREM. *Let R be a fir, then every full matrix can be written as a product of matrix atoms, and the factors are unique up to order and stable association.*

4. How is U = k(X) formed from R? In the commutative case the element u of U is obtained as solution of the equation

$$au + b = 0, \quad \text{where } a,b \in R,\ a \neq 0.$$

The same works in general, replacing a by a full matrix. Let us write A for the matrix (a,b), then we must solve

$$\text{(1)} \qquad Au = 0, \quad \text{where } A = (A_0, A_1, \ldots, A_m),\ u = (1, u_1, \ldots, u_m)^T.$$

The element $p \in U$ is obtained as the last component, say $p = u_m$, of the solution. Given a system (1), write $A_* = (A_1, \ldots, A_{m-1})$ and A_∞ for A_m, then one has Cramer's rule ([2], p. 251):

$$(2) \qquad (A_0\ A_*) = (A_\infty\ A_*)\begin{pmatrix} -u_m & 0 \\ -u_* & I \end{pmatrix}.$$

We shall call $(A_0\ A_*)$ the <u>numerator</u> and $(A_\infty\ A_*)$ the <u>denominator</u> of $p = u_m$ in the system (1), and the matrix A is said to be <u>admissible</u> for p. Generally a matrix A is <u>admissible</u> if it is $m \times m+1$ for some m and the submatrix consisting of the last m columns is full; this is just the condition for (1) to have a unique solution.

It is clear that different admissible systems for p will in general have different numerators and denominators, but we can obtain an invariant as follows. Let P be a matrix atom for R; the class of matrices stably associated to P is called a <u>prime divisor</u> of R, written π, say. For any full matrix A over R we define $v_\pi(A)$ or simply $v(A)$ as the number of factors of type π in a complete factorization of A. Clearly we have

$$v(AB) = v(A) + v(B).$$

Our aim will be to define v on U by putting

$$(3) \qquad v(p) = m(A_0\ A_*) - v(A_\infty\ A_*),$$

if p is given by (2). Of course we have to show that the righthand side of (3) is independent of the choice of the system A.

To do this let us in the first instance define, for any admissible matrix A,

$$(4) \qquad V(A) = v(A_0\ A_*) - v(A_\infty\ A_*),$$

putting $v(C) = \infty$ if C is not full. If A is admissible for p and B is admissible for q, then the matrix

$$(5) \qquad A.B = \begin{pmatrix} 0 & 0 & A_0 & A_* & A_\infty \\ B_0 & B_* & B_\infty & 0 & 0 \end{pmatrix}$$

is admissible for pq, and it is clear that we have

$$V(A.B) = v\begin{pmatrix} 0 & 0 & A_0 & A_* \\ B_0 & B_* & B_\infty & 0 \end{pmatrix} - v\begin{pmatrix} 0 & A_0 & A_* & A_\infty \\ B_* & B_\infty & 0 & 0 \end{pmatrix} = V(A) + V(B).$$

If S is the monoid of admissible matrices with the multiplication defined by (5), then $V:S \to Z$ is a homomorphism and we have to show that V has the same value on all admissible matrices defining a given p.

If $A = (A_0\ A_*\ A_\infty)$ is admissible for p, then $\overline{A} = (A_\infty\ A_*\ A_0)$ is admissible for p^{-1}, and hence $A.\overline{A}$ is admissible for $pp^{-1} = 1$. Hence if A, B are any matrices admissible for the same p, then $A.\overline{B}$ is admissible for $pp^{-1} = 1$, and so

$$V(A) - V(B) = V(A) + V(\overline{B}) = V(A.\overline{B}).$$

So to prove that v is well-defined we need only show that $V(C) = 0$ for any admissible matrix C for 1.

To establish this fact it is convenient first to define determinantal sums. Let A, B be two $n \times n$ matrices which agree in all but the first column; say $A = (A_1,\ldots,A_n)$, $B = (B_1, A_2,\ldots,A_n)$, then their <u>determinantal sum</u> with respect to the first column is

$$A \nabla B = (A_1 + B_1, A_2,\ldots,A_n). \tag{6}$$

If we are in a commutative ring, we have $\det(A \nabla B) = \det A + \det B$, but (6) makes sense over any ring, where det may not be defined.

LEMMA. <u>Let</u> R <u>be a fir, and</u>

$$C = (A_1 + A_1', A_2,\ldots,A_n) = A \nabla B',$$

<u>then if</u> A' <u>is non-full</u>, $v(C) = v(A)$.

<u>Proof</u>. By hypothesis, $A' = PQ$, where P is $n \times n - 1$ and Q is $n - 1 \times n$. Denote the first column of Q by Q_1, so that $Q = (Q_1\ Q')$, where Q' is $n - 1 \times n - 1$. Then

$$A = (A_1, PQ') = (A_1, P)) \begin{pmatrix} 1 & 0 \\ 0 & Q' \end{pmatrix}. \tag{7}$$

Secondly we have

$$A' = PQ = (PQ_1, PQ') = (PQ_1, P) \begin{pmatrix} 1 & 0 \\ 0 & Q' \end{pmatrix}.$$

It follows that

$$(8) \qquad C = (A_1 + PQ_1,\ PQ') = (A_1 + PQ_1,\ P)\begin{pmatrix} 1 & 0 \\ 0 & Q' \end{pmatrix} = (A_1, P)\begin{pmatrix} 1 & 0 \\ Q_1 & I \end{pmatrix}\begin{pmatrix} 1 & 0 \\ 0 & Q' \end{pmatrix}$$

Now the result follows by comparing (7) and (8).

We can now show that the kernel of the homomorphism $V: S \to Z$ contains all admissible matrices for 1. Suppose that A is admissible for 1; the condition for this is that $(A_0 + A_\infty\ A_*)$ is nonfull. Put $F = A_0 + A_\infty$, then

$$(A_\infty\ A_*) = (F - A_0\ A_*) = (F\ A_*) \nabla (-\ A_0\ A_*),$$

and by the lemma, $V(A) = v(A_0\ A_*) - v(A_\infty\ A_*) = 0$, which is what we wished to show. It now follows that all admissible matrices for a given $p \in U$ lead to the same value $v(p)$.

5. As we have seen in 4., for each prime divisor π we have a homomorphism $v: U^* \to Z$, which is clearly surjective, for we may take any full matrix A over R and reduce it to diagonal form as in the computation of the Dieudonné determinant. Let I be the set of all prime divisors and D the free abelian group on I, then the v_π combine to give a homomorphism

$$f: U^* \to D.$$

Again we can check that f is surjective, hence we obtain

$$GL_n(U)^{ab} \cong U^{*ab} \cong D \times N,$$

where $N = \ker\ f$.

In order to determine N more precisely, let us write $GL(R)$ for the stabilized group $\lim GL_n(R)$ and $E(R) = \lim E_n(R)$. As is well known, $E(R) = GL(R)'$ (cf. [1], p. 229). Consider any full matrices A and B over R; we have $f(A) = f(B)$ if and only if A and B have the same atomic factors, up to order and stable association. Suppose that $f(A) = f(B)$; take a complete factorization of A:

$$A = P_1 \ldots P_r$$

and let B be a product (in some order) of $Q_1, \ldots, Q_r$, where Q_i is stably associated to P_i. On replacing A by $\begin{pmatrix} A & 0 \\ 0 & I \end{pmatrix}$ for a suitably chosen unit matrix, and doing likewise for B, we may take each Q_i to be associated to

P_i (not just stably associated), say $Q_i = U_i P_i V_i$, where $U_i, V_i \in GL(R)$. Hence up to the order of the factors we have $B = P_1 \ldots P_r U_1 \ldots U_r V_1 \ldots V_r = AF$, where $F \in GL(R)$. Hence (remembering that $GL(U)' = E(U)$), we have

$$B \equiv AF \quad (\text{mod } E(U)).$$

It follows that

$$N = GL(R) \,/\, [GL(R) \cap E(U)].$$

When $R = k\langle X\rangle$ is the free algebra, one can show that $GL(R) \cap E(U) = E(R)$, hence we then have $N = GL(R) / E(R) \cong k^*$, and so

(9) $$k\langle X\rangle^{*ab} \cong D \times k^* ,$$

where D is a free abelian group. This solves our problem completely.

For general firs one obtains the formula

(10) $$U^{*ab} \cong D \times GL(R) \,/\, [GL(R) \cap E(U)],$$

and the second factor can in some cases be shown to be just $GL(R)^{ab}$ as above, and in others to be a proper homomorphic image. More precise results have been obtained by G. Révész [14], who also derives (9), by a slightly different route.

6. The scope of these results can still be extended somewhat. For any matrix A let us define the <u>inner rank</u> rk A as the least r such that $A = PQ$, where P has r columns. <u>E.g.</u>, a full matrix is just an $n \times n$ matrix of inner rank n. A ring R is said to be a <u>Sylvester domain</u> if for any $A \in {}^mR^r$, $B \in {}^rR^n$ such that $AB = 0$, we have

$$\text{rk } A + \text{rk } B \leq r.$$

This condition can easily be shown to be equivalent to Sylvester's law of nullity: $\text{rk } A + \text{rk } B \leq r + \text{rk } AB$. These rings were introduced by Dicks and Sontag [7], who also showed that Sylvester domains are the precise class of rings possessing a universal field of fractions inverting all full matrices. Thus every fir is a Sylvester domain; other examples are $k[x,y]$, the polynomial ring in two variables over a field (but not in more than two variables), and $Z\langle X\rangle$, the free algebra over Z. The full matrices over R, regarded as being contained in $GL(U)$, form a monoid $F(R)$, and we can form the universal abelian group $F(R)^{ab}$. Now one can show that for any Sylvester domain R with universal field of fractions U,

$$U^{*ab} \cong F(R)^{ab} . \tag{11}$$

This is proved by constructing two mutually inverse homomorphisms, one from left to right using Cramer's rule, and one from right to left using the Dieudonné determinant (cf. [5]). There is also another proof, by Revesz [14].

One can still enlarge the class of rings by replacing the inner rank by the <u>stable rank</u>, defined as $\lim \{\mathrm{rk}\begin{pmatrix} A & 0 \\ 0 & I_s \end{pmatrix} - s\}$. The weakly finite rings satisfying Sylvester's law of nullity for the stable rank are called <u>pseudo-Sylvester rings</u>; they again have a universal field of fractions, inverting all "stably full" matrices (cf. [6]), and for these rings one can again prove (11). They include for example the Weyl algebra $A_2(k)$, where k is commutative, $A_1(D)$ for any skew field, and $D[x,y]$.

REFERENCES

[1.] H. Bass, Algebraic K-theory, Benjamin (New York 1968).

[2.] P. M. Cohn, The embedding of firs in skew fields, Proc. London Math. Soc. (3) 23 (1971), 193-213.

[3.] P. M. Cohn, Free rings and their relations, LMS monographs No. 2, Academic Press (London, New York 1971).

[4.] P. M. Cohn, Skew field constructions, LMS Lecture Notes No. 27, Cambridge University Press (Cambridge 1977).

[5.] P. M. Cohn, The divisor group of a fir, to appear.

[6.] P. M. Cohn and A. H. Schofield, The law of nullity, to appear.

[7.] W. Dicks and E. D. Sontag, Sylvester domains, J. Pure and Appl. Algebra 13 (1978), 243-275.

[8.] J. Dieudonné, Les déterminants sur un corps non-commutatif, Bull. Soc. Math. France 71 (1943), 27-45.

[9.] J. L. Fisher, Embedding free algebras in skew fields, Proc. Amer. Math. Soc. 30 (1971), 453-458.

[10.] G. Frobenius and L. Stickelberger, Über Gruppen von vertauschbaren Elementen, J. reine u. angew. Math. 86 (1879), 217-262.

[11.] N. Jacobson, Pseudo-linear transformations, Ann. Math. 38 (1937), 484-507.

[12.] N. Jacobson, Theory of rings, Amer. Math. Soc. (Providence 1943).

13. T. Nakayama, A note on the elementary divisor theory in non-commutative domains, Bull. Amer. Math. Soc. 44 (1938),719-723.

14. G. Révész, On the abelianized group of universal fields of fractions, to appear.

15. O. Teichmüller, Der Elementarteilersatz für nichtkommutative Rings, Sitz.-ber. Preuss. Akad. Wiss. (1937),169-177.

16. J. H. M. Wedderburn, Non-commutative domains of integrity, J. reine u. angew. Math. 167 (1932),129-141.

Bedford College
Regent's Park
London NW1 rNS
ENGLAND

Contemporary Mathematics
Volume 13, 1982

WEAK COHOMOLOGY

by Moss E. Sweedler*

Dedication

This article is dedicated to the matchmaker, with gratitude.

Around 1965 Harry Allen and I were instructors at M.I.T. Jake suggested that we work together on his conjectured classification of the Witt Lie algebras. He figured that Harry knew Lie algebras and their Galois descent and that I knew Hopf algebras, which might provide a replacement for the Galois group for purely inseparable descent. Jake was right and in [1] we wrote.

Even though the subject of the present article is a survey of other work I am glad for the opportunity to publicly acknowledge my thanks to Jake.

§1. INTRODUCTION

In [3] Darrell Haile, Richard Larson and I describe a variation of the classical Amitsur and Galois cohomology theories which uncovers new invariants even for the most classical field extensions such as $\mathbb{C}$ over $\mathbb{R}$ and finite field extensions of finite fields. In [3] the cohomology theory is developed, the equivalence classes of algebras which the theory classifies is presented and many examples are given. Further development of the theory can be found in [4].

One aspect of the theory is that the cohomology groups are <u>not groups</u>. They are monoids in which the classical cohomology is the subgroup of invertible elements. For example for $\mathbb{C}$ over $\mathbb{R}$ the second cohomology is the multiplicative monoid $\{-1,0,+1\}$.

* The writing of this survey article supported by the John Simon Guggenheim Memorial Foundation.

0271-4132/82/0516/$03.75

In monoids there are often idempotent elements other than the unit. The multiplicative monoid $\{-1,0,+1\}$ for $\mathbb{C}$ over $\mathbb{R}$ has the idempotent 0 in addition to the unit 1. For field extensions with Galois group G the idempotents in the second cohomology monoid are in one correspondence with certain G sets. The classification of the idempotents in terms of G sets appears in [3,§7] and is the main topic of this survey. The sets on which the Galois group acts are partially ordered and the group action partially respects the partial order as explained at (3.4). The posets on which the Galois group acts are "photogenic" and some of their pictures appear at (3.9)-(3.13).

In section two of this survey the crossed product construction is used to motivate the definition of <u>weak two cocycle</u>. The weak two cocycle conditions are just what is needed to make the crossed product associative and unitary. The cohomology monoid consists of equivalence classes -- called cohomology classes -- of weak cocycles. In section two we hardly do more than define the weak cohomology since the aim of this survey is to present the classification of the idempotents.

The last section is a brief mention of the role played by a computer in this research. Many of our early examples were found by computer and noticing common features of these examples led to conjectures which we were able to prove.

§2. WEAK COHOMOLOGY

The definition of weak Galois two cocycles is motivated by the crossed product [2, p. 79]. Although the development was more 'round about -- involving Amitsur cohomology and multiplication alteration [5] -- we present it here the way it might have been.

Suppose A is a Galois field extension of R with Galois group G. (By Galois we mean Galois and $\dim_R A < \infty$.) To construct the crossed product of A by G form $^{\#}G$ copies of A

2.1 $$A \times G = \{A \times g\}_{g \in G}$$

and take their direct sum

2.2 $$A\#G = \bigoplus_{g \in G} A \times g$$

The product on $A\#G$ involves a two cochain i.e., a map $\sigma : G \times G \to A$. Component wise the product is given by

$$(a \times g)(b \times h) = (ag(b)\sigma(g,h)) \times gh$$

Since this depends on σ we write $A \#_\sigma G$. Experimenting with the product leads to the conclusion that is is associative if and only if

2.3 $$(1 \times g)[(1 \times h)(1 \times \ell)] = [(1 \times g)(1 \times h)](1 \times \ell)$$

for all $g,h,\ell \in G$ and this is satisified if and only if

2.4 $$[g(\sigma(h,\ell))][\sigma(g,h\ell)] = [\sigma(gh,\ell)][\sigma(g,h)]$$

Some further experimenting with the product shows that if there is an identity element it must be of the form $\lambda \times 1$ where $0 \neq \lambda \in A$ and 1 is the identity of G. For $\lambda \times 1$ to be the identity of $A \#_\sigma G$ considering

2.5 $$(\lambda \times 1)(1 \times g) = (1 \times g) = (1 \times g)(\lambda \times 1)$$

shows that σ must satisfy for all $g \in G$

2.6 $$\lambda\sigma(1,g) = 1 = g(\lambda)\sigma(g,1).$$

Hence $\lambda = \sigma(1,1)^{-1}$ and

2.7 $\sigma(1,g)$ and $\sigma(g,1)$ are invertible for all $g \in G$.

Of course <u>invertibility</u> of $\sigma(1,g)$ is equivalent to $\sigma(1,g) \neq 0$ since A is a field. For generalization to a broader setting invertibility is the proper condition.

2.8 DEFINITION: $\sigma : G \times G \to A$ is a weak two cocycle if (2.4) and (2.7) are satisified.

As we have seen σ is a weak two cocycle if and only if $A \#_\sigma G$ is an associative unitary algebra. However even when σ is merely a two cochain $A \#_\sigma G$ has a product as indicated.

Here are two examples:

2.9 σ is the constant function 1 i.e., $\sigma(g,h) = 1$ for all $g,h \in G$.

2.10 $$\sigma(g,h) = \begin{cases} 1 & \text{if } g \text{ or } h \text{ is the identity of } G \\ 0 & \text{otherwise} \end{cases}$$

In example (2.9) the equations (2.4) and (2.7) are obviously satisfied. There is a tricky natural isomorphism of $A \#_\sigma G$ with $\mathrm{End}_R A$. Hence $A \#_\sigma G$ is isomorphic to $n \times n$ matricies over R where $n = {}^\# G$.

In example (2.10) equation (2.7) is obviously satisfied and it is only a small effort to show that both sides of (2.4) are zero unless at least two of g,h,ℓ are the identity of G in which case both sides of (2.4) are 1.

Recall that a Galois two cocycle is a two cochain satisfying (2.4) and which is invertible with respect to pointwise product. Hence the Galois two cocycles are precisely the invertible weak two cocycles.

In a certain complex the two maps

2.11 $$\begin{aligned} &G \to A \\ &g \to \sigma(1,g) \\ &g \to \sigma(g,1) \end{aligned}$$

arise as "degeneracies" of σ. This illustrates the fact that the notion of weak cocycle is derived from the notion of cocycle by relaxing the condition that the cocycle be invertible but still requiring that the degeneracies of the cocycle be invertible. Of course the cocycle must have trivial coboundary but this can be expressed without invertibility simply by moving the terms of the coboundary usually involving the inverse to the other side of the equation. Equation (2.4) does this for two cochains

Back to one and two cochains. Suppose $\gamma: G \to A$ i.e., γ is a one cochain and suppose that γ is invertible i.e., $\mathrm{Im}\,\gamma \in A-\{0\}$. The coboundary of γ is denoted $\delta\gamma$ and is the map

2.12 $$\begin{aligned} G \times G &\longrightarrow A \\ (g,h) &\longrightarrow [g(\gamma(h))][\gamma(gh)^{-1}][\gamma(g)]. \end{aligned}$$

As is well known $\delta\gamma$ is a Galois two cocycle. Suppose σ is a two cochain and τ is the two cochain $[\sigma][\delta\gamma^{-1}]$; here as elsewhere we use the pointwise product of maps from $G \times G \to A$. The map

2.13
$$\begin{aligned} A \#_\sigma G &\longrightarrow A \#_\tau G \\ a \times g &\longrightarrow a\gamma(g) \times g \end{aligned}$$

is bijective, it is multiplicative and preserves some additional structure. Two two cochains σ and τ are called <u>cohomologous</u> if there is an inverrible one cochain γ where

2.14
$$\tau = [\sigma][\sigma\gamma^{-1}] .$$

Being cohomologous is an equivalence relation and yields equivalence classes -- called cohomology classes -- of two cochains. The pointwise product structure on two cochains induces a product structure on the cohomology classes by which they form a monoid. $M^2(G,A)$ denotes the submonoid of equivalence classes of weak two cocycles and $H^2(G,A)$ denotes the submonoid of equivalence classes of Galois two cocycles. A two cochain σ is a weak two cocycle if and only if every two cochain cohomologous to σ is a weak two cocycle and σ is invertible if and only if every two cochain cohomologous to σ is invertible. $H^2(G,A)$ is the largest subgroup of the monoid $M^2(G,A)$ in that $H^2(G,A)$ is the set of invertible elements of $M^2(G,A)$. The "old fashioned" second Galois cohomology group of the extension A over R is precisely $H^2(G,A)$.

The crossed product construction provides a correspondence between the cohomology and the equivalence classes of algebras classified by the cohomology. Multiplication alteration provides another means of obtaining the correspondence between the cohomology and the equivalence classes of algebras classified by the cohomology. Probably descent theory does too. Rather than going further along these lines we wish to present the classification of idempotent weak two cocycles.

§3. IDEMPOTENT WEAK TWO COCYCLES

Notice that the examples of weak two cocycles (2.9) and (2.10) are functions from $G \times G$ to A which only take the values 0 and 1. Hence they are idempotent under pointwise multiplication. Clearly the idempotent cochains are the ones which only take the values 0 and 1.

It is not hard to show that if two idempotent weak two cocycles are cohomologous then they are equal and that a cohomology class in $M^2(G,A)$ is idempotent if and only if it contains a -- necessarily unique -- idempotent weak two cocycle. Hence the idempotent weak two cocycles are in one to one correspondence with the idempotent cohomology classes in $M^2(G,A)$. As indicated in the last section weak two cocycles give rise to algebras via the crossed product. Even for the complex numbers over the reals (2.10) gives an idempotent weak two cocycle other than the identity.

Suppose $\sigma: G\times G \to A$ is an idempotent weak two cocycle. Then actually $\sigma: G \times G \to \{0,1\}$ and (2.7) becomes

3.1 $$\sigma(1,g) = 1 = \sigma(g,1) \quad \text{for all} \quad g \in G.$$

Since 0 and 1 lie in the fixed field of the Galois group the group action in (2.4) disappears and (2.4) becomes for all $g,h,\ell \in G$

3.2 $$[\sigma(h,\ell)][\sigma(g,h\ell)] = [\sigma(gh,\ell)][\sigma(g,h)]$$

Notice that the group action on the field has completely gone in the sense that if G is a finite group and σ is a map from $G \times G$ to the multiplicative monoid $\{0,1\}$ satisfying (3.1) and (3.2) then σ gives an idempotent weak two cocycle σ' for any field extension A over R for which G is the Galois group. σ' is the composite

3.3 $$G \times G \xrightarrow{\sigma} \{0,1\} \hookrightarrow A$$

Moreover all idempotent weak two cocycles from $G \times G$ to A arise in this manner.

The classification of idempotent weak two cocycles is in terms of actions of the group G on partially ordered sets.

3.4 DEFINITION: A partially ordered set (poset) P is <u>rooted</u> if it has a unique minimal element ρ which is called the <u>root</u> of P. Suppose P is a rooted poset with root ρ and P is a left G set for the group G. P is a <u>lower subtractive G poset</u> if the following condition is satisfied for all $g,h,\ell \in G$ with $g\cdot\rho \le \ell\cdot\rho$

* $$g\cdot\rho \le h\cdot\rho \le \ell\cdot\rho \quad \text{if and only if} \quad g^{-1}h\cdot\rho \le g^{-1}\ell\cdot\rho$$

P is a lower subtractive G <u>orbit</u> poset if in addition $P = G\cdot\rho$ i.e. P is a G orbit.

For pictures of lower subtractive G orbit posets look at (3.9)-(3.13).

If P is a lower subtractive G poset define $e_p:G \times G \to \{0,1\}$ by

3.5 $$e_p(g,h) = \begin{cases} 1 & \text{if } g\cdot\rho \le gh\cdot\rho \\ 0 & \text{otherwise} \end{cases}$$

where ρ is the root of P. The key result is that e_p satisfies (3.1) and (3.2). Hence if G is the Galois group of A over R then (3.5) gives an idempotent weak two cocycle where in (3.5) use 1_A for 1 and 0_A for A. Moreover all idempotent weak two cocycles from $G \times G$ to A arise in this fashion. Shortly we will give an indication of how to obtain the lower subtractive G orbit poset from the idempotent weak two cocycle.

In (3.5) we never get out of $G\cdot\rho$ which is why lower subtractive G *orbit* posets are of interest. Also if two lower subtractive G orbit posets are in bijective correspondence by a map preserving the G set structure and poset structure they are called equivalent and will give the same idempotent weak two cocycle. The classification theorem is

3.6 THEOREM: *Suppose A is Galois over R with Galois group G. There is a bijective correspondence between the idempotent weak two cocycles from $G \times G$ to A and the equivalence classes of lower subtractive G orbit posets. If $[P]$ is an equivalence class of lower subtractive G orbit posets the corresponding idempotent weak two cocycle is e_p.*

Here is a brief sketch of how to get the lower subtractive G orbit poset from the idempotent weak two cocycle $\sigma:G \times G \to A$. Define a relation "<<" on G by

3.7 $$g \ll h \quad \text{if} \quad \sigma(g,g^{-1}h) = 1$$

This relation is reflexive and transitive and satisfies (3.4,*) with 1 the unit of G replacing ρ. In general there is $g \neq h \in G$ with $g \ll h$ and $h \ll g$. Define

3.8 $$H_\sigma = \{g \in G : g \ll 1\}$$

This is a subgroup. Let P be the left G set G/H_σ, cosets of the form gH. Then "<<" induces a partial order relation on P_σ by which P_σ is a lower subtractive G orbit poset with root $1H_\sigma$. This is the lower subtractive G orbit poset which corresponds to the idempotent weak two cocycle σ.

Now to the pictures of lower subtractive G orbit posets.

For examples (3.9)-(3.12) G is a cyclic group of order n with generator g and G acts on itself by left translation. Then in (3.9) G is a lower subtractive G orbit poset with the partial order

3.9

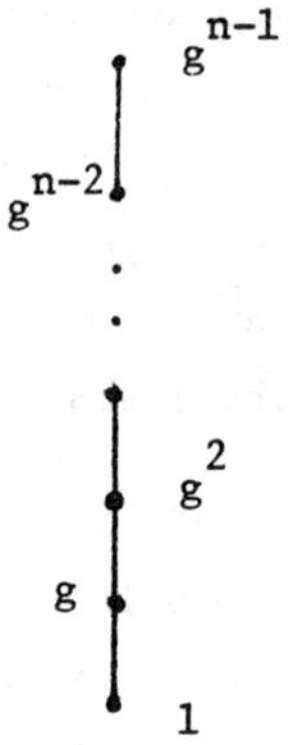

In the next example suppose $n = 6$ and that G is a lower subtractive G orbit poset with the partial order

3.10

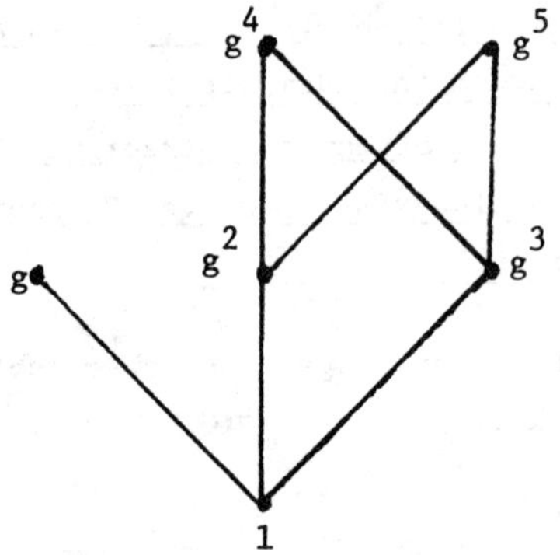

In the next two examples suppose $n = 4$ and that G is a lower subtractive G orbit poset with either of the two partial orders

3.11

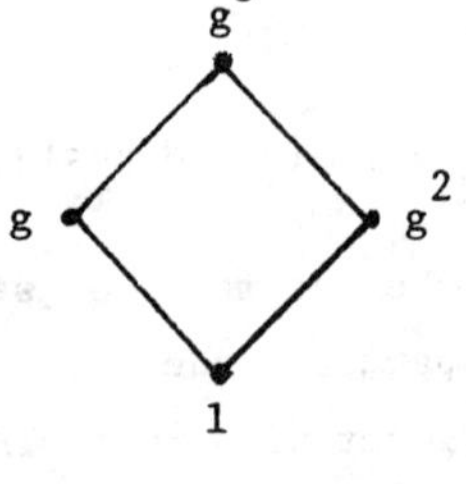

3.12

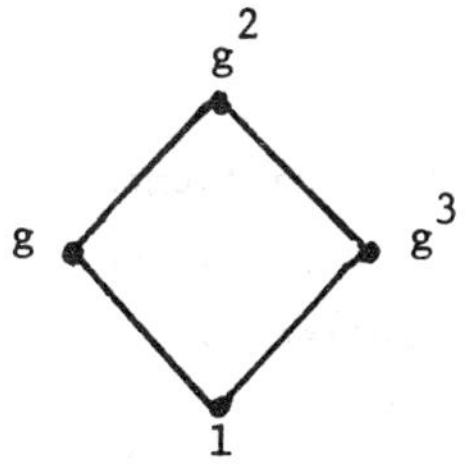

Notice that (3.11) and (3.12) differ by more than choice of generator since the top element g^3 in (3.11) is a generator of G but the top element g^2 in (3.12) is not a generator of G.

As a last example let G be any group with subgroup H. Let P be the set of cosets of the form gH so that P has a natural left G set structure. Give P the rooted poset structure where 1H = H is the root -- i.e. $1H \leq gH$ for all $g \in H$ -- and all other distinct pairs of elements of P are incomparable. P looks like

3.13

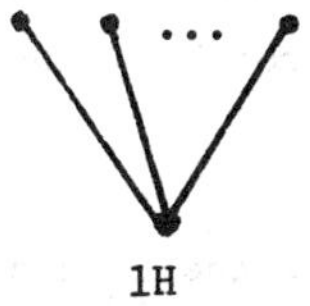

This example is easily verified to be a lower subtractive G orbit poset since (3.4,*) is trivially satisfied because P has no length three chains.

How does one <u>easily</u> verify that a rooted poset P which is a G set is a lower subtractive G orbit poset? Section eight of [3] is devoted to presenting techniques which simplify the verification. The elements of P which lie just above the root play an important role.

So now we can start with a picture of a lower subtractive G orbit poset, get an idempotent weak two cocycle from (3.5) and then form the crossed product algebra using the idempotent weak two cocycle. What can we tell about the algebra from the picture? In [3, 7.28,e] the length of chains in the original picture is related to the degree of nilpotence of the radical. In [4] Darrell Haile shows that if the poset is a tree -- has no loops -- then a certain map of cohomology is an isomorphism. More work needs to be done relating the posets to the algebras. In fact [3] begs for continued research in several different areas.

§4. THE ROLE OF THE COMPUTER

As mentioned earlier, finding idempotent weak two cocycles is equivalent to finding maps from $G \times G$ to $\{0,1\}$ satisfying (3.1) and (3.2). Before we knew anything about lower subtractive G orbit posets we had a computer produce such maps for various small groups G. We kept noticing distinguished subgroups and eventually realized that for the $\sigma:G \times G \to \{0,1\}$ in our print-out the distinguished subgroup could be characterized as

$$\{g \in G: \ \sigma(g,g^{-1}) = 1\} \tag{4.1}$$

Once we saw this it was not hard to prove in general for $\sigma:G \times G \to \{0,1\}$ satisfying (3.1) and (3.2) that (4.1) gives a subgroup of G.

Notice that (4.1) is (3.7) and (3.8) put together but we did not have them separately at that point. In fact we had no partial orders or anything about lower subtractive. The only invariant we had for idempotent weak two cocycles was this subgroup (4.1) and many different idempotent weak two cocycles can have the same subgroup. Nevertheless this already seemed like a good computer inspired result and we did not know then that we would be able to do better some day.

It is impossible to say how the idea of working with an order relation was conceived. But the original order relation was (3.7) with the quick realization that (3.8) then gives the same subgroup as (4.1) and the eventual realization that "$\ll$" induces a poset structure on G/H_σ. From that point the axiomatization of lower subtractive G orbit poset was straight forward.

I am sure that somehow knowing the "$\sigma(g,g^{-1}) = 1$" condition in (4.1) helped guide the experimentation which produced (3.7) and eventually the classification of idempotent weak two cocycles by lower subtractive G orbit posets. So the entire classification is in some measure inspired by observation of computer produced examples.

BIBLIOGRAPHY

1. H. P. Allen and M. E. Sweedler, A theory of linear descent based upon Hopf algebraic techniques, Journal of Algebra, 12, (1969), 242-294.

2. E. Artin, C. J. Nesbitt and R. Thrall, Rings with Minimum Condition, University of Michigan Press, 1944.

3. D. E. Haile, R. G. Larson and M. E. Sweedler, A new invariant for C over R, to appear.

4. D. E. Haile, On crossed product algebras arising from weak cocycles, Journal of Algebra, to appear.

5. M. E. Sweedler, Multiplication alteration by two-cocycles, Illinois J. Math. 15, (1971) 302-323.

Cornell University
Ithaca, NY 14853

Contemporary Mathematics
Volume 13, 1982

APPLICATIONS OF THE CLASSIFICATION OF FINITE SIMPLE GROUPS TO BRAUER GROUPS

Murray Schacher*

To Professor Jacobson, with gratitude

This note concerns joint work of mine with Burton Fein and William Kantor (see [2]) involving an interesting connection between the finite simple groups and Brauer groups. The end result is a theorem about the relative Brauer group of a pair of number fields, and an equivalent arithmetic property about pairs of number fields, whose proof at this time requires the classification of finite simple groups. We rely on the theorem, established during the last year, that this classification is complete. The derivation of our arithmetic property from the classification is accomplished in [2] and so will not be reproduced here; rather I hope to give a fuller exposition of the interconnections between group theory and number theory involved in [2].

Let $L \supset K$ be fields. We let $B(L/K)$, the relative Brauer group of L over K, denote those Brauer classes of finite-dimensional central simple K-algebras split by L. This investigation began with the natural question: if $L \supsetneq K$ are algebraic number fields, is $B(L/K) \neq 0$?

In all that follows, p will denote a prime integer, and $|A|$ will denote the order of a finite set A. Suppose $[L:K] = n$. One knows that $B(L/K)$ is an abelian group with n-torsion ([5], Theorems 28.5 and 29.22). If $p|n$ we set $B(L/K)_p$ = the elements of $B(L/K)$ of order p^a, some integer $a \geq 0$; this is the "p-primary subgroup" of $B(L/K)$. Then $B(L/K) \cong \bigoplus_{p|n} B(L/K)_p$. It is clear that $B(L/K)$ is non-0 (resp. infinite) $\Leftrightarrow B(L/K)_p$ is non-0 (resp. infinite) for some $p|n$. If G is a finite group, a p-element of G is an element whose order is a power of p.

With this notation we will prove:

THEOREM A: _The following are equivalent:_

(1) _For all number fields_ $L \supsetneq K$, $B(L/K) \neq 0$

(2) _For all number fields_ $L \supsetneq K$, $B(L/K)$ _is infinite._

(3) _Suppose_ G _is a finite group acting transitively on a set_ Ω .

---

*This paper was supported in part by NSF Grant No. MCS 78-27583.

0271-4132/82/0517/$02.25

<u>If, for all</u> $p \mid |G|$ <u>every</u> p-<u>element in</u> G <u>has a fixed point on</u> Ω, <u>then</u> $|\Omega| = 1$.

(4) <u>Suppose</u> G <u>is a finite simple group and</u> H <u>a maximal subgroup</u>. <u>Then</u> $\exists\, p \mid |G|$ <u>and a</u> p-<u>element</u> $g \in G$ <u>so that</u>: <u>no conjugate of</u> g <u>is in</u> H.

(5) <u>Suppose</u> G <u>is a finite group and</u> $H \subsetneq G$ <u>a subgroup</u>. <u>Then</u> $\exists\, p \mid |G|$ <u>and a</u> p-<u>element</u> $g \in G$ <u>so that</u>: <u>no conjugate of</u> g <u>is in</u> H.

Before proving Theorem A, certain remarks are in order.

<u>Remark 1</u>: Theorem A is neither stated nor proved in [2]. In [2] we prove (4) directly using the classification of finite simple groups, and then derive (2), (3) and (5) as consequences. I feel this would be a good place to show (1)-(5) are actually equivalent.

<u>Remark 2:</u> The conclusion in (5) does not in general hold for all p which divide $|G|$. If $G = A_4$ and $H = \langle(12)(34)\rangle$, then one can easily check: every 2-element in G has a conjugate in H. From the viewpoint of (3): every 2-element in G has in fact two fixed points when acting on the left cosets of H by left translation. We use this later to give examples of situations where $p \mid [L:K]$ but $B(L/K)_p = 0$. By class field theory, however, this could never happen when $L \supset K$ is normal.

<u>Remark 3:</u> Item (2) in Theorem A is still valid if $L \supsetneq K$ are function fields in one variable over a finite field (even if L/K is inseparable). In the function field case (1) through (5) are still equivalent, and the proof we are about to give will still hold, if all extensions $L \supsetneq K$ are assumed to be separable.

<u>Remark 4:</u> One cannot in (3) or (5) replace p-elements with elements of order p. Examples given in [2] show both statements then become false.

The proof of Theorem A will require the description of finite dimensional central division algebras over a number field K by Hasse invariants; the standard reference here will be to [1]. For such a division ring D, write $\mathrm{inv}_{\mathfrak{p}} D$ for the invariant of D at a prime $\mathfrak{p}$ of K; this is an element of the group Q/Z. If $L \supsetneq K$, then [1, Chapter 7] shows: L splits D $\Leftrightarrow$ for every $\mathfrak{p}$ with $0 \neq \mathrm{inv}_{\mathfrak{p}} D = \frac{a}{m}$, $(a,m) = 1$, every extension of $\mathfrak{p}$ to L must have local degree divisible by m. Using this and the fact that Hasse invariants must sum to 0 in Q/Z, we obtain:

(i) $B(L/K)_p \neq 0 \Leftrightarrow$ there are two primes $\mathfrak{p}_1$ and $\mathfrak{p}_2$ of K so that: every extension of $\mathfrak{p}_i$ to L has local degree divisible by p, $i = 1,2$.

(ii) $B(L/K)_p$ is infinite $\Leftrightarrow$ there are infinitely many primes $\mathfrak{p}_1, \mathfrak{p}_2, \ldots, \mathfrak{p}_i, \ldots$ of K so that: every extension of $\mathfrak{p}_i$ to L has local degree divisible by p.

Thus items (i) and (ii) ask for arithmetic properties about number fields which are equivalent to (1) and (2) in Theorem A; it is curious that, at least to date, one needs the classification to show such well-behaved primes exist.

We will show first that (3) ⇒ (2) in Theorem A. Suppose $L = K(\alpha)$ with α a primitive element of L/K, $f(x) \in K[x]$ the irreducible polynomial of L/K, and M is the splitting field of $f(x)$ over K. Set $G = G(M/K)$, the Galois group of M over K, and let $\Omega = \{\alpha = \alpha_1, \ldots, \alpha_n\}$ the roots of $f(x)$ in M. Then G acts transitively on Ω ; if $H = G(M/L)$ then this action is equivalent to the action of G on the set of left cosets G/H of H in G by left translation (where group elements act on the left). An element $g \in G$ has a fixed point on Ω (or on G/H) ⇔ $g\,a\,H = aH$ some a ⇔ $a^{-1}gaH = H$ ⇔ $a^{-1}ga \in H$ ⇔ some conjugate of g is in H. Using (3), choose some $\sigma \in G$ a p-element having no fixed points on Ω. If P is an unramified prime of L we write $\phi(P)$ for the Frobenius of L/K at P. By the Cebotarev density theorem there are infinitely many primes $P_1, \ldots, P_i, \ldots$ of L so that $\phi(P_i) = \sigma$. We can arrange so that $P_i \cap K = p_i$ and $p_1, p_2, \ldots, p_i, \ldots$ are distinct primes of K. By [4, Prop. 2.8] the local degrees of extensions of p_i to L are given by the cycle lengths of σ on Ω ; the cycles are of length p^s since σ is a p-element, and none of them is 1, by (3) and the choice of σ. Thus every extension of p_i to L has degree p^s, $s > 0$, and so (ii) gives that $B(L/K)_p$ is infinite. Thus (3) ⇒ (2). Clearly (2) ⇒ (1).

We show next that (3) ⇒ (5). Suppose G is a finite group and $H \subsetneq G$ a subgroup. Let $\Omega = G/H$ be the set of left cosets of H in G; G acts transitively on Ω by left translation. By (3), $\exists$ a p-element $g \in G$ so that $g\,aH \neq aH$ all a, and so no conjugate of g is in H.

Clearly (5) ⇒ (4). To show (4) ⇒ (3), assume (4) and suppose (G,Ω) gives a minimal counter-example to (3). For any $x \in \Omega$ set H = stabilizer of x in G. We may then assume $\Omega = G/H$ with the action left translation. If N is the set of elements of G acting trivially on Ω, then $N \triangleleft G$, and in fact $N = \bigcap_{g \in G} gHg^{-1}$ = the largest normal subgroup of G contained in H. By minimality $(G/N,\Omega)$ cannot be a counter-example to (3), so we may assume $N = \{1\}$ and that G acts faithfully on Ω. Then:

(I) We may assume H is maximal in G.

If H is not maximal, then $H \subsetneq H_1$ with H_1 maximal in G. As every p-element g in G has a conjugate in H, then g has a conjugate in H_1 and $(G, G/H_1)$ is a counter-example to (3).

(II) G is simple.

If $\{1\} \neq N \lhd G$, then NH is a subgroup of G, and so $G = NH$ by maximality of H as $N \not\subset H$. Now N acts transitively on G/H. If $|N| < |G|$, then $\exists$ a p-element $n \in N$ with no fixed point on G/H by minimality of $|G|$; this forces $N = G$.

Now (I) and (II) say a minimal counter-example to (3) looks like (4), and so (4) $\Rightarrow$ (3). We have now that (3), (4), (5) are equivalent and (3) $\Rightarrow$ (2) $\Rightarrow$ (1). We finish by showing (1) $\Rightarrow$ (5). Suppose then G is a finite group and $H \subsetneq G$ a subgroup. First find a pair of number fields $M_0 \supset K_0$ with $G = G(M_0/K_0)$; this is possible since S_n is a Galois group over Q and $G \subset S_n$ for some n. By [3] one can extend K_0 to K so that, for $M = M_0K$, $G = G(M/K)$ and M/K is everywhere unramified. Let L = fixed field of H, so $H = G(M/L)$. By (1) $B(L/K)_p \neq 0$ for some p, and so $\exists$ a prime $\mathfrak{p}$ of K with every extension of $\mathfrak{p}$ to L of local degree divisible by p. Choose P a prime of L with $P \cap K = \mathfrak{p}$ and set $\sigma = \phi(P)$; this makes sense as $P/\mathfrak{p}$ is unramified. By [4, Prop, 2.8], every cycle of σ on G/H has length divisible by p. If the order of σ is $p^a q$ with $(p,q) = 1$, then $g = \sigma^q$ is a p-element of G with no fixed point on G/H, and so no conjugate in H. This finishes the proof of Theorem A.

Example: Construct, as in (1) $\Rightarrow$ (5), a pair of number fields $M \supset K$ with $G(M/K) = A_4$ and M/K everywhere unramified. Let L = fixed field of H where $H = \langle(12)(34)\rangle$. Then $[L:K] = 6$, but by Remark 2 $B(L/K)_2 = 0$, while $B(L/K)_3$ is infinite by Theorem A. One can actually make such an example with $K = Q$, but we will not do it here.

This sort of pathology is extended in [2] in the following way: Suppose $\{p_1, \ldots, p_N\}$ is any set of N primes. Then $\exists$ a pair of number fields $L \supsetneq K$ so that: $p_i \mid [L:K]$ for $i = 1,2, \ldots, N$ but $B(L/K)_p = 0$ for $p = p_i$, $i = 1,2, \ldots, N$. Of course then $B(L/K)_q$ is infinite for some other prime q dividing $[L:K]$. Predicting what q might be given $\{p_1, \ldots, p_N\}$ will involve interesting (although perhaps monstrously difficult) problems in group theory.

The analogue of Theorem A for function fields is an open problem. In fact, one can ask: If $L \supsetneq K$ are function fields in one variable over Q, is $B(L/K) \neq 0$? The methods of [2] give an affirmative answer when L/K is normal.

REFERENCES

[1] E. Artin and J. Tate, Class Field Theory, Benjamin, New York, 1968 .

[2] B. Fein, W. Kantor, and M. Schacher, Relative Brauer groups II, Crelle's Journal, to appear.

[3] A. Fröhlich, On non-ramified extensions with prescribed Galois group, Mathematica 9 (1962), 133-134.

[4] G. Janusz, Algebraic Number Fields, Academic Press, New York, 1973.

[5] I. Reiner, Maximal Orders, Academic Press, New York, 1975.

University of California, Los Angeles
Los Angeles, California 90024

Contemporary Mathematics
Volume 13, 1982

GENERIC STRUCTURES AND FIELD THEORY

by

David J. Saltman*

In this paper we will outline some results concerning various generic constructions in Galois theory and the theory of simple algebras. We introduce the notion of a retract of a purely transcendental extension. A relationship is traced between these retracts, Noether's Problem, the generic division algebras, and certain lifting and approximation problems (as in the Grunwald-Wang theorem). For example, we can show that Swan's counterexample to Noether's Problem is a retract of a purely transcendental extension of Q, though not, of course, purely transcendental itself. Some of the results we will mention here have appear in [6], others will appear elsewhere in greater detail.

To fix some notation, F will always be an infinite field. All algebras, domains etc. will be associative F algebras. All maps will be F algebra homomorphisms. If $\varphi: R \to S$ is such a map, then S is an R module via φ. As φ may vary, it is useful to write the tensor product $M \otimes_R S$ as $M \otimes_\varphi S$. To say that F has characteristic prime to an integer n will mean that F has characteristic 0 or characteristic p, and p does not divide n. For such F and n, we will use $\rho(n)$ to denote a primitive n^{th} root of one. A field K, containing F, we call a rational F field if K is a purely transcendental extension of F.

The generic structures referred to in the title can best be described using an example. Let n be a positive integer and F a field of characteristic prime to n, containing a primitive n^{th} root of one. Consider fields $L \supseteq K \supseteq F$ such that L/K is Galois with group C_n, the cyclic group of order n. Kummer theory says that all such L/K can be described as follows. Set $R = F[x](1/x)$, the polynomial ring localized at x. Set $S = R[y]/(y^n - x)$. S/R is a Galois extension of commutative rings with group C_n. Furthermore,

*The author is grateful for support from the Sloan Foundation and the N.S.F.

0271-4132/82/0518/$03.00

if L/K is as above, then there is a $\varphi:R \to K$ such that $L \simeq S \otimes_{\varphi} K$. Note that if $K \supseteq F$ is any field, and $\varphi:R \to K$ is an F algebra map, then $S \otimes_{\varphi} R$ may not be a field, but it is Galois over K, and so is a direct sum of fields. In this paper, a Galois extension of a field will always mean an extension of this slightly more general type.

The above example motivates the following definition (see [6]). For a field F and a finite group G, we say a commutative ring extension S/R is a generic Galois extension for G over F if and only if:

1a) $R = F[x_1,\ldots,x_n](1/s)$ for some n and some $0 \neq s \in F[x_1,\ldots,x_n]$.

1b) S/R is Galois with group G.

1c) If K is a field containing F and L/K is Galois with group G then there is a $\varphi:R \to K$ such that $L \cong S \otimes_{\varphi} K$, the isomorphism preserving the action of G.

We call rings of the form 1a) localized polynomial rings. The fact that $L \cong S \otimes_{\varphi} K$ we express by saying that φ realizes L/K.

In [6], generic Galois extensions were constructed for various groups G and fields F. To mention one case, suppose G is abelian of exponent n, n is prime to the characteristic of F, and the cyclotomic extension $F(\rho(q))/F$ is cyclic, for q any prime power divisor of n. Then there is a generic Galois extension for G over F. Note that the above conditions hold if, for example, n is odd or F has nonzero characteristic.

Generic structures can also be defined for simple algebras. To be completely general, we consider classes of simple algebras as follows. Let $\mathcal{C}$ be a class of simple F algebras. $\mathcal{C}$ is called an F class of simple algebras if two properties hold. First of all, there is an integer n such that all elements of $\mathcal{C}$ have degree n (dimension n^2 over their centers). Secondly, if $A \in \mathcal{C}$ has center K and L is a field containing K, then $A \otimes_K L \in \mathcal{C}$. As examples of such classes, we can let $\mathcal{C}$ be all simple F algebras of degree n or all cyclic F algebras of degree n.

If $\mathcal{C}$ is an F class of simple algebras, a pure generic algebra for $\mathcal{C}$ is an Azumaya algebra A with center R such that

2a) R is a localized polynomial ring.

2b) If K is an F field and φ $R \to K$ is an F map, then $A \otimes_{\varphi} K \in \mathcal{C}$.

2c) If $B \in \mathcal{C}$ has center K then there is a $\varphi:R \to K$ such that $B \cong A \otimes_{\varphi} K$.

As an example of the existence of pure generic algebras, we quote the following.

THEOREM 1: Suppose n is an integer, F has characteristic prime to n, and $F(\rho(Q))/F$ is cyclic for q any prime power divisor of n. Let be the F class of cyclic algebras of degree n. Then there is a pure generic algebra for C.

The pure generic algebra of Theorem 1 is constructed as follows. Let S'/R' be a generic Galois extension for C_n over F. Set $R = R'[x](1/x)$, set $S = S' \otimes_{R'} R$, and set A to be the cyclic Azumaya R algebra $(S/R,x)$. Then A/R is the desired pure generic algebra.

For convenience sake, we use the term F class if C is either an F class of simple algebras or the class of all Galois extensions with a fixed group G. A generic structure for C is then either a pure generic algebra or a generic Galois extension.

The existence of generic structures for F classes is closely related to certain lifting problems. To be precise, we say a group G has the lifting property over F if the following holds. Let R be a local F domain with maximal ideal M. Assume L is a Galois extension of R/M with group G. Then there is a Galois extension T/R such that $T \otimes_R R/M \approx L$.

In a similar manner, an F class C of simple algebras has the lifting property if for such R and N, and for any $A \in C$ with center R/M, there is an Azumaya R algebra B such that $B \otimes_R R/M \approx A$ and $B \otimes_R K \in C$, where K is the quotient field of R.

We now turn to the relationship between lifting properties and generic structures.

THEOREM 2: Let C either be the class of G Galois extensions for a fixed group G or the class of simple algebras of fixed degree n. Then C has a generic structure if and only if it has the lifting property.

The proof of the Galois theory part of Theorem 2 appears in [6]. The algebra part is done in an analogous way. Two constructions appear in these proofs which we have not yet touched on. In either the Galois theory case, or in the algebra case, there is a natural candidate for a generic structure. In the case of simple algebras, start by considering the generic division algebra $UD(F,n,r)$ (see, e.g. [2]). Recall that $UD(F,n,r)$ is the central quotient ring of the ring, $M(F,n,r)$, of generic $n \times n$ matrices in r variables. It has been observed ([1] or [8]) that if the center, $Z(F,n,r)$, of $UD(F,n,r)$ is a rational F field, and if F has enough roots of unity, then every simple F algebra of degree n is similar (in the Brauer group) to a product of cyclic algebras. In fact, one can say something which turns out to be formally stronger. Namely, if $Z(F,n,r)$ is a rational F field then there is a domain $R \subset Z(F,n,r)$ and an Azumaya R algebra A such that A/R is a pure generic algebra for the class of simple algebras of degree n, and $A \otimes_R Z(F,n,r) \cong UD(F,n,r)$.

An analogous observation can be made in the case of Galois extensions. Instead of generic division algebras, one considers the construction of Noether's Problem. That is, let G be a finite group and set L to be the purely transcendental extension $F(x_g | g \in G)$. The group G acts on L by setting $h(x_g) = x_{hg}$. It was shown in [6] that if the fixed field L^G is purely transcendental over F, then one can use L^G to construct a generic Galois extension. All together, we have the following result.

THEOREM 3: a) *If L^G is a rational F field, then there is a generic Galois extension for G over F.*

b) *If $Z(F,n,r)$ is a rational F field, then there is a pure generic algebra for the class of all simple F algebras of degree n.*

Note that the converse of Theorem 3a) is not true. Swan ([9]) showed that if $G = C_{47}$ and $F = Q$, then L^G is not purely transcendental over Q. But a generic Galois extension does exist for C_{47} over Q. It turns out that pure transcendence is not quite the right property to ask for here. Instead, one has to define retracts of purely transcendental extensions.

Let K be a field finitely generated over F. We define K to be a retract rational F field if there are domains R and S and a split surjection $f:R \to S$ such that R is a localized polynomial ring and S has quotient field K. Note that these retracts are almost as "good" as rational F fields. For example, a variant of the Hilbert Irreducibility Theorem applies to them (trivially). For our purposes, their most important property is the following result. It shows that these retracts are, in some sense, projective objects. The proof of this next theorem closely parallels the usual proof that a projective module is a direct summand of a free module.

THEOREM 4: *The following are equivalent:*

a) *K is a retract rational F field.*

b) *K is the quotient field of a finitely generated F algebra S with the following property. Suppose R is a local F domain with maximal ideal M and canonical map $\tau:R \to R/M$. Assume $\varphi:S \to R/M$ is an F map. Then φ lifts to map $\varphi':S \to R$ such that $\varphi = \tau \circ \varphi'$.*

The next result shows how using these retracts one can improve Theorem 3.

THEOREM 5: a) _Let_ F _be a field,_ G _a group, and_ $L = F(x_g | g \in G)$. _There is a generic Galois extension for_ G _over_ F _if and only if_ L^G _is a retract rational_ F _field._

b) Let C be the F class of simple algebras of degree n. There is _a pure generic algebra for_ C _if and only if_ $Z(F,n,r)$ _is a retract rational_ F _field._

We will indicate a proof of the harder direction of 5b). The argument for 5a) is analogous. Unfortunately, we must assume the reader has some familiarity with the algebras $UD(F,n,r)$ (see [2]). Suppose that a pure generic algebra exists for all algebras of degree n. By Theorem 2, the simple F algebras of degree n have the lifting property. Suppose R is a local F domain with maximal ideal M. Set C to be the center of the ring, $M(F,n,r)$, of generic matrices. By, for example, [5], p. 70, there is a $0 \neq s \in C$ such that $A = M(F,n,r)(1/s)$ is Azumaya over $S = C(1/s)$. If $\varphi : S \to R/M$ is an F algebra map, set $B = A \otimes_{\varphi} R/M$. Also use φ to denote the induced map from A to B. By the lifting property, B lifts to an Azumaya R algebra B'. If $X_i \in M_n(F,n,r)$ are the generic matrices, set $y_i = \varphi(X_i)$, and choose $y_i' \in B'$ to be preimages of the y_i. The y_i' define a map $\varphi' : A \to B'$ where $\varphi'(X_i) = y_i'$. Restricted to S, φ' is a lifting for φ. Now apply Theorem 4.

A consequence of 5a) is that the field L^G for $F = Q$ and $G = C_{47}$ is a retract rational Q field but is not itself rational.

Our next step is to try to show that, sometimes, generic structures do not exist. To do this, it is natural to try to construct a counter example to the lifting property, and then to use Theorem 2. It turns out, however, that it is easier to investigate a related approximation problem. The key result is Theorem 6 below.

Let K be an F field with Henselian valuation v. Let Γ be the (linearly ordered) value group of v, where v need not be discrete. We choose to write v and Γ so that $v(a+b) \geq \min(v(a), v(b))$. Note that this means that basic open sets of K have the form $\{a | v(a-b) > \eta\}$. Suppose $L \subset K$ is a dense subfield. Assume C is an F class of simple algebras. We say C has the approximation property if for all K and L as above, and all $A \in C$ with center K, there is an $A' \in C$ with center L such that $A' \otimes_L K \cong A$. We also make the analogous definition for classes of Galois extensions.

THEOREM 6: *If an F class C has a generic structure then it has the approximation property.*

We will sketch a proof of Theorem 6. We begin by considering briefly the Galois theory case (see [6] for details). So assume S/R is a generic Galois extension for G over F. Write $R = F[x_1,\ldots,x_n](1/s)$. Suppose K, L are as above, and N/K is a Galois extension with group G. Choose $\varphi:R \to K$ such that $N \approx S \otimes_\varphi K$. We claim that maps "close" to φ define isomorphic extensions. More exactly, we claim that there is an $\eta \in \Gamma$ such that if $v(b_i - \varphi(x_i)) > \eta$, then setting $\varphi'(x_i) = b_i$ defines a map $\varphi':R \to K$ such that $S \otimes_{\varphi'} K \approx L$. We do not prove this claim, except to say that it is a restatement of Krasner's Lemma for Henselian valuations (e.g. [4], p. 195-8).

The proof of the Galois theory part of Theorem 6 is now straightforward. Let N/K, φ be as above, and choose such an η. Since $L \subseteq K$ is dense, there are $b_i \in L$ such that $v(b_i - \varphi(x_i)) > \eta$. Define $\varphi':R \to K$ by setting $\varphi'(x_i) = b_i$ and notice that $\varphi'(R) \subset L$. We can set $N' = S \otimes_{\varphi'} L$. Clearly $N' \otimes_L K \approx S \otimes_{\varphi'} K \approx N$.

Having proved the Galois theory part of Theorem 6, we turn to the simple algebra case. It is immediate that a parallel argument can be made, given a "Krasner's Lemma" for simple algebras. Such a result can be proved, and is of sufficient interest to state separately. Suppose $R = F[x_1,\ldots,x_n](1/s)$ is a localized polynomial ring, and A is an Azumaya algebra over R.

THEOREM 7: *Let K be an F field with valuation $v:K \to \Gamma$ such that K is Henselian. If $\varphi:R \to K$ is an F map, then there is an $\eta \in \Gamma$ such that if $v(b_i - \varphi(x_i)) > \eta$, then setting $\varphi'(x_i) = b_i$ defines a map $\varphi':R \to K$ and $A \otimes_{\varphi'} K \approx A \otimes_\varphi K$.*

We use Theorem 6 to show that in some cases no generic structure can exist. This example derives from the counter example to Grunwald's Theorem due to Wang ([10]). Namely, let Q_2 be the 2-adic completion of the rational field Q. Suppose n is an integer divisible by 8, and L/Q_2 is the unramified (cyclic) extension of degree n. Then there is no Galois extension L'/Q such that $L \approx L' \otimes_Q Q_2$. This immediately implies, using Theorem 6 and 5:

THEOREM 8: *Let G be a cyclic group of order a multiple of 8.*

a) *There is no generic Galois extension for G over Q.*

b) *If $L = Q(x_g | g \in G)$, then L^G is not a rational Q field, nor even a retract rational Q field.*

Thus the counterexamples to Noether's Problem (e.g. [9] and [3]) fall into two distinct nonempty categories, those which are and are not retract rational fields. One aim of this paper is to convince the reader that this is an important distinction.

In the talk presented at this conference, this author claimed that UD(Q,n,r) could be shown not to be retract rational over Q, for n a multiple if 8. The argument underlying this claim has a gap, and so the claim is withdrawn.

As the last result we exhibit a relation between the structures of UD(F,n,r) and Z(F,n,r). It is a corollary of Theorems 6 and 1.

Corollary 9: *Let* p *be an odd prime. If* UD(F,p,r) *is a cyclic algebra, then* Z(F,p,r) *is retract rational over* F.

We should note that with further work, Corollary 9 can be improved a bit. Namely, we need only assume that UD(F,p,r) is split by a Galois extension of degree pr where $r|p-1$.

Finally, we end by mentioning another application of the ideas of this paper. In [6], a version of Theorem 6 is used to prove a Grunwald-Wang type theorem for arbitrary valued fields and groups with generic Galois extensions.

Note: Since I wrote the above paper I discovered that Theorem 7 is closely related to a previous result of Zassenhaus which is the main result of [11]. The result of [11] is not in the same form, and is more general.

REFERENCES

1. Formanek, E., The center of the ring of 3×3 generic matrices. Lin. and Mult. Alg. 7 (1979), 203-212.

2. Jacobson, N., P. I. Algebras. Lecture Notes in Mathematics, no. 441, Springer-Verlag, Berlin/Heidelberg/New York, 1975.

3. Lenstra, H. W., Rational Functions Invariant under a Finite Abelian Group. Inv. Math. 25 (1974), 299-325.

4. Ribenboim, P., Theories des Valuations. Les Presses de L'Universite de Montreal, Montreal 1964.

5. Rowen, L., Polynomial Identities in Ring Theory. Academic Press, New York, 1980.

6. Saltman, D., Generic Galois Extensions and Problems in Field Theory, to appear, Adv. in Math.

7. Serre, J.-P., Local Fields. Springer-Verlag, Berlin/Heidelberg/New York, 1979.

8. Snider, R. L., Is the Brauer Group generated by cyclic algebras? In Ring Theory, Waterloo 1978, Lecture Notes in Mathematics no. 734. Springer-Verlag, Berlin/Heidelberg/New York. 1979, 279-301.

9. Swan, R., Invariant Rational Functions and a Problem of Steenrod. Inv. Math. 7 (1969), 148-158.

10. Wang, S., A Counterexample to Grunwald's Theorem. Annals of Math. 49 (1948), 1008-1009.

11. Zassenhaus, H., On Structural Stability, Comm. in Alg., 8(19), 1799-1844, (1980).

Yale University
Box 2155, Yale Station
New Haven, CT 06520

Contemporary Mathematics
Volume 13, 1982

BALANCED SPACES OF FORMS AND SKEW CYCLIC TRANSFORMATIONS

by
Darrell E. Haile*

Dedicated to Professor Nathan Jacobson

1. INTRODUCTION

Let V be a vector space of dimension m over a field F, which we assume throughout is of characteristic zero. In [3] the idea of a nondegenerate space of functionals on $V^n (= \overbrace{V \otimes_F V \otimes \ldots \otimes_F V}^{n})$ was introduced to generalize the one dimensional space generated by a single nondegenerate symmetric or skew bilinear form (See the next section for definitions). In addition to being of possible independent interest, these spaces lead to certain linear maps on $End_F(V^{n-1})$ we call skew cyclic transformations, in the same way that a nondegerate symmetric or skew bilinear form on V leads to an involution of $End_F(V)$. The study of these maps for arbitrary central simple algebras of given exponent was begun in [2].

The group GL(V) acts on $(V^n)^*$ and preserves the set of nondegenerate subspaces of $(V^n)^*$. In this paper we initiate a study of this action. To do this it is useful to restrict attention to a distinguished subset of the set of all nondegenerate subspaces -- the "balanced" ones of the title. The group GL(V) preserves this subset and seems to act much more coherently on it than on the whole collection. As evidence of this we prove (Theorem 3.1) that if F is algebraically closed and n = m = 3 (in which case there are two "types" of nondegenerate subspaces of $(V^3)^*$, determined by the possible representation types of the symmetric group S_3 acting on the subspaces), then the balanced spaces of a given type form a single orbit under the action of GL(V). This is analogous to the fact that on a given vector space over an algebraically closed field there is, up to the usual notion of equivalence, exactly

*This work was partially supported by NSF Grant MCS-7904165

0271-4132/82/0519/$04.75

one nondegenerate symmetric bilinear form and, if the dimension of the space is even, one nondegenerate skew form. The action of $GL(V)$ on the whole collection of nondegenerate subspaces seems chaotic.

In addition we examine the meaning of balance for skew cyclic transforma- in general and, as an application of the theorem stated above, determine all the balanced skew cyclic transformations on $M_3(F) \otimes M_3(f)$.

2. BALANCED SPACES OF FORMS

We first recall the definition of a nondegenerated subspace (See Section 2 of [3]). Let F be a field and let V be an m dimensional F-vector space. Let n be an integer, $0 < n \leq m$. For each element f in $(V^n)^*$ let $\phi_f : V^{n-1} \to V^*$ be given by $\phi_f(v_1 \otimes \ldots \otimes v_{n-1})(v) = f(v_1 \otimes \ldots \otimes v_{n-1} \otimes v)$ for all $v_1, \ldots, v_{n-1}$, v in V. A subspace W of $(V^n)^*$ is called <u>nondegenerate</u> if it satisfies the following properties:

1. $[W : F] = m^{n-2}$
2. W is stable under the canonical action of S_n on $(V^n)^*$.
3. $\bigcap_{f \in W} \operatorname{Ker} \phi_f = 0.$

Let $r = m^{n-2}$. Given condition (1), condition (3) is equivalent to the following statement: For every basis $f_1, \ldots, f_r$ of W, the map $\phi : V^{n-1} \to \underbrace{V^* \oplus \ldots \oplus V^*}_{r}$ given by $\phi(v) = \sum_{i=1}^{r} \phi_{f_i}(v)$ is an isomorphism of vector spaces. In the case $n = 2$, W is the span of a single nondegenerate symmetric or skew bilinear form.

It is shown in [3] that if $m = n^k$ for some k, then $(V^n)^*$ contains a nondegenerate subspace. Also the S_{n-1} module structure on a nondegenerate subspace W of $(V^n)^*$ (where S_{n-1} is viewed as the subgroup of S_n of permutations fixing the last letter) is uniquely determined and in fact $W \simeq U^{n-1} \otimes X$ where U and X are F vector spaces, $[U:F] = [X:F] = n$, and S_{n-1} acts in the standard way on U^{n-1} and trivially on X. The representation type of S_n on W is called the <u>type</u> of W. For example if $\dim V = 3$, then any nondegenerate subspace of $(V^3)^*$ is isomorphic as an S_2-module to $U_1 \oplus U_1 \oplus U_2$ where U_1 is the trivial representation and U_2 is the non-trivial representation of S_2. There are two ways of extending this S_2 action to an S_3 action and there is a nondegenerate subspace of $(V^3)^*$ of each of these types: Explicitly letting $\{v_1, v_2, v_3\}$ be a basis of V and

v_1^*, v_2^*, v_3^* the corresponding dual basis of V^*, the subspaces

$$W_0 = \langle \sum_{i=1}^{3} v_i^* \otimes v_i^* \otimes v_i^*, \sum_{\sigma\in S_3} v_{\sigma(1)}^* \otimes v_{\sigma(2)}^* \otimes v_{\sigma(3)}^*, \sum_{\sigma\in S_3} \mathrm{sn}(\sigma) v_{\sigma(1)}^* \otimes v_{\sigma(2)}^* \otimes v_{\sigma(3)}^* \rangle \quad \text{and}$$

$$W_1 = \langle v_1^* \otimes v_1^* \otimes v_2^* + v_2^* \otimes v_2^* \otimes v_3^* + v_3^* \otimes v_3^* \otimes v_1^*,$$
$$v_2^* \otimes v_1^* \otimes v_1^* + v_3^* \otimes v_2^* \otimes v_2^* + v_1^* \otimes v_3^* \otimes v_3^*,$$
$$v_1^* \otimes v_2^* \otimes v_1^* + v_2^* \otimes v_3^* \otimes v_2^* + v_3^* \otimes v_1^* \otimes v_3^* \rangle$$

are nondegenerate. The subspace W_0 is isomorphic to $U_1 \oplus U_1 \oplus U_2$ where U_1 is the trivial representation of S_3 and U_2 is the non-trivial one dimensional representation. The space W_1 is isomorphic to $U_1 \oplus U_3$, where U_3 is the two dimensional irreducible representation of S_3.

Again let W be a nondegenerate subspace of $(V^n)^*$, $[V:F] = m$. Then associated to W there is a skew cyclic transformation $\tau : \mathrm{End}_F(V^{n-1}) \to \mathrm{End}_F(V^{n-1})$ characterized by the formula $f((1 \otimes B)(x)) = f((\tau(B) \otimes 1)(x))$ for all x in V^n, B in $\mathrm{End}_F(V^{n-1})$ and f in W (see Section 2 of [3]).

The group $GL(V)$ acts on V^n (diagonally) and hence on $(V^n)^*$. We will write this action on the right, so if $f \in (V^n)^*$ and $g \in GL(V)$ then $f\cdot g(x) = f(gx)$ for all $x \in V^n$. If W is nondegenerate and g is in $GL(V)$ then $Wg = \{f\cdot g | f \in W\}$ is also nondegenerate. Since the action of S_n on $(V^n)^*$ commutes with that of $GL(V)$ we see in fact that $GL(V)$ stabilizes the set of nondegenerate subspaces of a given type. As noted in the introduction, if $n = 2$ and F is algebraically closed each type represents a single orbit. If $n > 2$ the situation is more complicated and an argument on dimension shows that there must be an infinite number of orbits. To improve matters we need to restrict our attention to a certain subset of nondegenerate subspaces.

Let $W \subseteq (V^n)^*$ be a nondegenerate subspace and let $f_1,\ldots,f_r$ be a basis of W, where $r = m^{n-2}$. Let $\phi : V^{n-1} \to \underbrace{V^* \oplus\ldots\oplus V^*}_{r}$ be the corresponding isomorphism. There is an induced isomrophism ${}^*\phi^{-1} : (V^*)^{n-1} \to \underbrace{V \oplus\ldots\oplus V}_{r}$ (the contragredient -- we have identified V and V^{**}). We obtain then r projections from $(V^*)^{n-1}$ to V and hence r elements $x_1,\ldots,x_r$ of V^n

(again identifying V^n and $(V^n)^{**}$). In fact $x_1,\ldots,x_r$ are characterized as follows: x_i is the unique element of V^n such that $f_j((1\otimes\alpha)(x_i)\otimes v) = \delta_{ij}(v)$ for all $v \in V$ and $\sigma \in V^*$ where $(1\otimes\alpha)(x_i)$ means $\sum_k \alpha(z_k)y_k$ if $x_i = \sum_k y_k \otimes z_k$, for $y_k \in V^{n-1}$, $z_k \in V$. In particular $f_j(x_i) = \delta_{ij}m$ (recall $m = [V:F]$).

Let W' be the subspace of V^n generated by $x_1,\ldots,x_r$. The characterization of $x_1,\ldots,x_r$ given above shows that W' is independent of the choice of basis for W. It is easy to see that W' satisfies conditions (1) and (3) for a nondegenerate subspace of V^n. In the case $n = 2$, W' is the span of the usual dual form on V^* associated to a given generating form of W on V and this dual form is symmetric or skew corresponding to the type of W. In particular W' is S_2 invariant. However examples show that if $n > 2$ then W' is not necessarily S_n invariant. We therefore make the following definition.

DEFINITION 2.1. A nondegenerate subspace of W of $(V^n)^*$ is called _balanced_ if the corresponding subspace W' of V^n is S_n invariant.

The remarks before the definition show that if W is balanced, then W' is a nondegenerate subspace of V^n. Moreover the dual relationship between the bases of W and W' show that if W is balanced then the representation of S_n on W' is naturally contragredient to the representation of S_n on W (and so $W \simeq_{S_n} W'$ since each representation of S_n is equivalent to its contragredient). If $n = 2$ then we have seen that all nondegenerate subspaces are balanced. A computation shows that both of the examples given above in the case $m = n = 3$ are balanced; in fact W_0' and W_1' are given explicitly as follows:

$$W_0' = \langle \sum_{i=1}^{3} v_i \otimes v_i \otimes v_i,\ \sum_{\sigma\in S_3} v_{\sigma(1)} \otimes v_{\sigma(2)} \otimes v_{\sigma(3)},$$
$$\sum_{\sigma\in S_3} \mathrm{sn}(\sigma) v_{\sigma(1)} \otimes v_{\sigma(2)} \otimes v_{\sigma(3)} \rangle$$

$$W_1' = \langle v_1 \otimes v_1 \otimes v_2 + v_2 \otimes v_2 \otimes v_3 + v_3 \otimes v_3 \otimes v_1,$$
$$v_2 \otimes v_1 \otimes v_1 + v_3 \otimes v_2 \otimes v_2 + v_1 \otimes v_3 \otimes v_3,$$
$$v_1 \otimes v_2 \otimes v_1 + v_2 \otimes v_3 \otimes v_2 + v_3 \otimes v_1 \otimes v_3 \rangle.$$

The following is an example of nondegenerate subspace of

$(V^3)^*$ that is not balanced: $<\sum_{i=1}^{3} v_i^* \otimes v_i^* \otimes v_i^*$,

$\sum_{i=0}^{2} \phi^i(v_1^* \otimes v_1^* \otimes v_2^* + v_2^* \otimes v_2^* \otimes v_3^* + v_3^* \otimes v_3^* \otimes v_1^*)$,

$\sum_{\sigma \in S_3} \mathrm{sn}(\sigma) v_{\sigma(1)}^* \otimes v_{\sigma(2)}^* \otimes v_{\sigma(3)}^*>$, where $\phi : (V^3)^* \to (V^3)^*$

is given by $\phi(v_i^* \otimes v_j^* \otimes v_k^*) = v_k^* \otimes v_i^* \otimes v_j^*$.

If $g \in GL(V)$ and W is a balanced nondegenerate subspace of $(V^n)^*$ then Wg is also balanced and nondegenerate. In the next section we will investigate the action of $GL(V)$ on the set of balanced nondegenerate subspaces of a given type. Here we want to indicate the significance of balance for the corresponding skew cyclic transformation.

Let F be a field and let A be a central simple F-algebra such that A^n is a full matrix algebra. Assume there is a skew cyclic transformation $\tau : A^{n-1} \to A^{n-1}$ (see [3]). For the purposes of this discussion we will call τ a <u>right</u> skew cyclic transformation. By a <u>left</u> skew cyclic transformation on A^{n-1} we will mean a nonsingular transformation $\gamma : A^{n-1} \to A^{n-1}$ satisfying conditions (1) and (3) for a skew cyclic transformation (See the introduction of [3]) and the condition (2)'' : $\gamma(x \otimes y) = \bar{\gamma}(y)(x \otimes 1)$ for all $y \in A$, $x \in A^{n-2}$. In the original paper on this subject [2], we considered left skew cyclic transformations. The shift to the right in [3] was prompted by the connection with nondegenerate subspaces.

Given a right skew cyclic transformation τ on A^{n-1}, define $\tau' : A^{n-1} \to A^{n-1}$ by $\tau'(x \otimes y) = \bar{\tau}(y)(x \otimes 1)$ for $x \in A^{n-2}$, $y \in A$.

<u>Definition 2.2</u>. The (right) skew cyclic transformation τ is called <u>balanced</u> if the associated map τ' is a left skew cyclic transformation.

The example given at the beginning of Section 3 of [3] is balanced. In case $n = 2$ all skew cyclic transformations (that is all involutions) are balanced. In a moment we will show that the skew cyclic transformations arising from balanced nondegenerate subspaces are balanced. We first need the following result. As in [3], given a left or right skew cyclic transformation $\tau : A^{n-1} \to A^{n-1}$, we let $\bar{\tau} : A \to A^{n-1}$ be given by $\bar{\tau}(a) = \tau(\underbrace{1 \otimes \ldots \otimes 1}_{n-1} \otimes a)$.

<u>THEOREM 2.3</u>. <u>Let W be a nondegenerate subspace of $(V^n)^*$</u> $[V:F] = m$, $n \le m$. <u>Let $A = \mathrm{End}_F(V)$. Let τ be the (right) skew cyclic transformation on A^{n-1} corresponding to W. Let $\{e_{ij}\}$ be a set of matrix units for A and let</u> $e = \frac{1}{m} \sum_{i,j} \bar{\tau}(e_{ji}) \otimes e_{ij}$. <u>Then</u> e <u>has the following properties:</u>

(1) $e^2 = e$

(2) $e(\overline{1} \otimes a) = e(\overline{\tau}(a) \otimes 1)$ <u>and</u> $(\overline{1} \otimes a)e = (\overline{\tau}(a) \otimes 1)e$ <u>for all</u> $a \in A$ (where $\overline{1} \otimes a$ means $\underbrace{1 \otimes \ldots \otimes 1}_{n-2} \otimes a$).

(3) <u>The canonical maps</u> $A^{n-1} \to e(A^{n-1} \to 1)$ <u>and</u> $A^{n-1} \to (A^{n-1} \otimes 1)e$ <u>are injections</u>.

(4) <u>The element</u> e <u>is independent of the matrix units chosen</u>.

(5) $(V^n)^* e = W$ <u>and</u> $eV^n = W'$. (Here A^n acts on the right on $(V^n)^*$ via $(f \cdot a)(x) = f(ax)$ for all $a \in A^n$, $f \in (V^n)^*$ and $x \in V^n$).

<u>REMARK</u>: The first three properties show that e is almost an associated idempotent for τ, in the sense of [3]. However e is not necessarily fixed by S_n.

<u>PROOF</u>: The first three properties are straightforward calculations, so we will verify only parts of them. To see the first half of property (2) for example it suffices to show $e(\overline{1} \otimes e_{ij}) = e(\overline{\tau}(e_{ij}) \otimes 1)$ for all i,j. But $e(\overline{1} \otimes e_{ij}) = \frac{1}{m}(\sum_{r,s} \overline{\tau}(e_{sr})(\overline{1} \otimes e_{ij})) = \frac{1}{m}\sum_r \overline{\tau}(e_{ir}) \otimes e_{rj}$ and $e(\overline{\tau}(e_{ij}) \otimes 1) =$ $\frac{1}{m}(\sum_{r,s} \overline{\tau}(e_{sr}) \otimes e_{rs})(\overline{\tau}(e_{ij}) \otimes 1) = \frac{1}{m}\sum_{r,s} \overline{\tau}(e_{sr})\overline{\tau}(e_{ij}) \otimes e_{rs} =$ $\frac{1}{m}\sum_r \overline{\tau}(e_{ir}) \otimes e_{rs} = \frac{1}{m}\sum_r \overline{\tau}(e_{ir}) \otimes e_{rj}$, as desired.

For the second half of property (3), if $(x \otimes 1)e = 0$, $x \in A^{n-1}$, then $\sum_{i,j} x\overline{\tau}(e_{ji}) \otimes e_{ij} = 0$. Hence $x\overline{\tau}(e_{ji}) = 0$ for all i and j, so $x = 0$.

Property (4) is a consequence of (2): If $\{f_{ij}\}$ is another set of matrix units, then there is an invertible element a in A such that $ae_{ij}a^{-1} = f_{ij}$ for all i and j. Hence $\frac{1}{m}\sum_{i,j} \overline{\tau}(f_{ji}) \otimes f_{ij} =$ $\frac{1}{m}\sum_{i,j} \overline{\tau}(a)^{-1}\overline{\tau}(e_{ji})\overline{\tau}(a) \otimes ae_{ij}a^{-1} = (\overline{\tau}(a)^{-1} \otimes a)e(\overline{\tau}(a) \otimes a^{-1}) = e.$

We proceed to property (5). It is clear that if $f \in W$, then $f \cdot e = f$: $(f \cdot e)(z) = f(\frac{1}{m}(\sum_{i,j} \overline{\tau}(e_{ji}) \otimes e_{ij})z) = \frac{1}{m}\sum_{i,j} f(\overline{\tau}(e_{ji})\overline{\tau}(e_{ij})z) =$ $\frac{1}{m}\sum_{i,j} f(\overline{\tau}(e_{ii})z) = \frac{1}{m}\sum_j f(z) = f(z)$ for all $z \in V^n$. Moreover properties (2) and (3) determine the rank of e, thus it follows that $(V^n)^* e = W$.

To show $eV^n = W'$ it suffices to show $ex_i = x_i$ for all i, where $\{x_i\}$ is the basis of W' determined by the basis $\{f\}$ of W in the usual way. We have seen that x_i is characterized by $f_j((1 \otimes k)(x_i) \otimes v) = \delta_{ij}k(v)$ for all $k \otimes V^*$, $v \otimes V$ and all j. It suffices then to show

$f_1((1 \otimes k)(ex_i) \otimes v) = \delta_{ij}k(v)$. This can be verified by choosing a basis for V, expressing k in terms of the dual basis for V^* and e in terms of the corresponding matrix units. The calculation is left to the reader. □

As we noted before e is not necessarily S_n-invariant. The following corollary shows the significance of this extra condition.

<u>COROLLARY 2.4</u>. <u>The following conditions are equivalent.</u>

(1) W <u>is balanced</u>

(2) τ <u>is balanced</u>

(3) e <u>is fixed by</u> S_n.

<u>PROOF</u>: We first prove (1) and (3) are equivalent. If e is S_n-invariant, then $W' = eV^n$ is S_n-invariant, so W is balanced. Now assume W is balanced and let $U = \{z \in V^n | (z) = 0$ for all $f \in W\}$. Then $U = (1 - e)V^n$. Moreover U is S_n-invariant. Since $V^n = W' \oplus U = eV^n \oplus (1 - e)V^n$ and both W' and U are S_n-invariant, it follows that e is fixed by S_n.

We now show (2) and (3) are equivalent. If e is S_n-invariant then referring to the remarks after the statement of the theorem, we have e is an associated idempotent for τ. Since $(\overline{1} \otimes a)e = (\overline{\tau}(a) \otimes 1)e$ for all $a \in A$, it follows that $(1 \otimes x)e = (\tau'(x) \otimes 1)e$ for all $x \in A^{n-1}$. The conditions on e then say that e is an associated idempotent on the left as well as the right. In particular τ' is a left skew cyclic transformation, so τ is balanced.

Now assume τ is balanced. Then τ is a right skew cyclic transformation and as in the introduction of [3] we have an exact sequence of right A^n-modules

$$0 \to J \to A^n \xrightarrow{\mu_\tau} A^{n-1} \to 0$$

where μ_τ is given by $\mu_\tau(x \otimes a) = \overline{\tau}(a)x$ for $a \in A$ and $x \in A^{n-1}$, and J is the kernel of μ_τ. In a similar way we have an exact sequence of left A^n-modules

$$0 \to K \to A^n \xrightarrow{\eta_{\tau'}} A^{n-1} \to 0$$

where $\eta_{\tau'}$ is given by $\eta_{\tau'}(x \otimes a) = x\overline{\tau}(a)$ for $a \in A$ and $x \in A^{n-1}$, and K is the kernel of μ_τ. From the properties of e it follows that

$J = (1-e)A^n$ and $K = A^n(1-e)$. If $f \in A^n$ is another idempotent such that $J = fA^n$ and $K = A^n f$, then it easily follows that $f = 1 - e$. By Theorem 2.1 of [2], we know J is S_n-invariant. Similarly K is S_n-invariant. Hence if $\sigma \in S_n$ then $J = (1-\sigma(e))A^n$ and $K = A^n(1-\sigma(e))$. Thus $\sigma(e) = e$. Since σ was chosen arbitrarily in S_n, we have e is S_n-invariant. □

COROLLARY 2.5. *Let B be F-central simple, B^n a full matrix algebra and $[B:F] \geq n^2$. Let $\tau : B^{n-1} \to B^{n-1}$ be a balanced skew cyclic transformation. Then there is a unique idempotent f in B^n which is an associated idempotent for both τ and τ'.*

PROOF: The proof of uniqueness follows as in the second half of the proof of the previous corollary. To prove existence consider again the sequences

$$0 \to J \to B^n \xrightarrow{\mu_\tau} B^{n-1} \to 0$$

$$0 \to K \to B^n \xrightarrow{\eta_{\tau'}} B^{n-1} \to 0$$

If we pass to $\overline{B} = B \otimes_F \overline{F}$, where $\overline{F}$ is an algebraic closure of F, then $\tau \otimes 1 : \overline{B}^{n-1} (= B^{n-1} \otimes_F \overline{F}) \to \overline{B}^{n-1}$ is a balanced skew cyclic transformation. Moreover $J \otimes_F \overline{F} = \ker \mu_{\tau\otimes 1}$ and $K \otimes_F \overline{F} = \ker \eta_{\tau'\otimes 1}$. Let e be as described in the theorem for the skew cyclic transformation $\tau \otimes 1$ (note that $\overline{B}$ is split). Then $JK \otimes_F \overline{F} = (J \otimes_F \overline{F})(K \otimes_F \overline{F}) = (1-e)(\overline{B}^n)(1-e)$. Hence $JK \otimes_F \overline{F}$ is a central simple $\overline{F}$-algebra. Thus JK is a simple algebra, so in particular JK has an identity element f. An easy calculation shows $1 - f$ is an idempotent with the desired properties.

We now proceed to a result that will be useful in the next section. If $\tau : A^{n-1} \to A^{n-1}$ is a skew cyclic transformation, we let C be the centralizer of $\overline{\tau}(A)$ in A^{n-1}. Then $A^{n-1} \cong \overline{\tau}(A) \otimes C$ as F-algebras.

THEOREM 2.6. *Let A be F-central simple, A^n a full matrix algebra,* $[A:F] \geq n^2$ and let $\tau, \gamma : A^{n-1} \to A^{n-1}$ be balanced skew cyclic transformtions. *There is an invertible element x in $(A^{n-1})^{S_{n-1}}$ and an invertible element c in C such that $\tau(x) = cx$, $\tau'(x^{-1}) = x^{-1}c^{-1}$* and $\overline{\gamma}(a) = x^{-1}\overline{\tau}(a)x$ *for all* a A. If $\overline{\tau}(A) = \overline{\gamma}(A)$ then we may take x in $\overline{\tau}(A)$. *Conversely if x in $(A^{n-1})^{S_{n-1}}$ is invertible and $\tau(x) = cx$, $\tau'(x^{-1}) = x^{-1}c^{-1}$ for some c in C_τ, then the transformation δ determined by $\overline{\delta}(a) = x^{-1}\overline{\tau}(a)x$ for $a \in A$ is skew cyclic and balanced.*

PROOF: The proof is similar to the proofs of Theorems 1.6 and 1.7 of [3]. Let $\tau, \gamma : A^{n-1} \to A^{n-1}$ be balanced skew cyclic transformations. As in the proof of Theorem 1.6 of [3], there is an invertible element x in $(A^{n-1})^{S_{n-1}}$ such that $\bar{\tau}(a) = x^{-1}\bar{\tau}(a)x$ for all a in A. Let e be the associated idempotent for τ. Consider the idempotent $f = (1 \otimes x^{-1})e(1 \otimes x)$. Let $0 \to J \to A^n \xrightarrow{\mu_\gamma} A^{n-1} \to 0$ and $0 \to K \to A^n \xrightarrow{\eta_{\gamma'}} A^{n-1} \to 0$ be the usual sequences. The calculations in the proof of Theorem 1.6 show that $J = (1 - f)A^n$ and $K = A^n(1 - f)$. It follows that f is the unique idempotent associated to γ. In particular $f \in (A^n)^{S_n}$, so $(1 \otimes x^{-1})e(1 \otimes x) = (x^{-1} \otimes 1)e(x \otimes 1)$. This leads to $\tau(x) = cx$ and $\tau'(x^{-1}) = x^{-1}r$ for some invertible elements c and r in C_τ. But since $(\tau(x^{-1}) \otimes 1)e(\tau(x) \otimes 1) = (x^{-1} \otimes 1)e(x \otimes 1)$, we conclude $(r \otimes 1)e(c \otimes 1) = e$. Hence $r = c^{-1}$ as desired.

If $\bar{\tau}(A) = \bar{\gamma}(A)$ then $\tau^{-1}\gamma : A \to A$ is an F-algebra automorphism, hence inner. Then $\bar{\tau}(a) = \bar{\tau}(d)^{-1}\bar{\tau}(a)\bar{\tau}(d)$ for some invertible element d in A, all $a \in A$. Now we can choose $x = \bar{\tau}(d)$.

The last part is proved using arguments quite similar to those of Theorem 1.7 in [3]. We omit the details. □

COROLLARY 2.7. *Let $\tau : A^{n-1} \to A^{n-1}$ be a balanced skew cyclic transformation and let $\gamma : A^{n-1} \to A^{n-1}$ be a right or left skew cyclic transformation.* Suppose $\bar{\tau}(A) = \bar{\gamma}(A)$. Then γ is balanced.

PROOF: Assume γ is a right skew cyclic transformation. (The argument for the left is similar). By the proof of the theorem there is an invertible element d in A such that $\tau(\tau(d)) = c\bar{\tau}(d)$ where c is invertible in C and $\bar{\tau}(a) = \bar{\tau}(d)^{-1}\bar{\tau}(a)\bar{\tau}(d)$ for all $a \in A$. But for any $y \in A$, $\tau(\bar{\tau}(y)) = y \otimes \bar{1} = \tau'(\bar{\tau}'(y))$. Hence $\tau'(\bar{\tau}(d)^{-1}) = \tau'(\bar{\tau}'(d^{-1})) = \tau(\bar{\tau}(d)^{-1}) = \bar{\tau}(d)^{-1}c^{-1}$. By the last part of the theorem, γ is balanced. □

3. ACTION OF GL(V)

In this section we begin an investigation of the action of $GL(V)$ on the balanced nondegenerate subspaces of $(V^n)^*$. Our results are quite special, so we can only claim to have scratched the surface of this subject. However the results are perhaps encouraging and justify the consideration of other cases. After the main theorem we derive some consequences for the behavior of balanced skew cyclic transformations on $M_3(F)$.

THEOREM 3.1. Let F be algebraically closed (characteristic zero). Let $[V:F] = 3$. Let U_1 and U_2 be balanced nondegenerated subspaces of $(V^3)^*$. Then U_1 and U_2 belong to the same orbit under the action of $GL(V)$ if and only if U_1 and U_2 are of the same type (that is, if and only if U_1 and U_2 are isomorphic as S_3-modules).

PROOF: Let v_1, v_2, v_3 be a basis of V and let v_1^*, v_2^*, v_3^* be the dual basis of V^*. Let W_0 and W_1 be as described previously. We have seen that if W if balanced, nondegenerate then W is isomorphic as an S_3-module to W_0 or W_1. We will show that if $W \simeq_{S_3} W_i$ then W is $GL(V)$ conjugate to W_i, $i = 0, 1$. First note that both W_0 and W_1 contain symmetric forms, so W contains symmetric forms. There is a linear isomorphism between the space $S((V^3)^*)$ of symmetric forms in $(V^3)^*$ and the space of cubic forms in the polynomial ring $F[X_1, X_2, X_3]$ given by the the symmetrization map $X_iX_jX_k \to \frac{1}{6} \sum_{\sigma\in S_3} v^*_{\sigma(i)} \otimes v^*_{\sigma(j)} \otimes v^*_{\sigma(k)}$. The group $GL(V)$ acts on each of these spaces as usual and this isomorphism is a $GL(V)$-map. Let $f \in W$ be symmetric, $f \neq 0$. Since we may replace W by any of its $GL(V)$ conjugates we may replace f by any of its conjugates. Passing to the corresponding cubic form $\overline{f}$ we may replace $\overline{f}$ by any cubic form projectively equivalent to it. Recall that since $f \in W$, the corresponding map $\phi_f : V^2 \to V^*$ is onto (see the Proof of Theorem 2.2 in [3]). It is then easy to list the possibilities for $\overline{f}$. If $\overline{f}$ is non-singular then $\overline{f}$ is projectively equivalent to $X_1^3 + X_2^3 + X_3^3 + \lambda X_1X_2X_3$ where $\lambda^3 \neq -27$. If $\overline{f}$ is irreducible and singular then either $\overline{f}$ has a node, in which case it is projectively equivalent to $X_2^2X_3 - X_1^3 - X_1^2X_3$ or $\overline{f}$ has a cusp in which case it is projectively equivalent to $X_2^2X_3 - X_1^3$. If $\overline{f}$ is reducible, then using that ϕ_f is onto it follows that $\overline{f}$ is projectively equivalent to $X_1(X_2^2 + X_1X_3)$ or $X_1X_2X_3$. (See [1], pp. 117 and 106 and [4], p. 27). So we conclude that up to projective equivalence $\overline{f}$ is one of the following:

(1) $X_1^3 + X_2^3 + X_3^3 + \lambda X_1X_2X_3$, $\lambda^3 \neq -27$

(2) $X_2^2X_3 - X_1^3 - X_1^2X_3$

(3) $X_2^2X_3 - X_1^3$

(4) $X_1(X_2^2 + X_1X_3)$

(5) $X_1X_2X_3$

Now assume $W \simeq_{S_3} W_0$. Let $f \in W$, f symmetric, $f \neq 0$. We may assume f corresponds to one of the cubic forms listed above. We will show that f cannot be of type (2), (3) or (4) and if f is of type (1) or (5) then $W = W_0$. That will finish this half of the proof.

Let $W = \langle f_1 = f, f_2, f_3 \rangle$ where f_2 is symmetric and f_3 is skew. It follows that $f_3 = \alpha \sum_{\sigma \in S} sn(\sigma) v^*_{\sigma(1)} \otimes v^*_{\sigma(2)} \otimes v^*_{\sigma(3)}$ for some $\alpha \in F^*$. Let $W' = \langle x_1, x_2, x_3 \rangle$ where the x_i are obtained as described before Definition 2.1. Since $f_i(x_j) = 3\delta_{ij}$ and W' is S_3-stable, it follows that x_1, x_2 are symmetric and x_3 is skew. Now assume f_1 is of type (2), (3) or (4). The calculations for each of these cases are similar, so we consider only type (2). Suppose then that f_1 corresponds to the cubic form $X_2^2X_3 - X_1^3 - X_1^2X_3$. We have the relations $f_1((1 \otimes v_i^*)(x_2) \otimes v_j) = 0$ for all i, j. From this and the symmetry of x_2 it follows that $x_2 = 0$, a contradiction: Letting $j = 2$, the only possible nonzero contribution to $f_1((1 \otimes v_i^*)(x_2) \in v_2)$ comes about when $(1 \otimes v_i^*)(x_2) \in F(v_2 \otimes v_3) + F(v_3 \otimes v_2)$. Letting $i = 1$ (and $j = 2$), we infer that the coefficient of $\sum_{\sigma \in S_3} v^*_{\sigma(1)} \otimes v^*_{\sigma(2)} \otimes v^*_{\sigma(3)}$ in x_2 must be zero. Letting $i = 2$, we infer that the coefficient of $v_2 \otimes v_2 \otimes v_3 + v_2 \otimes v_3 \otimes v_2 + v_3 \otimes v_2 \otimes v_2$ in x_2 must be zero and letting $i = 3$ we infer the coefficient of $v_2 \otimes v_3 \otimes v_3 + v_3 \otimes v_2 \otimes v_3 + v_3 \otimes v_3 \otimes v_2$ in x_2 is zero. Continuing in this way for $j = 1$ and 3 results in $x_2 = 0$.

Now suppose f_1 corresponds to a cubic form $X_1^3 + X_2^3 + X_3^3 + \lambda X_1X_2X_3$, $\lambda^3 \neq -27$. An examination of the relations $f_1((1 \otimes v_i^*)(x_k) \otimes v_j) = \delta_{ik}\delta_{ij}$ for $k = 1, 2$ (in conjunction with the symmetry of x_1 and x_2) leads to the conclusion that

$$x_1, x_2 \in \langle \sum_{i=1}^{3} v_i \otimes v_i \otimes v_i, \sum_{\sigma \in S_3} v_{\sigma(1)} \otimes v_{\sigma(2)} \otimes v_{\sigma(3)} \rangle \subseteq W_0'$$

Since x_3 is determined up to scalar multiple by being skew, it follows that $W' = W_0'$. Hence $W = W_0$.

Similarly if f_1 corresponds to $X_1X_2X_3$ then the usual relations show $x_1, x_2 \in W_0'$ (in fact x_2 is a scalar multiple of $\sum_{i=1}^{3} v_i \otimes v_i \otimes v_i$), so $W = W_0$. This finishes the first half of the proof.

Now assume $W \approx_{S_3} W_1$. Let $\alpha = (1,2)$, $\beta = (1, 2, 3)$ in S_3. Then we can write $W = \langle f_1, f_2, f_3\rangle$ where f_1 is symmetric and $\alpha f_2 = f_3$, $\beta f_2 = - f_3$ and $\beta f_3 = f_2 - f_3$. Let $W' = \langle x_1, x_2, x_3\rangle$ as usual. The relations $f_i(x_j) = 3\delta_{ij}$ now lead to $\alpha x_2 = x_3$, $\beta x_3 = x_2$, $\beta x_2 = -(x_2 + x_3)$ and x_1 is symmetric. We may assume f_1 corresponds to one of the five types listed above.

We first <u>claim</u> that f_1 must be of type (1), that is non-singular. For example, if f_1 corresponds to $x_2^2X_3 - X_1^3 - X_1^2X_3$ then the relations above together with $f_1((1 \otimes v_i^*)(x_k) \otimes v_j) = 0$ for $k = 1, 2$ lead to $x_2 = x_3 = 0$, a contradiction. The cases (3), (4) and (4) follow similarly.

Suppose then that f_1 is of type (1). This time the relations $f_1((1 \otimes v_i^*)(x_k) \otimes v_j) = 0$ for $k = 1, 2$ lead to the conclusion $\lambda = 0$ or $\lambda^3 = 216$. However these four cubic forms are all projectively eqivalent: For example if ξ is a primitive third root of unity and $a^3 = 1/3$, then $X_1^3 + X_2^3 + X_3^3$ is projectively equivalent to $X_1^3 + X_2^3 + X_3^3 + 6X_1X_2X_3$ via the transformation that sends X_1 to $a(X_1 + X_2 + X_3)$, X_2 to $a(X_1 + X_2 + {}^2X_3)$ and X_3 to $a(X_1 + {}^2X_2 + X_3)$. The equivalences between $X_1^3 + X_2^3 + X_3^3$ and the other two forms are similar.

We may assume then that $\lambda = 0$, so f_1 corresponds to $X_{1_3}^3 + X_2^3 + X_3^3$. The usual relations now show that $W = \langle \sum_{i=1}^{3} v_i^* \otimes v_i^* \otimes v_i^*, g, \alpha g\rangle$ where $g = v_1^* \otimes v_2^* \otimes v_3^* + \xi v_2^* \otimes v_3^* \otimes v_1^* + \xi^2 v_3^* \otimes v_1^* \otimes v_2^* - v_1^* \otimes v_3^* \otimes v_2^* - \xi v_2^* \otimes v_1^* \otimes v_3^* - \xi^2 v_3^* \otimes v_2^* \otimes v_1^*$, ξ a primitive third root of unity and $\alpha = (1,2)$ as before. (Using ξ^2 in place of ξ gives a space easily seen to be $GL(V)$ conjugate to the one given.) But this space is $GL(V)$ conjugate to W_1 under the transformation that sends v_1^* to $v_1^* + v_2^* + v_3^*$, v_2^* to $b^2v_1^* + b^5v_2^* + b^8v_3^*$ and v_3^* to $bv_1^* + b^7v_2^* + b^4v_3^*$, where b is a primitive ninth root of unity. This finishes the proof. □

In the following corollaries we show that this theorem allows the complete determination of the balanced skew cyclic transformations on $M_3(F)^2$, F algebraically closed, characteristic zero.

COROLLARY 3.2. *Let* $\tau: M_3(F)^2 \to M_3(F)^2$ *be a balanced skew cyclic transformation. Let* $S = \{x \in ((M_3(F))^2)^{S_2} \mid x$ *is invertible,* $\tau(x) = x$ *and* $\tau'(x^{-1}) = x^{-1}\}$. *Then* $S = \{c\bar{\tau}(g)(g \otimes g) \mid g \in GL_3(F),\ c \in (C_\tau)^{S_2}$ *and invertible, and* $\tau(c) = c\}$.

PROOF: First note that $\tau|_{C_\tau} = \tau'|_{C_\tau}$: If $c \in C_\tau = C_{\tau'}$ is invertible then $(1 \otimes c)e(1 \otimes c^{-1}) = e$ where e is the idempotent associated to τ. Hence $(\tau(c) \otimes 1)e(\tau'(c^{-1}) \otimes 1) = e$, so $\tau(c) = \tau'(c^{-1})^{-1} = \tau'(c)$. A computation then shows that elements of the form $c\bar{\tau}(g)(g \otimes g)$ as described above are in S. (See Lemma 1.5 of [3]). Conversely, suppose $x \in S$. Let $W = (V^3)^*e$, the nondegenerate space corresponding to τ. Then $W(x \otimes 1)$ is another balanced nondegenerate subspace, of the same type as W. By the theorem there is an element g in $GL_3(F)$ such that $W(x \otimes 1) = W(g \otimes g \otimes g)$. But $W(g \otimes g \otimes g) = We(g \otimes g \otimes g) = We(\bar{\tau}(g)(g \otimes g) \otimes 1) = W(\bar{\tau}(g)(g \otimes g) \otimes 1)$. Hence $\bar{\tau}(g)(g \otimes g)x^{-1}$ is in C_τ. Hence $x = c\bar{\tau}(g)(g \otimes g)$ for some c in C_τ. Then c is in $(C_\tau)^{S_2}$ and since $\tau(x) = x$, it follows that $\tau(c) = c$. □

COROLLARY 3.3. *Let* $\tau, \gamma : M_3(F)^2 \to M_3(F)^2$ *be balanced skew cyclic transformations of the same type. Then there is an element* g *in* $GL_3(F)$ *such that* $\bar{\gamma}(a) = (g \otimes g)^{-1}\bar{\tau}(g)^{-1}\bar{\tau}(a)\bar{\tau}(g)(g \otimes g)$ *for all* a *in* $M_3(F)$.

PROOF: By Theorem 2.6 and the previous Corollary there is an element g in $GL_3(F)$ and an invertible element c in C_τ such that $\bar{\gamma}(a) = (g \otimes g)^{-1}\bar{\tau}(g)^{-1}c^{-1}\bar{\tau}(a)c\bar{\tau}(g)(g \otimes g)$. But this latter is just $(g \otimes g)^{-1}\bar{\tau}(g)^{-1}\bar{\tau}((a)\bar{\tau}(g)(g \otimes g)$, so we are done. □

We will say a skew cyclic transformation τ on $M_3(F)^2$ is of *type zero* if its corresponding nondegenerate subspace W is S_3-isomorphic to W_0 and *type one* if W is S_3-isomorphic to W_1.

COROLLARY 3.4. *Let* $\tau : M_3(F)^2 \to M_3(F)^2$ *be balanced skew cyclic. There are exactly nine skew cyclic transformations* γ *such that* $\bar{\gamma}(M_3(F)) = \bar{\tau}(M_3(F))$ *and all of these are balanced. More precisely, there is one of type zero and there are eight of type one.*

PROOF: Let $A = M_3(F)$. By Corollary 2.7, if τ is balanced then so is every right or left skew cyclic transformation $\gamma : A^2 \to A^2$ such that $\bar{\gamma}(A) = \bar{\tau}(A)$. □

Let τ_0 be the skew cyclic transformation associated to W_0. We claim it suffices to show there are exactly nine skew cyclic transformations γ such that $\bar{\gamma}(A) = \bar{\tau}_0(A)$ and all except τ_0 are of type one: To see the claim, suppose τ is balanced skew cyclic on A^2. Since both possible types occur on $\bar{\tau}_0(A)$, we know there is a skew cyclic transformation γ such that $\bar{\gamma}(A) = \bar{\tau}_0(A)$ and an element g in $GL_3(F)$ such that $(g \otimes g)^{-1}\bar{\gamma}(g)^{-1}\bar{\gamma}(a)\bar{\gamma}(g)(g \otimes g) = \bar{\tau}(a)$ for all a in A. In particular $(g \otimes g)^{-1}\bar{\gamma}(A)(g \otimes g) = \bar{\tau}(A)$. Now let δ be any skew cyclic transformation such that $\bar{\delta}(A) = \bar{\tau}_0(A)$. By Corollary 3.2 and Theorem 2.6, we can obtain another balanced skew cyclic transformation β of the same type as δ determined by $\bar{\beta}(a) = (g \otimes g)^{-1}\bar{\delta}(g)^{-1}\bar{\delta}(a)\bar{\delta}(g)(g \otimes g)$. But $\bar{\beta}(A) = (g \otimes g)^{-1}\bar{\delta}(A)g \otimes g = \bar{\tau}(A)$. In this way we produce nine skew cyclic transformations on $\bar{\tau}(A)$ of the desired types and another application of Corollary 3.2 and Theorem 2.6 shows there can be no others.

It suffices then to understand the situation on $\bar{\tau}_0(A)$. We have seen (Theorem 2.6) that if $\bar{\gamma}(A) = \bar{\tau}_0(A)$ then there must exist an element d in $GL_3(F)$ such that $\tau_0(\bar{\tau}_0(d)) = c\bar{\tau}_0(d)$ for some c in C_{τ_0} and

$\bar{\gamma}(a) = \bar{\tau}_0(d)^{-1}\bar{\tau}_0(a)\bar{\tau}_0(d) = \bar{\tau}_0(dad^{-1})$ for all a in A. Let $X = \begin{pmatrix} 0 & 0 & 1 \\ 1 & 0 & 0 \\ 0 & 1 & 0 \end{pmatrix}$ and let $y = \begin{pmatrix} 1 & 0 & 0 \\ 0 & \xi & 0 \\ 0 & 0 & \xi^2 \end{pmatrix}$ where is a primitive third root of unity. Then A is is generated as an F-algebra by x and y, and we have $x^3 = y^3 = 1$ and $yx = \xi xy$. Using W_0 we see that $\bar{\tau}_0(x) = x^{-1} \otimes x^{-1}$ and $\bar{\tau}_0(y) = y^{-1} \otimes y^{-1}$. The F-algebra C_{τ_0} is generated by $x^2 \otimes x$ and $y^2 \otimes y$, and $\tau_0|_{C_{\tau_0}} =$ identity. Hence it follows from $\tau_0(\bar{\tau}_0(d)) = c\bar{\tau}_0(d)$ that $c = d^{-1} \otimes d$, and so $\bar{\tau}_0(d)(1 \otimes d) = d^2 \otimes 1$. Expanding d in terms of the basis $\{1, x, x^2, y, y^2, xy, xy^2, x^2y, x^2y^2\}$ of A and using this last relation, we see that d must be of the form αu where $\alpha \in F^*$ and u is one of the nine basis elements given above. This gives the nine skew cyclic transformations. Moreover one can compute explicitly the corresponding balanced non-degenerate subspaces -- they are the spaces $W_0(1 \otimes 1 \otimes u)$ where u runs over the nine basis elements. By inspection one sees that W_1 is one of these (letting $u = x$) and that $W_0(1 \otimes 1 \otimes u) \simeq_{S_3} W_1$ unless $u = 1$. □

BIBLIOGRAPHY

[1] W. Fulton, Algebraic Curves, Benjamin, Reading, Mass., 1969.

[2] D. E. Haile, On central simple algebras of given exponent, J. Algebra 57 (1979), 449-465.

[3] D. E. Haile, On central simple algebras of given exponent II, J. Algebra 66 (1980), 205-219.

[4] J. G. Semple and G. T. Kneebone, Algebraic Curves, Oxford University Press, London, 1959.

Indiana University
Bloomington, Indiana 47405

Contemporary Mathematics
Volume 13, 1982

RINGS WITH BOUNDED INDEX OF NILPOTENCE

by Abraham A. Klein

FOR NATHAN JACOBSON

A ring R is said to be of bounded index (of nilpotence) if there is a positive integer n such that $a^n = 0$ for any nilpotent element $a \in R$. The least such integer is called the index of R, and we denote it by $i(R)$.

Rings with bounded index have nice properties; e.g. one-sided inverses are two-sided. For if $a,b \in R$, $ab = 1$ and $ba \neq 1$, then by a trick of Jacobson [1, p. 32], R contains an infinite number of matric units e_{ij}, $i,j=1,2,\ldots$ and therefore it contains nilpotent elements of arbitrary index.

The question that we would like to consider is: under what extensions the property of being with bounded index is preserved. We consider the following extensions: the matrix ring $M_k(R)$, $k \geq 2$, and the polynomial ring $R[T]$, T a set of commuting variables.

The general question with respect to $M_k(R)$ has a negative answer. An example of Shepherdson [2] shows that there exists a domain R (which may be chosen to be an algebra over the rationals) such that $M_2(R)$ contains two matrices A,B satisfying $AB = 1$, $BA \neq 1$. So $M_2(R)$ cannot have bounded index.

Let us consider the ring of polynomials $R[T]$. Here the answer is positive at least when R is assumed to be an algebra over an infinite field and $R[T]$ has the same index as R. The problem becomes difficult if no assumptions are imposed on the additive structure of R. We answer the problem affirmatively for PI rings. The basic result is:

0271-4132/82/0520/$02.00

THEOREM 1. <u>If</u> R <u>is a nil ring with bounded index and</u> C <u>is any commutative ring, then</u> $R \otimes C$ <u>has bounded index. Moreover, if</u> $i(R) \leq n$ <u>and</u> $n = 1,2$ <u>then</u> $i(R \otimes C) \leq n$ <u>and if</u> $n \geq 3$ <u>then</u> $i(R \otimes C) \leq \frac{2}{3}n!$

One easily observes that it suffices to prove the theorem when C is the free commutative ring, namely the ring of polynomials in a set T of commuting variables over the integers.

As in the proof of the theorem of Nagata-Higman [1] we consider the ideal S in R generated by $\{a^{n-1} | a \in R\}$. Since $(R/S)[T] = R[T]/S[T]$ and $i(R/S) \leq n-1$, the result will follow by induction if we prove that $S[T]$ is nil of index $\leq n$. This is the most difficult part of the proof. We are going to sketch the proof here and the full details will appear elsewhere.

We introduce the following non-commutative polynomials. Given n and $i_1,\ldots,i_{n-1}$ satisfying $0 \leq i_j \leq n$, $i_1 +\ldots+ i_{n-1} = n$, we let $P_{i_1\ldots i_{n-1}}(x_1,\ldots,x_{n-1})$ be the polynomial which is the sum of all the monomials of degree i_j in x_j. Ir will be convenient to write only the i_j's which are ≥ 1. These polynomials arise when x^n is multilinearized. For example we have:

$$(1) \qquad (x + y)^n - x^n - y^n = P_{n-1,1}(x,y) + P_{n-2,2}(x,y) +\ldots+ P_{1,n-1}(x,y).$$

An important step in the proof is the following result:

PROPOSITION. <u>If</u> R <u>satisfies all the polynomial identities</u> $P_{i_1\ldots i_{n-1}}$, $0 \leq i_j \leq n$, $i_1 +\ldots+ i_{n-1} = n$, <u>and</u> C <u>is any commutative ring then</u> $R \otimes C$ <u>is nil with index</u> $\leq n$.

The main trick in the proof of this result is to show that $(\sum_{i=1}^{k} x_i \otimes c_i)^n$ can be expressed as a combination with integral coefficients of the elements $(\sum_{j=1}^{n} x_{i_j} \otimes c_{i_j})^n$, $r \leq n - 1$.

The proof that S satisfies all the identities $P_{i_1\ldots i_{n-1}}$ is by induction on the set $\{(i_1,\ldots,i_{n-1}) | 0 \leq i_j \leq n,\ i_1 +\ldots+ i_{n-1} = n\}$ with the lexicograph order. We prove here that S satisfies $P_{n-1,1}$.

We first show that R satisfies $x^{n-1}yx^{n-1} = 0$. Since R is nil of index $\leq n$ we have

$$0 = (x^{n-1}y + x)^n = \sum_{j=1}^{n-1} (x^{n-1}y)^j x^{n-j} = x^{n-1}yx^{n-1}(1 + z)$$

with $z \in R$. Since z is nilpotent, $1+z$ is invertible so $x^{n-1}yx^{n-1} = 0$. Now we prove that S satisfies $P_{n-1,1}$. In fact we show that $P_{n-1,1}(x,s) = 0$ for all $x \in R$ and $s \in S$. Since $P_{n-1,1}$ is linear in its second variable and s is a sum of elements of the form $ua^{n-1}v$ with $a \in R$ and $u,v \in R \cup \{1\}$, it suffices to show that $P_{n-1,1}(x,ua^{n-1}v) = 0$. But for $i = 2,\ldots,n-1$, $P_{n-i,i}(x,ua^{n-1}v) = 0$ since in each term $ua^{n-1}v$ occurs at least twice, so it is 0 since $a^{n-1}ba^{n-1} = 0$ for any $a \in R$ and $b \in R \cup\{1\}$. Now using (1) we get $P_{n-1,1}(x,ua^{n-1}v) = 0$.

How we get $i(R \otimes C) \leq \frac{2}{3}n!$ when $n \geq 3$? It can be proved that if R is nil with index 3 then $R[T]$ is nil with index ≤ 4 and using this, the induction hypothesis is $i((R/S)[T]) \leq \frac{2}{3}(n-1)!$ and the resûlt follows since $i(S[T]) \leq n$.

Our result in the case $n = 3$ is the best possible one. Indeed, let R be the commutative algebra generated by two elements x,y over $GF(2)$ with the relations

$$x^3 = y^3 = x^2y^2 = 0, \quad x^2y = xy^2.$$

It is easy to show that R is nil with index 3 and the polynomial ring $R[t]$ in one variable has index 4.

It has been pointed out by Armendariz that Theorem 1 yields

THEOREM 2. <u>If</u> R <u>is a PI-ring with bounded index then</u> $R[T]$ <u>has bounded index</u>.

The proof follows using the fact that $R/B(R)$ ($B(R)$ is the prime radical) is embeddable in a matrix ring $M_k(C)$, C a direct product of fields.

Now let us consider commutative rings with bounded index. Such rings have been considered by Armendariz and Dulin, and using completely different methods they have proved that $R[t]$ has bound index (unpublished). They have obtained a very large bound for the index of $R[t]$. Our result is:

THEOREM 3. <u>Let</u> R <u>be a commutative ring which is nil with index</u> $\leq n$. <u>If</u> C <u>is any commutative ring then</u> $i(R \otimes C) \leq 1 + \frac{1}{2}n(n-1)$.

The result of Theorem 3 is the best possible one for $n \le 3$. This is clear for $n = 1,2$ and for $n = 3$ consider the above example. The result can be improved for $n = 4$; the bound given in Theorem 3 is 7 but we can prove that the bound is 5.

Finally we have:

THEOREM 4. *If* R *is a commutative ring with bounded index* $\le n$ *then* $R[T]$ *has bounded index* $\le 1 + \frac{1}{2}n(n-1)$.

REMARK. We have seen that the general problem concerning $M_k(R)$ has a negative answer. It may be that if R is assumed to be as in our Theorems then the answers are positive. If as in Theorem 1 we assume that R is nil with bounded index we do not know whether $M_k(R)$ has bounded index ($M_k(R)$ is nil and even locally nilpotent since R is locally nilpotent). We can show that if R is nil and $i(R) \le 2$ then $i(M_k(R)) \le k^2 + 1$ and if $i(R) \le 3$ then $i(M_k(R)) \le (k^2 + 1)(3k^2 + 1)$. With respect to Theorem 2, the result is positive for any index n for which the answer to the previous question is positive. If R is commutative it is easy to get positive answers. If R is nil with index $\le n$ then $i(M_k(R)) \le 1 + (n-1)k^2$, and if R is only assumed to have index $\le n$ then $i(M_k(R)) \le k(1 + (n-1)k^2)$.

Acknowledgement. I am indebted to Armendariz for his interest and remarks.

REFERENCES

1. Jacobson, N., Structure of Rings, Amer. Math. Soc. Colloq. Publ., 37 (1964), Providence.

2. Shepherdson, J. C., Inverses and zero divisors in matrix rings, Proc. London Math. Soc., 1 (1951), 71-85.

The University of Texas at Austin
and
Tel Aviv University

Added in proof (August 6, 1981): Now we can prove that if R is nil with bounded index then $M_k(R)$ has bounded index, but we do not have a bound for $i(M_k(R))$ when $i(R) \ge 4$.

Contemporary Mathematics
Volume 13, 1982

ON THE JACOBSON DENSITY THEOREM

J. M. Zelmanowitz
University of California
Santa Barbara, California

The area we wish to survey is the classification and structure theory of subrings of full linear rings[1]. Posed in this fashion, the problem is certainly too broad, as this class of rings includes many nilpotent rings such as rings of strictly lower triangular matrices. We will therefore modify our investigation in the spirit of two well-known results which should be regarded as primary models.

CHEVALLEY-JACOBSON [7; p. 28]. A ring R is (isomorphic to) a subring of the ring T of linear transformations of a vector space V_Δ with the property that given any $\tau \in T$ and finite dimensional subspace $U_\Delta \subseteq V$ there exists $r \in R$ with $r|_U = \tau|_U$ if and only if R has a faithful simple module.

GOLDIE [5]. A ring R is (isomorphic to) a subring of a ring T of linear transformations of a finite dimensional vector space V_Δ with the properties that (a) every non-zero-divisor in R is a unit in T, and (b) every element of T is of the form $r^{-1}s$ for some $r,s \in R$ if and only if R is a prime ring satisfying the ascending chain condition on annihilator left ideals[2] and closed left ideals[3].

A ring R of linear transformations with the property stated in Jacobson's theorem is called a Jacobson dense ring of linear transformations on V_Δ, or a primitive ring. A ring R of linear transformations with the property stated in Goldie's theorem will be called a Goldie order on V_Δ.

Primitive rings and Goldie orders provide two basic instances of what should be regarded as "large" subrings of full linear rings. The spirit of the two quoted theorems is quite distinct, howèver. Primitive rings are characterized by a representation-theoretic requirement (the existence of a faithful

[1] A full linear ring is the ring of linear transformations of a right vector space over a division ring.

[2] A left ideal L of R is an annihilator if $L = \ell_R(X) = \{r \in R | rX = 0\}$ for some $X \subseteq R$.

[3] A left ideal L of R is closed if ${}_RL$ has no essential extensions in R. A ring satisfying the ascending chain condition on closed left ideals is called finite dimensional.

0271-4132/82/0521/$03.00

simple module), while Goldie's theorem supplies an internal ring-theoretic characterization. Jacobson's theorem is of further interest because it permits us to reverse our perspective, and ask what restrictions the existence of a "special" faithful representation places on the structure of a ring? Thus we take the point of view that a model theorem would read: R is a "large" subring of a full linear ring if and only if it has a "special" faithful module; with appropriate definitions of the words "large" and "special".

HISTORY AND MOTIVATION

By virtue of a sequence of contributions by many investigators ([8], [11], [6], [1], [17]) it became possible to develop a theory which simultaneously embraced both the Jacobson theorem (and theory) and Goldie's theorem. We will describe these accomplishments, as they will condition our later discussion. The terminology used is that of the survey article [16]. The principal theorem which emerged from these developments reads as follows.

THEOREM A [17; Theorem 2.2]. A ring R is (isomorphic to) a weakly dense[4] subring of a full linear ring if and only if R has a faithful critically compressible[5] module.

The preceding theorem specializes to the Jacobson theorem in case the critically compressible module is actually simple; and the following module-theoretic version of the Goldie theorem may be regarded as the finite dimensional specialization of this result.

THEOREM B [17; Corollary 2.3]. R is a Goldie order if and only if R has a critically compressible module M with $\ell_R(m_1,\ldots,m_t) = 0$ for some $m_1,\ldots,m_t \in M$.

One structural property of Goldie orders is known as the Faith-Utumi theorem. It reads as follows.

FAITH-UTUMI [4]. If R is a Goldie order in the matrix ring $M_n(\Delta)$, Δ a division ring, then R contains a subring isomorphic to $M_n(D)$ for some order D in Δ.

A structural property of primitive rings is the following.

---

[4] A subring R of the ring of linear transformations T of a vector space V_Δ is weakly dense in T if it satisfies the following property: there exists an R-submodule M of V such that given $\tau \in T$ and finite dimensional subspaces U_Δ and $U' = \sum_{i=1}^{t} m_i\Delta$ $(m_i \in M)$ of V_Δ there exists $r,s \in R$ with $r\tau|_U = s|_U$ and with $r|_{U'}$ an automorphism. Such a ring is also called a weakly primitive ring.

[5] A module $_RM$ is critically compressible if for every nonzero submodule $_RN$ of M, $\mathrm{Hom}_R(M,N)$ contains a monomorphism while $\mathrm{Hom}_R(M,M/N)$ contains no monomorphisms.

THEOREM C [7; p. 33]. If R is a Jacobson dense ring of linear transformations on V_Δ, then either $R \cong M_n(\Delta)$ for some positive integer n, or else for each positive integer n, $M_n(\Delta)$ is a homomorphic image of a subring of R.

With the advent of the theory of weakly primitive rings it became possible to simultaneously embrace both of the preceding results under one umbrella.

THEOREM D [17, Theorem 3.3]. If R is a weakly dense ring of linear transformations on the vector space V_Δ, then either V_Δ is finite dimensional and R contains a subring isomorphic to $M_n(D)$ for some positive integer n and some order D of Δ; or else for each positive integer n there exists an order D of Δ such that $M_n(D)$ is a homomorphic image of a subring of R.

Moreover, the weakly dense rings of linear transformations on V_Δ which possess linear transformations of finite rank on V_Δ turned out to be a class of rings which had been extensively investigated, and which has a rich structure theory which parallels and extends the theory of primitive rings with nonzero socle. The precise situation reads as follows.

THEOREM E [11], [1], [17]. The following conditions are equivalent for a ring R.

(i) R is a weakly dense ring of linear transformations on V_Δ which possesses a linear transformation of finite rank.

(ii) R has a faithful critically compressible left ideal.

(iii) R is a prime ring[6] with a nonsingular[7] uniform[8] left ideal.

(iv) R is a prime ring with a maximal annihilator left ideal and a maximal closed left ideal.

(v) R can be embedded in the ring of linear transformations S of a left vector space ${}_\Delta W$ in such a way that given any $\sigma \in S$ and finite dimensional subspace ${}_\Delta U \subseteq W$ there exists $r,s \in R$ with $r\tau = s$, $Wr \subseteq U$ and $Ur = U$.

(vi) There exists a Morita context (R,M,N,S) with S a left Ore domain[9], ${}_SN_R$ faithful, and such that $[n,M] = 0$ and $[N,m] = 0$ respectively imply that $n = 0$ and $m = 0$.

In summary then, what had emerged from these investigations was an extension of the Jacobson theory of primitive rings which was sufficiently broad to embrace and further illuminate the structure of Goldie orders. One success of this new theory was the first significant information on the structure of poly-

[6] A ring is prime if $AB = 0$ for ideals A and B implies that $A = 0$ or $B = 0$.

[7] An R-module M is nonsingular if $\{m \in M \mid \ell_R(m)$ is an essential left ideal of $R\} = 0$.

[8] An R-module is uniform if every pair of non-zero submodules has non-zero intersection.

[9] A left Ore domain is a Goldie order in a division ring.

nomial rings over primitive rings; such rings are always weakly primitive [13]. The ladder of successive generalization suggested by this result ends with the first rung; for a polynomial ring over a weakly primitive ring is again weakly primitive.

Before moving on we should mention that efforts in this general direction were also being made by other researchers. In [2], a notion of "k-primitive" rings was introduced, and some properties of these rings were discussed, but the authors were unable to describe the nature of the embedding of k-primitive rings in full linear rings, although they conjectured that such a description should be possible. A distinct concept of "K-primitive" rings also appears in [10].

All rings that have been discussed up to this point have been prime rings. There are, however, candidates for the title of "large" subrings of full linear rings which are not prime. The most elementary such examples are perhaps triangular matrix rings over division rings. The sense in which triangular matrix rings and their infinite dimensional analogues are "large" subrings of full linear rings can be made precise by the concept of maximal quotient rings[10].

For the rest of this discussion we will assume that all modules are cofaithful[11]; in particular, all rings are assumed to have zero right annihilator. $Q(R)$ will always denote the maximal quotient ring of R. For R cofaithful, $Q(R)$ exists and is unique up to isomorphism over R [3; p. 68].

The fact is that if $T_n(\Delta)$ denotes the ring of lower triangular $n \times n$ matrices over the division ring Δ, then $Q(R) = M_n(\Delta)$. Explicitly then, given any matrices $\tau, 0 \neq \tau' \in M_n(\Delta)$ there exists $r, s \in T_n(\Delta)$ with $r\tau = s$ and $r\tau' \neq 0$. In view of this we have a new question before us. For which rings R is it the case that $Q(R)$ is a full linear ring? The mathematical literature provides at least two relevant results.

SANDOMIERSKI [15]. $Q(R)$ is a semisimple artinian ring if and only if R is a finite dimensional nonsingular ring.

JOHNSON [9]. $Q(R)$ is a full linear ring if R is an irreducible[12] nonsingular ring such that every nonzero left ideal of R contains a nonzero uniform left ideal.

---

[10] An overring Q of R is called a maximal quotient ring of R if (a) $(R:x)y \neq 0$ for every $x, 0 \neq y \in Q$, where $(R:x) = \{r \in R \mid rx \in R\}$, and (b) Q is maximal with respect to this property.

[11] A module ${}_R M$ is cofaithful if $r_M(R) = \{m \in M \mid Rm = 0\} = 0$.

[12] A ring R is irreducible if $A \cap r_R(A) = 0$ for A a nonzero ideal of R implies that $r_R(A) = 0$. This terminology is due to R. E. Johnson [9].

It can readily be verified that the converse of Johnson's theorem holds (see [13]), and so there is in fact available an internal characterization of the rings in question.

RECENT RESULTS

We intend now to update the survey in [16] with a description of the most recent progress in this area. First of all we would like a representation-theoretic characterization of the rings R for which $Q(R)$ is a full linear ring. For this we need to begin by identifying a class of modules such that there is associated to each module in the class a division ring.

This search is simplified somewhat by the realization that critically compressible modules are monoform[13]. Furthermore, a module ${}_RM$ is monoform if and only if $\Delta = \mathrm{End}_R\bar{M}$ is a division ring, where ${}_R\bar{M}$ is the quasi-injective hull of ${}_RM$ (cf. [3; p. 24]). Δ can also be constructed as a direct limit of partial endomorphisms of M, but this need not yet concern us. Thus the class of rings possessing faithful monoform modules emerges as a natural candidate for study, and we can ask: What does the existence of a faithful monoform module imply about the structure of the ring?

The finite dimensional situation is completely settled by the following result.

THEOREM F [18]. Each of the following conditions is necessary and sufficient for $Q(R)$ to be isomorphic to $M_n(\Delta)$ for some positive integer n and division ring Δ.

(1) R has a monoform module M such that $\ell_R(m_1,\ldots,m_t) = 0$ for some $m_1,\ldots,m_t \in M$.

(2) R has a nonsingular uniform module M with $\ell_R(m_1,\ldots,m_t) = 0$ for some $m_1,\ldots,m_t \in M$.

(3) R is an irreducible, finite dimensional, nonsingular ring.

One can recognize in condition (3) an interweaving of the earlier results of Sandomierski and Johnson. Conditions (1) and (2) are the representation-theoretic characterizations we demanded.

This approach produces a dividend in the form of structural information about this class of rings. For, if $Q(R) \cong M_n(\Delta)$ then we may regard the right Δ-vector space $V = \Delta^{(k)}$ as a R-Δ-bimodule. A careful investigation of the action of R on V produces the following information.

THEOREM G [19]. Let $Q(R) = M_n(\Delta)$ with Δ a division ring. Then there exist matrix units $\{e_{ij} \mid 1 \le i,j \le n\}$ for $M_n(\Delta)$ and an order D in $\Gamma = \{a \in Q \mid ae_{ij} = e_{ij}a \text{ for all } i,j\}$ such that $\sum_{j=1}^{n} De_{1j} \subseteq R$.

[13] A module ${}_RM$ is monoform if every nonzero partial endomorphism of M is a monomorphism; i.e., every $0 \neq f \in \mathrm{Hom}_R(N,M)$ with ${}_RN \subseteq M$ is a monomorphism.

This settles in the affirmative a question first raised by C. Faith as Open Problem 13 in [3]. One should note that the stated condition is also sufficient for $Q(R)$ to equal $M_n(\Delta)$, and thus we have in hand an internal structural characterization of the subrings R of $M_n(\Delta)$ with $Q(R) = M_n(\Delta)$.

The study of the general infinite dimensional situation is both complicated and made attractive by the realization that rings with large nilpotent ideals have now entered the picture. Furthermore, the condition $\ell_R(m_1,\ldots,m_t) = 0$ in (1) and (2) of Theorem F is equivalent in this context to the requirement that $\bar{M}$ be finite dimensional as a vector space over $\mathrm{End}_R\bar{M}$. (This fact is a consequence of the deceptively modest "double annihilator lemma" of Jacobson [7; p. 27], a result which continues to play an important role in the results described herein.) Thus we must distinguish between two different infinite dimensional situations according as R has a faithful monoform module or a faithful nonsingular uniform module. Nonsingular uniform modules are monoform, and monoform modules are uniform, but there do exist monoform singular modules.

Observe that the requirement that $Q(R)$ be a full linear ring imposes a "global" relationship between the elements of R and those of $Q(R)$ in rough parallel to the statement of Theorem E(v). By contrast, a weakly dense subring of a full linear ring T admits in general only a "local" relationship between the elements of R and those of T. These considerations motivated the following theorems.

THEOREM H [18]. A ring R is isomorphic to an m-dense[14] subring of a full linear ring if and only if R has a faithful monoform module.

THEOREM I. $Q(R)$ is isomorphic to the ring of linear transformations of a left vector space ${}_\Delta W$ if and only if R has a faithful nonsingular uniform module.

SOME QUESTIONS

For the rings of Theorem H the question of when they possess linear transformations of finite rank is going to be more complicated than the situation described for weakly primitive (or primitive) rings. One might conjecture, on the basis of Theorem E, that what is required is, respectively, a fait-ful monoform left ideal or a faithful nonsingular uniform left

[14] A subring R of the ring of linear transformations T of a vector space V_Δ is m-dense in T if it satisfies the following property: given any finite dimensional subspace U_Δ of V and any $\tau \in T$ with $\tau|_U \neq 0$ there exists $r,s \in R$ with $r\tau|_U = s|_U \neq 0$.

ideal. That this is not the case is already evident from the example $R = \begin{pmatrix} 0 & 0 \\ \Delta & \Delta \end{pmatrix} = \{\begin{pmatrix} 0 & 0 \\ a & b \end{pmatrix} | a,b \in \Delta\}$, Δ a division ring. Here $Q(R) = M_2(\Delta)$, but R has no faithful nonsingular uniform (or even monoform) left ideal. However $M = \begin{pmatrix} \Delta \\ \Delta \end{pmatrix} = \{\begin{pmatrix} x \\ y \end{pmatrix} | x,y \in \Delta\}$ is a faithful nonsingular uniform R-module, consistent with Theorem I. Thus we are left with a set of problems.

QUESTION 1. Which of the rings described by THEOREM H possess linear transformations of finite rank?

QUESTION 2. Which rings have faithful nonsingular left ideals or faithful monoform left ideals?

QUESTION 3. Is there an infinite dimensional analogue of Theorem G (presumably along the lines of Theorem D)?

For rings with faithful monoform modules, satisfactory answers to Question 1 and Question 2 are already in hand. For instance, an m-dense ring of linear transformations contains a transformation of finite rank if and only if it has a faithful monoform module M such that $Z(M) \neq M$, where $Z(M)$ denotes the singular submodule of M. A full description of this situation is provided in [18].

A related and particularly intriguing question is the following. We do not have any conjecture to offer as to its resolution.

QUESTION 4. Provide a characterization (structural or representation-theoretic) of the subrings of $M_n(\Delta)$ which contain a full ring of triangular (or triangulable) matrices over an order in Δ.

Looking still further ahead, it would appear that advances in the structure theory of rings along the lines discussed above will eventually necessitate the study of rings which are "large" subrings of endomorphism rings of free modules over local rings[15]. Some but by no means all of the tools of investigation which have been employed to date are applicable in this wider setting. The starting points might be to examine the "large" subrings of $M_n(\Gamma)$ for Γ a local ring, and to classify the rings with faithful uniform modules.

REFERENCES

[1] S. A. Amitsur, Rings of quotients and Morita contexts, J. Algebra 17 (1971), 273-298.

[2] M. G. Deshpande and E. H. Feller, The Krull radical, Comm. in Algebra 3 (1975), 185-193.

[3] C. Faith, Lectures on Injective Modules and Quotient Rings, Lecture Notes in Math., vol. 49, Springer-Verlag, Berlin, 1967.

[4] C. Faith and Y. Utumi, On Noetherian prime rings, Trans. Amer. Math. Soc. 114 (1965), 53-60.

[15] A ring is local if its non-units form an ideal.

[5] A. W. Goldie, Semiprime rings with maximum condition, Proc. London Math. Soc. 10 (1960), 201-220.

[6] A. Heinicke, Some results in the theory of radicals of associative rings, Ph.D. Dissertation, University of British Columbia, 1969.

[7] N. Jacobson, Structure of Rings, Amer. Math. Soc. Colloq. Publ. vol. 37, Providence, R. I., 1964.

[8] R. E. Johnson, Representations of prime rings, Trans. Amer. Math. Soc. 74 (1953), 351-357.

[9] R. E. Johnson, Quotient rings of rings with zero singular ideal, Pac. J. Math. 4 (1961), 1385-1392.

[10] J. P. Kezlan, On K-primitive rings, Proc. Amer. Math. Soc. 74 (1979), 24-28.

[11] K. Koh and A. C. Mewborn, Prime rings with maximal annihilator and maximal complement right ideals, Proc. Amer. Math. Soc. 16 (1965), 1073-1076.

[12] K. Koh and A. C. Mewborn, A class of prime rings, Can. Math. Bull. 9, (1966), 63-72.

[13] J. Lambek and G. O. Michler, On products of full linear rings, Publ. Math. Debrecen 24 (1977), 123-127.

[14] W. K. Nicholson, J. F. Watters and J. M. Zelmanowitz, On extensions of weakly primitive rings, Canadian J. Math. 32 (1980), 937-944.

[15] F. L. Sandomierski, Semisimple maximal quotient rings, Trans. Amer. Math. Soc. 128 (1967), 112-120.

[16] J. M. Zelmanowitz, Dense rings of linear transformations, Ring Theory II: Proceedings of the Second Oklahoma Conference, Marcel Dekker, New York, 1977, 281-294.

[17] J. M. Zelmanowitz, Weakly primitive rings, Comm. in Algebra 9 (1981), 23-45.

[18] J. M. Zelmanowitz, Monoform representations of rings.

[19] J. M. Zelmanowitz, Large rings of matrices contain full rows, J. Algebra.

University of California
Santa Barbara, CA 93106

Contemporary Mathematics
Volume 13, 1982

DERIVATIONS OF PRIME RINGS HAVING POWER CENTRAL VALUES

I. N. Herstein[1]

(To Nathan Jacobson, with affection and esteem)

In a recent paper [2] we showed that if R is a prime ring with center Z, and $a \in R$, $a \notin Z$ such that $(ax - xa)^n \in Z$ for all $x \in R$, then R is an order in a 4-dimensional simple algebra. Put another way, if δ is the inner derivation defined by $\delta(x) = ax - xa$ for $x \in R$, and if $\delta \neq 0$ but $\delta(x)^n \in Z$ for all $x \in R$ then R is an order in a 4-dimensional simple algebra. It is natural to try to extend this result from inner derivations to arbitrary ones. We do that in this paper.

The theorem we shall prove is

THEOREM. *Let R be a prime ring with center Z and suppose that $d \neq 0$ is a derivation of R such that $d(x)^n \in Z$ for all $x \in R$. Then either R is commutative or is an order in a 4-dimensional simple algebra.*

We shall prove the theorem by making a series of reductions. We first assert that $Z \neq 0$, otherwise $d(x)^n = 0$ for all $x \in R$; however, a recent result by Giambruno and Herstein [1] shows that in this case $d = 0$. So we have that $Z \neq 0$.

Before starting the proof of the theorem itself we prove two rather special results about prime rings.

LEMMA 1. *Let R be a prime ring, Z its centroid, and $L \neq 0$ a left ideal of R. Suppose that $p(x_1,\dots,x_n) \neq 0$ is a multilinear homogeneous polynomial, with coefficients in Z, and that $a \neq 0 \in R$ is such that $ap(r_1,\dots,r_n) = 0$ for all $r_1,\dots,r_n \in L$. Then L satisfies a polynomial identity over Z.*

---

[1] This work was supported by an NSF grant, MCS78-01153, at the University of Chicago.

Proof: If $p(x_1,\ldots,x_n)$ vanishes on L then there is nothing to prove. So we may suppose that p does not vanish over L. Now $p(x_1,\ldots,x_n) = x_1q(x_2,\ldots,x_n) + h(x_1,\ldots,x_n)$ where x_1 is never the first variable in any of the monomials appearing in $h(x_1,\ldots,x_n)$. Let $x_1 = u \in L$, $u_2 = p(v_1^{(2)},\ldots,v_n^{(2)}),\ldots, u_n = p(v_1^{(n)},\ldots,v_n^{(n)})$, where all the $v_i^{(j)}$ are in L; thus $u_i \in L$ and $au_i = 0$ for $i = 2,\ldots,n$. Thus $0 = ap(u,u_2,\ldots,u_n) = auq(u_2,\ldots,u_n) + ah(u,u_2\ldots u_n)$. Since every term in $h(u,u_2,\ldots,u_n)$ begins with some u_k, $i \geq 2$, we have $ah(u,u_2,\ldots,u_n) = 0$. Therefore $aLq(u_2,\ldots,u_n) = 0$. Because $a \neq 0$ and R is prime, we conclude that $Lq(u_2,\ldots,u_n) = 0$. Remembering the form of the u_i we see that L satisfies $t(x,x_{21},\ldots,x_{2n},\ldots,x_{n1}, x_{n1},\ldots,x_{nn}) = xq(p(x_{21},\ldots,x_{2n}),\ldots,p(x_{n1},\ldots,x_{nn}))$. This proves the lemma.

We extend Lemma 1 slightly to

LEMMA 2: *Let $R, Z, p(x_1,\ldots,x_n)$ be as in Lemma 1, and suppose that for some non-zero left ideal L of R and some $a \neq 0 \in R$, $ap(r_1,\ldots,r_n) \in Za$ for all $r_1,\ldots,r_n \in L$. Then L satisfies a polynomial identity over Z.*

Proof: Let $u_1,\ldots,u_n, v_1,\ldots,v_n \in L$; then $ap(u_1,\ldots,u_n) = \lambda a$, $ap(v_1,\ldots,v_n) = \mu a$ where $\lambda,\mu \in Z$. Thus

$ap(u_1,\ldots,u_n)p(v_1,v_2,\ldots,v_n) = \lambda ap(v_1,\ldots,v_n) = \lambda\mu a = \mu\lambda a =$

$ap(v_1,\ldots,v_n)p(u_1,\ldots,u_n)$. Thus if $q(x_1,\ldots,x_n,y_1,\ldots,y_n) =$

$p(x_1,\ldots,x_n)p(y_1,\ldots,y_n) - p(y_1,\ldots,y_n)p(x_1,\ldots,x_n)$ then $aq(r_1,\ldots,r_{2n}) = 0$ for $r_1,\ldots,r_{2n} \in L$. By Lemma 1 we conclude that L satifies a polynomial identity.

We now turn our attention to the problem mentioned at the outset, namely, prime rings R where $d(x)^n \in Z$ for all $x \in R$. The approach depends on first settling the special case in which R is a division ring having a finite center.

LEMMA 3: *Let D be a division ring having a finite center Z, and suppose that d is a derivation of D such that $d(x)^n \in Z$ for all $x \in D$. If for some $y \in D$, $d^2(y) = 0$, then $d(y) = 0$.*

Proof: Suppose that $d^2(y) = 0$ but $d(y) \neq 0$. Then $d(yd(y)^{-1}) = 1 + yd(d(y)^{-1}) = 1 - yd(y)^{-1}d^2(y)d(y)^{-1} = 1$. So if $t = yd(y)^{-1}$, then $d(t) = 1$. Hence $d(t^p) = pt^{p-1} = 0$ where $p = \text{char } D$, and $d(t^{p+1}) = t^p$. Since $d(t^{p+1})^n \in Z$, $t^{pn} \in Z$, hence t is algebraic over Z, whence $t^{p^r} = t$ for some r. But then $d(t) = d(t^{p^r}) = p^r t^{p^r-1} = 0$, contradicting $d(t) = 1$.

We are now ready to settle the special case when R is a division ring having a finite center.

<u>THEOREM 1</u>. *Let D be a division ring having a finite center Z, and suppose that d is a derivation of D such that $d(x)^n \in Z$ for all $x \in D$. Then* $d = 0$.

<u>Proof</u>: Since $d(Z) \subset Z$ and Z is a finite field we must have $d(Z) = 0$. Suppose that $d(t) \neq 0$ for some $t \in D$; by Lemma 3 we have that $d^2(t) \neq 0$. Now $d(t)^n = \sigma \in Z$; if K is the algebraic closure of Z then d extends to a K-linear derivation of $R = D \otimes_Z K$. Moreover, R is a simple ring with 1, and for some $\sigma \in K$ such that $\sigma^n = \alpha$, $d(t) - \sigma \neq 0$ is a zero divisor in R.

While it may not be true that $d(x)^n \in Z(R) = K$ for all $x \in R$, however, because $\sum_{\sigma \in S_n} d(x_{\sigma(1)}) \ldots d(x_{\sigma(n)}) \in Z$ for $x_i \in D$ - where S_n is the symmetric group of degree n - we also have $\sum_{\sigma \in S_n} d(r_{\sigma(1)}) \ldots d(r_{\sigma(n)}) \in R$ for all $r_1, \ldots, r_n \in R$.

Let $u = d(t) - \sigma$ and let $L = x \in R | xd(u) = 0$. Since u is a zero divisor, $L \neq 0$. Also, since $d(u) = d^2(t) \neq 0 \in D$ is a unit, $Ld(u) \neq 0$.

Suppose $x_1, \ldots, x_n \in L$; thus $x_i u = 0$, and so $0 = d(x_i u) = d(x_i)u + x_i d(u)$. Now $\sum_{\sigma \in S_n} d(ux_{\sigma(1)}) \ldots d(ux_{\sigma(n)}) \in K$, therefore $\sum_{\sigma \in S_n} d(ux_{\sigma(1)}) \ldots d(ux_{\sigma(n)})u \in Ku$. But $d(ux_{\sigma(1)}) \ldots d(ux_{\sigma(n)})u =$ $d(ux_{\sigma(1)}) \ldots d(ux_{\sigma(n-1)})(d(u)x_{\sigma(n)} + ud(x_{\sigma(n)})u =$ $d(ux_{\sigma(1)}) \ldots d(ux_{\sigma(n-1)})ud(x_{\sigma(n)})u = \ldots = ud(x_{\sigma(1)})ud(x_{\sigma(2)}) \ldots ud(x_{\sigma(n)})u =$ $u(d(x_{\sigma(1)})u \ldots d(x_{\sigma(n)})u$. In this way we see that

$$(1) \qquad Ku \ni \sum_{\sigma \in S} d(ux_{\sigma(1)}) \ldots\ldots d(ux_{\sigma(n)})u$$
$$= u \sum_{\sigma \in S_n} d(x_{\sigma(1)})u \ldots ud(x_{\sigma(n)})u.$$

But $d(x_{\sigma(i)})u = -x_{\sigma(i)}d(u)$ since $x_{\sigma(i)}u = 0$. Thus (1) tells us that

$$u \sum_{\sigma \in S_n} (x_{\sigma(1)}d(u)) \ldots (x_{\sigma(n)}d(u)) \in Ku$$

for $x_i \in L$. Thus for $w_1, \ldots, w_n \in Ld(u) \neq 0$, $u \sum_{\sigma \in S_n} w_{\sigma(1)} \ldots w_{\sigma(n)} \in Ku$.

By Lemma 2 we conclude that the left ideal $Ld(u) \neq 0$ of R satisfies a polynomial identity over K. By a result of Martindale [3], R has a minimal left ideal Re where eRe is a finite dimensional division algebra over K, hence

$\bar{e}Re = Ke$ since K is algebraically closed. Since R is simple with 1 and has a minimal left ideal, we get that R is simple artinian and $R \approx K_m$ for some m.

Since $R = D \otimes_Z K \approx K_m$ is finite dimensional over K, D is finite dimensional over Z. Because Z is finite D must be finite, hence a finite field, whence $d = 0$. This proves the theorem.

Suppose now in all that follows that R is a prime ring, center $Z \neq 0$, $d(x)^n \in Z$ for all $x \in R$, where $d \neq 0$ is a derivation of R. We suppose that R is not commutative. We then assert that

LEMMA 4: *If* $\alpha \in Z$ *then* $d(\alpha) = 0$.

Proof: If Z is finite then $d(\alpha) = 0$ for all $\alpha \in Z$. So, suppose that Z is infinite.

If char $Z = 0$ and $d(\alpha) \neq 0$ for some $\alpha \in Z$, then $\frac{d(\alpha+m)}{\alpha+m} = \frac{d(\alpha)}{\alpha+m}$ takes on an infinite set of distinct values in the field of quotients of Z, as m runs over the integers.

If char $Z = p \neq 0$ then, for $\alpha \in Z$, $d(\alpha^p) = p\alpha^{p-1}d(\alpha) = 0$, so $\alpha^p \in Z_0$, where $Z_0 = \{\beta \in Z | d(\beta) = 0\}$. If $\alpha \notin Z_0$ then α is purely inseparable over Z_0, hence Z_0 is infinite. So $\frac{d(\alpha+\beta)}{\alpha+\beta} = \frac{d(\alpha)}{\alpha+\beta}$ takes on an infinite set of distinct values in the quotient field of Z as β runs over Z_0.

So we may assume that $\frac{d(\alpha)}{\alpha}$ takes on an infinite set of distinct values as α runs over Z. If $x \in R$ then $d(\alpha x)^n = (d(\alpha)x + \alpha d(x))^n \in Z$, so, in the localization of R_Z of R at Z, $(\frac{d(\alpha)}{\alpha}x + d(x))^n \in Z_1$, the center of R_Z. Since this is true for an infinity of distinct values $\frac{d(\alpha)}{\alpha} \in Z_1$ we get $x^n \in Z$ for all $x \in R$. By a result of ours and Kaplansky [4] we can conclude that R is commutative, a contradiction. Hence $d(\alpha) = 0$ for all $\alpha \in Z$.

Since $d(\alpha) = 0$ for $\alpha \in Z$, localizing R at Z and extending d by defining $d(\frac{x}{\alpha}) = \frac{d(x)}{\alpha}$ for $x \in R$, the conditions on R carry over to this localization. Since the center of the localization of R at Z is a field, *we may assume henceforth that* Z *is a field*.

With this assumption on Z we proceed to

LEMMA 5: *If* R *is a domain then* R *is a division ring*.

Proof: We show that R has no non-trivial left ideals. Let $L \neq 0$, R be a left ideal of R. Then, as is easily verified, $L + d(L)$ is also a left ideal of R. If $L + d(L) \neq R$ then, since Z is a field, $(L + d(L)) \cap Z = 0$. So, if $a \in L$ then $d(a)^n \in Z \cap (L + d(L)) = 0$, and since R is a domain, $d(a) = 0$. So $d(Ra) = 0$ for all $a \in L$, which yields $d(R)a = 0$. But then $d(R) = 0$; and so $d = 0$, a contradiction.

So we may suppose that $L + d(L) = R$; therefore $1 = a + d(b)$ for suitable $a,b \in L$. Thus $(1 - a)^n = d(b)^n = \alpha \in Z$. This tells us that a is algebraic over Z, and since Z is a field, R a domain, if $a \neq 0$ then a is invertible in R. But since $a \in L$, $L \neq R$, we cannot have a invertible in R, hence $a = 0$. Thus $1 = d(b)$ where $b \in L$. Therefore $d(b^2) = 2b$, $d(b^3) = 3b^2$ and so $(2b)^{2n} = d(b^2)^{2n} \in Z$ and $(3b^2)^n = d(b^3)^n \in Z$; these give us that $b^{2n} \in Z$, and so b is invertible. This again forces $L = R$. Therefore R has no non-trivial left ideals, so must be a division ring.

COROLLARY. *If R is a domain and Z is finite then R is commutative.*

Proof: This follows from the Lemma and Theorem 1.

We exploit the fact that Z is a field even further, to prove

LEMMA 6. *R is a simple ring.*

Proof: Suppose that $I \neq 0$, R is an ideal of R. Then $I^2 \neq 0$ is an ideal of R and $d(I^2) \subset I$, so, if $x \in I^2$, $d(x)^n \in I \cap Z = 0$ since Z is a field. By a result of Giambruno and Herstein [1], $d = 0$ follows, contradicting that $d \neq 0$. Thus R must be simple.

We digress now to prove a result of the "commutativity" type. Although we merely shall use this result in case R is simple, we prove it in a more general form.

THEOREM 2. *Let R be a ring having no non-zero nil ideals, with center Z, and suppose that for every $x \in R$, $x^{n(x)} = \lambda(x)x$ where $n(x) > 1$ and $\lambda(x) \in Z$. Then R satisfies the standard identity in 4 variables, and is a subdirect product of commutative integral domains and 2×2 matrix rings over fields algebraic over finite fields.*

Proof: Since R has no nil ideals, to prove the theorem it is enough to prove it for prime rings with the same hypotheses. So we suppose that R is prime. Since R has no nil ideals, $Z \neq 0$. Let F be the field of quotients of Z and let $Q = R \otimes_Z F$ be the localization of R at Z. Then Q is prime, the hypotheses on R carry over to Q and the center of Q is F, a field. So we may suppose that the center of R is a field.

Suppose that T is <u>not</u> commutative. If J is the Jacobson radical of R and $x \in J$ then $x^n = \lambda x$ so $(x^{n-1}-\lambda)x = 0$. If $\lambda \neq 0$, $x^{n-1}-\lambda$ is invertible, giving $x = 0$; so $\lambda = 0$ for $x \in J$. But then J is nil. In consequence, $J = 0$.

So we may suppose that R is semi-simple, and by structure theory, we may assume that R is primitive. If R is a division ring, $x^n = \lambda x$ implies $x^{n-1} \in Z$; by a result of Kaplansky [6], R is a field. If R is not a division ring, then, since R is primitive, if R, is not isomorphic to the 2×2 matrices over a division ring, then D_3, the 3×3 matrices over a division ring is a homomorphic image of a subring of R, which is an algebra over Z and so inherits the hypotheses of R. But if $x = \begin{pmatrix} 0 & 1 & 0 \\ 0 & 0 & 0 \\ 0 & 0 & 1 \end{pmatrix}$ then $x^n \neq \lambda x$ for any $n > 1$ or $\lambda \in Z$. So $R = D_2$. If $\alpha \in D$ then then $\begin{pmatrix} \alpha & 0 \\ 0 & 1 \end{pmatrix}^m = \lambda \begin{pmatrix} \alpha & 0 \\ 0 & 1 \end{pmatrix}$, for some $\lambda \in Z$, $m > 1$, which yields $\lambda = 1$, $\alpha^m = \alpha$. By a famous result of Jacobson [5] D is a field algebraic over a finite field.

<u>COROLLARY</u>. <u>If in the theorem R is simple then R is a field or is the ring of all 2×2 matrices over a field algebraic over a finite field</u>.

We now prove a very special result, with a peculiar hypothesis, for simple rings. Actually a much stronger result can be proved, which we state after the proof.

<u>LEMMA 7</u>. <u>Let R be any simple ring with 1, center Z, and suppose that d is a Z-linear derivation of R. If for some $a \neq 0 \in R$, $ad(r)a + ard(a) = 0$ for all $r \in R$, then d is an inner derivation of R.</u>

<u>Proof</u>: Since R is simple with 1 and $a \neq 0$, there exist $x_i, y_i \in R$ such that $\Sigma x_i a y_i = 1$. Thus $\Sigma x_i a d(y_i r) a + \Sigma x_i a (y_i r) d(a) = 0$ for all $r \in R$. This gives $d(r)a + ura + rd(a) = 0$ where $u = \Sigma x_i a d(y_i)$.

Therefore $\Sigma d(rx_i)ay_i + \Sigma urx_i ay_i + \Sigma rx_i d(a)y_i = 0$. On evaluating, using $\Sigma x_i a y_i = 1$, we get

$$d(r) + r\Sigma d(x_i)ay_i + ur + r\Sigma x_i d(a)y_i = 0,$$

hence $d(r) = rt_1 + t_2r$, t_1, t_2 fixed elements of R, for every $r \in R$. Since $0 = d(1) = t_1 + t_2$, we get $t_2 = -t_1$, hence $d(r) = rt_1 - t_1r$. In this way we see that d is inner.

Actually more can be proved for Z-linear derivations of a simple ring R with 1. Suppose that $\{\sum_1^k a_i d(r) b_i + \sum_1^m u_k r v_i \mid r \in R\}$ is finite-dimensional over Z, where $0 \neq a_1, \ldots, a_k$ are linearly independent over Z and $b_1 \neq 0$. Then d is inner. The proof is along the lines of the proof above, and reaches the situation of a result of Martindale [3].

LEMMA 8. *Suppose that R is simple with 1, $d \neq 0$ a derivation of R such that $d(x)^n \in Z$ for all $x \in R$. Suppose that $L \neq 0$, R is a left ideal of R such that $d(L) \subset L$. Then d is inner, hence R is a 4-dimensional simple algebra.*

Proof: We saw in Lemma 4 that we may assume that d is Z-linear. Since $d(L) \subset L$, if $T = \{x \in L \mid Lx = 0\}$ then T is an ideal of L and $d(T) \subset T$, so d induces a derivation $\bar{d}$ on $\bar{L} = L/T$. Also $\bar{L}$ is a simple ring. Because $L \cap Z = 0$, (since Z is a field), if $x \in L$ then $d(x)^n \in L \cap Z = 0$. Hence in $\bar{L}$ we have that $\bar{d}(x)^n = 0$ for all $\bar{x} \in L$. By the result of Giambruno and Herstein [1] quoted earlier, we have that $\bar{d} = 0$, which is to say, $d(L) \subset T$. Therefore $Ld(L) = 0$. So, if $a \neq 0 \in L$ then $ad(Ra) = 0$; this gives us that $ad(r)a + ard(a) = 0$ for all $r \in R$. By Lemma 7, d must be inner, so $d(x) = ux - xu$ for all $x \in R$; since $d \neq 0$, u cannot be in Z. Thus $(ux - xu)^n = d(x)^n \in Z$. By the main result of [2], R is an order in a 4-dimensional simple algebra.

We are now in the final stages of the proof of the theorem stated at the beginning. The last auxilliary piece that we need is given us by

THEOREM 3. *Suppose that R is a simple ring with center Z such that $d(x)^n \in Z$ for all $x \in R$, where $d \neq 0$ is a derivation of R. Suppose that R is not a domain. Then $R = F_2$, the ring of 2×2 matrices over a field F.*

Proof: Since $d \neq 0$ we know by [1] that $d(x)^n \neq 0$ for some x, hence $Z \neq 0$; since R is simple, Z is a field. Let $a \neq 0$ in R be a (right) zero-divisor, and let $L = \{x \in R \mid xa = 0\}$. If $d(L) \subset L$ then we are done by Lemma 8. So we may suppose that $d(L) \not\subset L$.

If $x \in L$ then $d(ax) = ad(x) = d(a)x$, hence $(d(a)x + ad(x))^n = \lambda(x) \in Z$. But then $x(d(a)x + ad(x))^n d(a) = \lambda(x)xd(a)$. However, $x(d(a)x + ad(x))^n d(a) = (xd(a))^{n+1}$ since $xa = 0$. So in $W = Ld(a)$ we have that $w^{n+1} = \lambda(w)w$, $\lambda(w) \in Z$, for all $w \in W$. Moreover, $W \neq 0$, for $W = Ld(a) = 0$ implies that $0 = d(La) = Ld(a) + d(L)a = d(L)a$, that is, $d(L) \subset L$. Clearly W is an algebra over Z.

Let $T = \{x \in W | Wx = 0\}$; then T is an algebra over Z and $A = {}^{W}/_{T}$ is a simple ring; since $w^{n+1} = \lambda(w)w$ for $w \in W$ this condition carries over to A. By the Corollary to Theorem 2, A is a field, or the 2×2 matrices over a field, F. This easily implies that R has a minimal left ideal - and since R is simple with 1, R must be simple artinian. So $R = D_m$, where D is a division ring. From the form of A, above, we have that $D = F$. So $R = F_m$, and since R is not a domain, $m > 1$.

Now we know that we may assume that d is Z-linear, hence on the finite-dimensional simple algebra, R, d must be inner. So $d(x) = ux - xu$ for all $x \in R$, and since $d \neq 0$, $u \notin Z$.

Since $(ux = xu)^n = d(x)^n \in Z$ we conclude from [2] that R is 4-dimensional over Z. Becuase R has zero-divisors, $R = F_2$ for the field $F = Z$.

We are now able to put the finishing touches on the proof of the Theorem mentioned at the outset.

Proof of Theorem. Let R be prime, center Z, and $d \neq 0$ a derivation of R such that $d(x)^n \in Z$. We saw, in Lemma 6, that we may assume R to be simple.

If Z is finite and R a domain then we saw, in the Corollary to Lemma 5, that R is commutative. On the other hand, if R is not a domain then, by Theorem 3, $R = F_2$, the 2×2 matrices over a field.

So we may suppose that Z is an infinite field. If R has zero divisors then we are done by Theorem 3. So we may assume that R is a domain; by Lemma 5, R is a division ring. Since $d(Z) = 0$, if $K \supset Z$ is an extension field, d extends to $R \otimes_Z K$ by defining $d(r \otimes k) = d(r) \otimes k$. If $x \in R \otimes_Z K$ then $d(x) = \sum_1^m d(x_i) \otimes k_i$ where $x_i \in R$, $k_i \in K$. Now, since Z is infinite and $(\sum_1^m d(x_i)\alpha_i)^n \in Z$ for all $\alpha_i \in Z$, we immediately have (by a van der Monde determinant argument) that $d(x)^n \in K$. So the conditions on R carry over to $R \otimes_Z K$. Pick K the algebraic closure of Z; since R has algebraic elements (namely, all $d(r)$ for $r \in R$), $R \otimes_Z K$ has zero divisors. By Theorem 3 we conclude that $R \otimes_Z K$ is a 2×2 matrix ring over a field, so satisfies the standard identity in 4-variables. Hence R also satisfies this identity, whence R is a 4-dimensional simple algebra. With this the proof of our main theorem is complete.

BIBLIOGRAPHY

1. A Giambruno and I. N. Herstein, Derivations of prime rings having nilpotent values, (to appear).

2. I. N. Herstein, Center-like elements in prime rings, Jour. of Algebra 60 (1979), 567-574.

3. I. H. Herstein, Rings with Involution, University of Chicago Press, 1976.

4. I. N. Herstein, Non-commutative Rings, Carus Monograph No. 15, Math. Assoc. Amer. 1968.

5. N. Jacobson, Structure theory for algebraic algebras of bounded degree, Ann. of Math. 46 (1945), 695-707.

6. I. Kaplansky, A Theorem on division rings, Canad. Jour. Math. 3 (1951), 290-292.

University of Chicago
Chicago, IL 60637

Contemporary Mathematics
Volume 13, 1982

LIE AND JORDAN MAPPINGS

Wallace S. Martindale, 3rd

1. In the setting of prime and semiprime rings, the study of Lie and Jordan mappings leads in a natural way to nonassociative prime and semiprime rings and to the notion of the extended centroid. Therefore in this opening section we shall touch upon some results in these areas obtained recently in collaboration with W. E. Baxter. Let A be an arbitrary nonassociative ring. We say A is _prime_ if for any ideals U and V, $UV = 0$ implies $U = 0$ or $V = 0$. A is _semiprime_ if for any ideal U, $U^2 = 0$ implies $U = 0$. The centroid Γ_A of a prime (semiprime) nonassociative ring A is easily seen to be a commutative domain (commutative ring without nilpotent elements). The _extended centroid_ C_A may be defined for a semiprime nonassociative ring A and is shown [3] to be a commutative von Neumann regular ring containing Γ_A which generalizes the prime case [5] where C_A is a field. We say a semiprime ring A is _closed_ if it is an algebra over C_A, i.e., $C_A = \Gamma_A$. Given a semiprime (prime) ring A one may form its central closure $Q_A = (A \otimes C)/M$ (M an appropriate ideal of $A \otimes C$), and one shows that Q_A is a closed semiprime (prime) algebra with (extended) centroid C_A [3].

The above notions had been studied some time ago in the case where $A = R$ was an associative prime [14] or semiprime [1] ring. (Henceforth in this article the term ring will mean an associative ring which is 2-torsion free). From the work of Herstein it is known that R^+, the Jordan ring via $x \circ y = xy + yx$, is a prime (semiprime) Jordan ring when R is prime (semiprime), and in [3] it is shown that $C_R = C_{R^+}$. If R is a ring with involution then the symmetric elements S form a Jordan ring under $xy + yx$ and the skew elements K form a Lie ring under $[x,y] = xy - yx$. For R semiprime with involution $*$, an involution is induced on $C = C_R$ and the subring C_* of fixed elements of C is called the _*-extended centroid_ of R. In [3] it is shown that $C_* = C_S$. With regard to Lie structure we are most interested in the case where R is a _*-prime_ ring (in the sense that for any two *-ideals U and V $UV = 0$ implies $U = 0$ or $V = 0$). In [4] we extend some of Herstein's Lie structure theory to *-prime rings and in particular establish

THEOREM A. Let R _be a *-prime ring (not satisfying the standard identity_ S_8 _if_ R _is prime and not satisfying_ S_4 _if_ R _is not prime), and let_ $L(R) = [K,K]/[K,K] \cap Z$ _(_Z _the center of_ R_). Then_

0271-4132/82/0523/$02.25

(1) $L(R)$ is a prime Lie ring.

(2) $C_{L(R)} = C_*$.

2. Let R and R' be rings and let J be a Jordan subring of R^+. An additive map $\phi: J \to R'$ is a Jordan homomorphism if $\phi(xy + yx) = \phi(x)\phi(y) + \phi(y)\phi(x)$. We first consider the situation in which conditions are imposed on the "image" ring R'. In some recent joint work with Baxter [2] we have extended a result of Herstein for R' prime [6] to R' being semiprime:

THEOREM B. *If ϕ is a Jordan homomorphism of a ring R onto a semi-prime ring R' then there is an essential ideal E of R such that $\phi = \sigma_1 \oplus \sigma_2$, $\sigma_1: E \to R'$ a homomorphism and $\sigma_2: E \to R'$ an antihomomorphism.*

THEOREM C. *If ϕ is a Jordan homomorphism of a ring R onto a closed semiprime ring R' then $\phi = \sigma_1 \oplus \sigma_2$, $\sigma_1: R \to R'$ a homomorphism and $\sigma_2: R \to R'$ an antihomomorphism.*

A simple example of Kaplansky [2], p. 458, shows the necessity of either shrinking the "domain" or requiring the "image" to be closed.

For the situation where conditions are imposed on the "domain" ring R we cite an old theorem of ours [13].

THEOREM D. *If R is a ring with involution containing three orthogonal symmetric idempotents e_1, e_2, e_3 whose sum is 1 and such that $Re_iR = R$, $i = 1,2,3$, then any Jordan homomorphism of S into a ring R' can be extended uniquely to a homomorphism of R into R' (i.e., R is the special universal envelope for S).*

This generalizes a result of Jacobson and Rickart [11], where $R = M_n(Q)$, $n \geq 3$, and is also applicable to von Neumann factors. An advantage to there being no conditions on R' is that the analogous theorem for a Jordan derivation $\delta: S \to R$ comes free of charge via the well known observation that

$$s \to \begin{bmatrix} s & 0 \\ \delta(s) & s \end{bmatrix} \quad \text{is a Jordan homomorphism.}$$

At this conference it was pointed out to us by Professor Jacobson that the Russian mathematician Zelmanov, in some important recent work on Jordan division algebras, has in fact established Theorem D for the case that R is an arbitrary division ring with involution. At this writing we have not gone through the details of his methods, but it appears hopeful that these methods may allow one to drop the idempotent assumptions when R is simple, or even prime. An account of Zelmanov's results will be provided by Jacobson in [10].

3. We again let R and R' be associative rings (often with involution) and we now consider Lie isomorphisms between certain Lie rings related to R and R'. Our point of view has been one in which R and R' do not satisfy finiteness conditions - a complete treatment of Lie isomorphisms in a finite dimensional setting is given in Jacobson's book [9], Chapter 10. In [7] Hua proved that any Lie automorphism ϕ of $R = R' = M_n(D)$, $n \geq 3$, D a division ring, was of the form $\phi = \sigma + \tau$, σ an automorphism or the negative of an antiautomorphism and τ an additive map of R into its center sending commutators to zero. In [15] we extended Hua's result as follows:

THEOREM E. *Let R and R' be prime rings with R containing idempotents e_1 and e_2 whose sum is 1. Then any Lie isomorphism ϕ of R onto R' is of the form $\sigma + \tau$, σ an isomorphism (or negative or an antiisomorphism) of R into R'C' (the central closure of R') and τ an additive map of R into C' sending commutators to zero.*

It was this work which led us to formalize the notion of extended centroid and thence to our results [14],[16] on prime rings satisfying a generalized polynomial identity. A simple example [18], p. 77, shows that the image of σ cannot always be accommodated by R' alone.

For rings with involution our most general result to date appears in [17]:

THEOREM F. *Let R and R' be closed prime rings with involutions of the first kind, with (extended) centroids C and C', such that*

(a) $(R:C) \neq 1,4,9,16,25,64$

(b) *R contains two orthogonal symmetric idempotents e_1, e_2 such that* $e_1 + e_2 \neq 1$

(c) *For $i = 1,2$, $e_i \in \overline{M}_i$, the associative subring generated by* $M_i = e_iRe_i \cap [K,K]$.

Then any Lie isomorphism of $L(R) = [K,K]$ onto $L(R') = [K',K']$ can be extended uniquely to an isomorphism of $\overline{[K,K]}$ onto $\overline{[K',K']}$.

We touch upon just one aspect of the proof of the theorem in order to illustrate the use made of the theory of generalized polynomial identities (GPI's). Specifically we consider the case when neither R nor R' satisfies a GPI (over C or C' respectively), set $N_i = \phi(M_i)$, $i = 1,2,3$, and aim to show $N_iN_j = 0$ for $i \neq j$. Indeed, pick $m_i \neq 0 \in M_i$, set $a = \phi(m_1)$, $b = \phi(m_2)$, $c = \phi(m_3)$ and note that (1) $[[[n,a],b],c] = 0$ for all $n \in [K',K']$. Using [16] and the assumption that R' does not satisfy a GPI we conclude that (1) must be "trivial" and so (2) $[[[y,a],b],c] = 0$ for all $y \in R'$. Using a result of ours [15] that R' closed over C' implies $R' \otimes_{C'} (R')^{\circ} \cong R'_r R'_\ell$ (right and left multiplications) we may translate (2) into a tensor product equation, whence a short argument finally produces $ab = ac = bc = 0$.

If R and R' are simple rings (hence automatically closed) condition (c) is not required and ϕ can be lifted to an isomorphism of R onto R'.

Analogous results on Lie derivations were established in [12] and [8] - undoubtedly these are capable of generalization to prime rings.

At this writing a student of ours, Mary Rosen, is in the process of generalizing Theorem F to the setting where R and R' are *-prime rings and where the involutions may be of either the first or second kind. The mild simplicity condition on idempotents will still be kept. Here the assumption will be that ϕ is a Lie isomorphism of $L(R)$ onto $L(R')$. A helpful first step in "lifting" ϕ has already been taken, namely, it follows immediately from Theorem A that the *-extended centroids of R and R' are isomorphic. This "theorem" (as yet not spelled out) should include all previous results of this nature. In particular, a non-involution result such as Theorem E should follow by forming the *-prime ring $R \oplus R^{\circ}$ (with $*$ given by $(x,y) \to (y,x)$) and noting that R is Lie isomorphic to K via $x \to (x, -x)$.

4. The reader may be interested in a more detailed and leisurely account of Lie and Jordan mappings which we gave in [18], though that account (1976) is somewhat outdated. In closing, we wish to acknowledge our indebtedness to Professors Jacobson and Herstein for the helpful advice and encouragement they have provided over the years on questions involving Lie and Jordan structure.

REFERENCES

1. S. A. Amitsur, On rings of quotients, Instituto Nazionale di Alta Matematica, Symposia Matematica, vol. 8 (1972).

2. W. E. Baxter and W. S. Martindale, 3rd, Jordan homomorphisms of semiprime rings, J. Algebra 56 (1979), 457-471.

3. W. E. Baxter and W. S. Martindale, 3rd, Central closure of semiprime non-associative rings, Com., in Algebra 7 (1979), 1103-1132.

4. W. E. Baxter and W. S. Martindale, 3rd, The extended centroid in *-prime rings, submitted.

5. T. S. Erickson, J. M. Osborn, and W. S. Martindale, 3rd, Prime non-associative algebras, Pac. J. Math. 60 (1975), 49-63.

6. I. N. Herstein, Jordan homomorphisms, TAMS 81 (1956), 331-351.

7. L. Hua, A theorem on matrices over an s-field and its applications, J. Chinese Math. Soc. (N.S.) 1 (1951), 110-163.

8. D. R. Jacobs, Lie derivations on the skew elements of simple rings with involution, Thesis, University of Massachusetts, 1973.

9. N. Jacobson, Lie algebras, Interscience tracts in pure and applied math., no. 10, Interscience, New York, 1962.

10. N. Jacobson, Notes on Jordan algebras, University of Arkansas, 1981, to appear.

11. N. Jacobson and C. E. Rickart, Homomorphisms of Jordan rings of self-adjoint elements, TAMS 72 (1952), 310-322.

12. W. S. Martindale, 3rd, Lie derivations of primitive rings, Mich. Math. J., 11 (1964), 182-187.

13. ____________________, Jordan homomorphisms of the symmetric elements of a ring with involution, J. of Algebra, 5 (1967), 232-249.

14. ____________________, Prime rings satisfying a generalized polynomial identity, J. of Algebra, 12 (1969), 576-584.

15. ____________________, Lie isomorphisms of prime rings, Trans. Amer. Math. Soc., 142, (1969), 437-455.

16. ____________________, Prime rings with involution and generalized polynomial identities, J. Algebra, 22 (1972), 502-516.

17. ____________________, Lie isomorphisms of the skew elements of a prime ring with involution, Communications in Algebra, 4 (1976), 929-977.

18. ____________________, Lie and Jordan mappings in associative rings, Proceedings of Ohio University conference on ring theory, May, 1976, Marcel Dekker.

University of Massachusetts
Amherst, Massachusetts 01002

Contemporary Mathematics
Volume 13, 1982

SOME PROPERTIES OF RINGS REFLECTED IN ENDOMORPHISM RINGS OF FREE MODULES

by

Barbara L. Osofsky

Dedicated to Nathan Jacobson with much respect and gratitude for his many contributions to algebra and mathematical life.

Let R be a ring with 1, C a set, F_C the free right R-module on C, and $\Lambda_C = \mathrm{End}_R\,(F_C)$. There are various properties of the ring R and its category of unital right modules $\mathcal{M}_R$ that are reflected in properties of the ring C. We will indicate some such properties in this paper.

If C is finite, R and Λ_C are Morita equivalent and their categories of right modules are equivalent. So we will be concerned here with the case that the cardinality of C, Λ_C , is infinite. Λ_C has many properties independent of R. For example, it contains a Boolean algebra of idempotents corresponding to the Boolean algebra of subsets of C. This implies that Λ_C contains a right ideal of projective dimension k where $2^{|C|} = \aleph_k$ (see [6]). It also contains two sided ideals analogous to ideals of linear transformations of rank $\leq \aleph$ in the case that R is a division ring. But the results listed here are properties of Λ_C equivalent to properties of the ring R, and not properties due solely to the fact that F_C is free.

We start with the basic observation that since $F_C \simeq \oplus_1^n F_C$ and so $\Lambda_C \simeq \oplus_1^n \Lambda_C$, finitely generated Λ_C-modules are cyclic. For $\lambda \in \Lambda$, properties of $\lambda\Lambda_\Lambda$ and $\lambda F \subseteq F_R$ are closely related. This relationship is basic to our discussion.

We next list some elementary propositions whose proofs are the level of exercises in a graduate algebra course.

PROPOSITION 1. <u>Λ_C is right coherent iff $|C|$ - generated submodules of</u> F (or equivalently of R) <u>are $|C|$ - related.</u>

0271-4132/82/0524/$01.75

PROPOSITION 2. Λ_C *is von Neumann regular iff* R *is semisimple Artin.*

PROPOSITION 3. Λ_C *is right semi-hereditary iff* $|C|$ - *generated submodules of* F (or equivalently of R) *are projective.*

Proposition 1 through 3 are proved using basic properties of projective modules. But F_C has properties other than projectivity. One of the major ones is that F_C is a generator. In addition, there are deep results on Λ_R itself which enable us to get additional connections between Λ_C and R.

PROPSITION 4. Λ_C *is right self injective iff* R *is quasi-Frobenius.* (See [1]).

Remarks on the proof. Using the fact that F_C is a projective generator, one can show that X_R is injective iff $[\mathrm{Hom}_R(F,X)]_{\Lambda_C}$ is injective. Thus Λ_C is injective iff F_R is injective. Now a result due to Faith [3] about R itself applies, namely an infinitely generated free R-module is injective iff R is quasi-Frobenius. Properties of annihilator ideals in R are crucial in the proof of this latter result.

It is not difficult to show that Λ_C is never left self-injective. Indeed, one would not particularly expect to be able to connect properties of R with properties of Λ_C on the left. However, we have

PROPOSITION 5. Λ_C *is left coherent for all infinite* C *iff* R *is right perfect and left coherent* (see [5]).

Remarks on the proof. By Chase [2], a ring is left coherent iff any product of flat modules is flat, and right perfect plus left coherent iff any product of a "sufficiently large" number of copies of the ring is projective. In [5], Menal obtains a technical criterion for Λ_C to be left coherent, which for sufficiently large C, forces a product of enough copies of R to be projective. One of the interesting consequences of Menal's result is that for $R = \mathbb{Z}$ = the integers, $\Lambda_{\aleph_0}$ is left coherent, but Λ_C for $|C| \geq \aleph_1$ is not.

The last property of Λ_C I wish to mention is actually two-sided and phrased somewhat differently from those above.

PROPOSITION 6. The weak global dimension of Λ_C does not exceed sup {projective dimension (M) | M a |C| - presented R-module}, and equality holds if R is right coherent (see [7]).

Remarks on the proof. Proposition 6 is a generalization of both propositions 2 and 3. The key property of F_C used is the property that F_C is |C| - generated and isomorphic to a coproduct of |C| copies of itself. This property plus the projectivity of F_C connects a form of "|C|-flatness" of R-modules M with the ordinary flatness of Λ_C-modules $\mathrm{Hom}_R(F_C, M)$, and this is extended down a projective resolution.

The results mentioned in this survey are basically homological in nature. There are other aspects of Λ_C, e.g. conditions on annihilators, its Jacobson radical, that have also been studied. Many aspects still remain to be explored.

REFERENCES

[1] G. Brodskii, Endomorphism rings of free modules, Math. USSR Sbornik 23 (1974), 215-230.

[2] S. Chase, Direct products of modules, Trans. Amer. Math. Soc. 97 (1960), 457-473.

[3] C. Faith, Rings with ascending condition on annihilators, Nagoya Math. J. 27 (1966), 179-191.

[4] N. Jacobson, Structure of rings, Amer. Math. Soc., Providence, R.I. 1956.

[5] P. Menal, On the endomorphism ring of a free module, to appear.

[6] B. Osofsky, Projective dimension of "nice" directed unions, J. Pure and Applied Alg. 13 (1978), 179-219.

[7] B. Osofsky, Weak global dimension of endomorphism rings of free modules, J. Pure and Applied Alg, to appear.

Rutgers University
New Brunswick, N.J. 08903

Contemporary Mathematics
Volume 13, 1982

ON THE GALOIS THEORY OF COMMUTATIVE RINGS I: DEDEKIND'S THEOREM ON THE INDEPENDENCE OF AUTOMORPHISMS REVISITED

Carl Faith[1]

FOR JAKE

INTRODUCTION

The Galois Theory of a commutative field K contains Dedekind's theorem on the (linear) independence over K of automorphisms $g_1,\dots,g_n$ as functions [2] $K \to K$, i.e., it is impossible for elements $k_i \in K$, $i=1,\dots,n$ to exist that satisfy the identity

(R K) $\quad \Sigma_{i=1}^{n} k_i g_i = 0$ on K with k_i not all zero,

or, equivalently, impossible that

(R'K) $\quad \Sigma_{i=1}^{n} k_i g_i(x) = 0 \ \forall x \in K$ with some $k_i \neq 0$.

(This implies one inequality, $[K:K^G] \geq n$, needed for the important dimension relation $[K:K^G] = (G:1) = |G|$. See [1].) Redoing the proof of Artin [1] for an arbitrary commutative ring yields the Dependence Theorem [3] for a commutative ring K: Every dependence relation (R K) implies

(DR K): $\quad k(g-1) = 0$ on K

for some $0 \neq k \in K$ and $1 \neq g$ belonging to the group G generated by $g_1,\dots,g_n$. Expressed otherwise, g induces the identity in the factor ring $K/k^{\perp}$, where the exponent denotes annihilation. It follows that automorphisms belonging to a group G are independent over a local ring K if $g = 1$ is the only $g \in G$ that induces the identity (id.) in the residue field $\bar{K}$.

We show for any commutative ring K that when G is a torsion group, then (DR K) implies that K is n-radical over $K^{g,k} = K^g + k^{\perp}$ in the sense that $x^n \in K^{g,k}$, where $n = |(g)|$; and, dually, K is n-torsion over $K^{g,k}$. Thus, denying the conclusion, (e.g., by requiring that G be a torsion group with $|(g)|$ a unit) yields independence theorems for automorphisms over K. Furthermore, by a theorem of Kaplansky on division rings (Can. J. Math. 3 ((1951)), if K is a local ring which is not extended from its radical J by some Galois subring K^g with $1 \neq g \in G$, then (DR K) implies that $\bar{K} = K/J$ has prime characteristic p, that $|(g)| = p^e$, and that $\bar{K}$ is purely inseparable over $\overline{K^g}$ of exponent e. (Finiteness of $\bar{K}$, another possibility offered by Kaplansky's theorem, is ruled out by the fact that $\bar{K}$ is p^e-radical

0271-4132/82/0525/$03.50

over $\overline{K^g}$.)

Similar theorems hold for the structure of the classical quotient ring $Q = Q_c(K)$ of K, e.g., if $|(g)|$ is a non zero-divisor of K, then (DR K) implies $Q = Q^{g^{ex},k}$, where ex denotes the "extension of." If Q is a local ring, meaning that the zero divisors of K form an ideal zd K, and if no $g \neq 1$ induces id. on K/(zd K), then the automorphisms are independent. Furthermore, if $(\text{zd } K)^{\perp} \neq 0$, e.g., when Q is a local PIR, then the converse holds, as the Dependence Theorem and (DR K) readily show, since the radical of Q is annihilated by some $0 \neq k \in K$.

Another type of dependence of an automorphism group occurs when some K^g for $g \neq 1$ contains a nonzero ideal I of K. But, then, if $0 \neq k \in I$, we have

(FI) $$kg(x) = g(kx) = kx \qquad \forall x \in K$$

so (DR K) holds. Conversely, for a reduced ring K, (DR K) implies that K^g contains an ideal $\neq 0$ (Proposition 8.0); in fact, K^g contains $(N_g(k))$, the norm of k with respect to (g). Thus, a reduced ring K has dependent automorphism group G iff some K^g with $1 \neq g \in G$ contains a nonzero ideal of K. Furthermore, if a reduced ring K is n-power radical over $K^{g,k}$, where K has prime characteristic n, then (DR K) holds when $g \neq 1$, and $k \neq 0$ (Corollary 4.4).

A third kind of dependence of a group G occurs when G is a torsion group and there exists $1 \neq g \in G$ such that the K^g-submodule $Y(g)$ consisting of all elements with zero (g)-trace contains a nonzero ideal. In fact, by Theorem 8.2, G is dependent iff K^g or $Y(g)$ contains a nonzero ideal, for some $1 \neq g \in G$.

We can easily classify dependent automorphism groups generated by translations or rotations of a polynomial ring $K = F[x]$ in a single variable, e.g., if g is the translation sending $x \mapsto x+a$ where $0 \neq a \in F$, then (g) is dependent iff some nonzero multiple ma is a zero divisor.[4] (Similarly for a rotation $x \mapsto ax$.) Significantly, the switching automorphisms of $K = F[x_1,\ldots,x_n]$ generate independent automorphism groups.

Similar results hold for automorphisms of a power series ring $K = R[[x]]$. It is known that an automorphism g that leaves the elements of R fixed maps x onto $\Sigma_{i=0}^{\infty} a_i x^i$, where a_1 is a unit, and a_0 belongs to the ideal $I_c(R)$ consisting of all $a \in R$ for which there exists a homomorphism $R[[x]] \to R$ sending x onto a (see [8]). The structure of Aut $R[[x]]$ is unknown in general, even when R is a domain (see, e.g., [9]). However, when R contains the rational number field, and G is a torsion subgroup of $\text{Aut}_R R[[x]]$, then by [8] G is conjugate to a subgroup of the circle group, where the circle group consists of all "rotations" of finite order, that is, automorphisms sending $x \to cx$ for an n-th root of unity c. Thus, a necessary and sufficient condition for G to be independent is that some primitive n-th root of unity c, for some $n > 1$, be such that $c^m - 1$ is a zero divisor in R, for $1 < m < n$.[4]

In the original version of this paper I proved a normal basis theorem for a local ring K with finite independent automorphism group G. Since then S. Endo showed me a more general theorem: K has a normal basis $g_1(u),\ldots,g_n(u)$ over K^G iff K is free of rank n over K^G, where $G = (g_1,\ldots,g_n)$. (In fact, K may be semilocal if it is also assumed that $t_G(K) = K^G$, where $t_G: K \to K^G$ is the trace map. (Letter of July 1980).)

Furthermore, under the hypothesis of the Independence Theorem (i.e., no $g \neq 1$ induces id. in $\bar{K}$), K is actually Galois over K^G by [10], and hence has a normal basis over K^G by [3].[4,5]

THE DEPENDENCE THEOREM

We begin this section with the Dependence Theorem stated in the introduction.

1. DEPENDENCE THEOREM. *When* (R K) *holds, then there is a "shortest" relation of the form*

(SR K) $$k(g_i - g_j) = 0 \quad \text{for some } i \neq j \text{ and } 0 \neq k \in K$$

and, in fact, there is a relation of the form

(DR K) = (DR K)(k,g) $$k(g-1) = 0 \quad \text{for some } 1 \neq g \in G, \text{ and } 0 \neq k \in K,$$

that is,

(DR'K) = (DR'K)(k,g) $$kg(x) = kx \quad \forall x \in K$$

Proof. We note that (SR K) is obtained from (R K) by the familiar technique of replacing x in (R'K) by $x = uy$ for a fixed $u \in K$. First allowing y to range over K, we obtain another (R K) of the form

($R\ K_u$) $$\Sigma_{i=1}^n k_i g_i(u) g_i = 0 \quad \text{on } K.$$

Next, multiply (R K) by $g_{i_0}(u)$, where i_0 is any i such that $k_i \neq 0$, and assuming that some $k_i g_{i_0}(u) \neq 0$, subtract the result from ($R\ K_u$) to obtain a "shorter" relation inasmuch as the coefficient of g_{i_0} is now $= 0$. Thus, in this case, all coefficients must $= 0$, that is, for all i,

($DR'K_u$) $$k_i g_i(u) = k_i g_{i_0}(u)$$

holds. If this is true for all u, then

$$k_i(g_i - g_{i_0}) = 0 \quad \text{on } K,$$

so (SR K) holds as asserted. Then (DR K) follows by letting $k = g_{i_0}^{-1}(k_i)$ and $g = g_{i_0}^{-1}g_i$. Since the only assumption made on u was that some $k_i g_{i_0}(u) \neq 0$, for some i_0 for which $k_i \neq 0$, then denying the assumption yields $k_i g_j(u) = 0$ for all i,j, which shows that $DR'K_u$ holds without restriction completing the proof.

The <u>trivial crossed product</u> or <u>skew group ring</u> $K*G$ of K and G is the ring consisting of all finite linear combinations $\Sigma_{i=1}^n k_i g_i$, with $k_i \in K$, and $g_i \in G$, with multiplication defined by the formula

$$(kg)(ph) = kg(p)gh \qquad \forall k,p \in K \text{ and } g,h \in G$$

and its implications. There is a canonical homomorphism

$$h: K*C \to \mathrm{End}_F K, \quad \text{where } F = K^G.$$

$$\sigma = \Sigma_{i=1}^n k_i g_i \to \sigma', \quad \text{where } \sigma'(x) = \Sigma_{i=i}^n k_i g_i(x) \quad \forall x \in K.$$

The next result gives equivalent conditions for h to be a monomorphism. As stated, a group of automorphisms of K will be called <u>dependent</u> (<u>independent</u>) accordingly as its elements are linearly dependent (independent) over K.

1.2. <u>COROLLARY</u>. <u>For a group</u> G <u>of automorphisms of</u> K, <u>the following conditions are equivalent</u>:

1.2.1. K <u>is faithful as a canonical left</u> $K*G$ <u>module</u>.

1.2.2. $K*G \hookrightarrow \mathrm{End}\, K_F$, <u>canonically, where</u> $F = K^G$.

1.2.3. G <u>is independent over</u> K.

1.2.4. <u>Every cyclic subgroup of</u> G <u>is independent over</u> K.

1.2.5. <u>Every pair</u> $\{1,g\}$ <u>is linearly independent over</u> K, $\forall 1 \neq g \in G$.

<u>When this is so, then</u> K <u>is a faithful module over the group ring</u> $R = K^G G$ <u>of the ring</u> K^G <u>and the group</u> G.

<u>Proof</u>. (1) ⇔ (2) by the preceding remarks, (2) ⇔ (3) is obvious, and both (3) ⇔ (4) and (3) ⇔ (5) by the Dependence Theorem. The last statement follows, since $R \hookrightarrow K*G$ canonically.

We record the following curiosity.

1.3. <u>PROPOSITION</u>. <u>If</u> G <u>is a torsion group</u>, H <u>a finite normal subgroup, and if</u> G <u>induces a dependent group of automorphisms of</u> $L = K^H$, <u>then</u> G <u>is dependent over</u> K.

<u>Proof</u>. Let $\bar{G}$ denote the group of automorphisms of L induced by $G(\bar{G} \approx G/H$ canonically), and suppose $(DR\ L)(\bar{g},k)$ holds, for some $0 \neq k \in L$, so

$$k\bar{g}(y) = ky \qquad \forall y \in L.$$

Then

$$kgT_{(g)}(x) = kT_{(g)}(x) \qquad \forall x \in K$$

defines a dependence relation (R K) over K where $T_{(g)}(x) = x+g(x) +\cdots+ g^{n-1}(x)$ is the (g)-trace of $x \in K$.

2. REMARK. Regarding Theorem 1: (DR K) is equivalent to the statement that the ideal $((1-g)K)$ generated by the set $(1-g)K - \{g(x)-x \mid x \in K\}$ has nonzero annihilator. Note: in general $(1-g)K$ is not an ideal!). Thus, G is independent iff

$$((1-g)K)^{\perp} = 0 \qquad \forall 1 \neq g \in G. \tag{2.1}$$

This happens, e.g., if

$$((1-g)K) = K \qquad \forall 1 \neq g \in G. \tag{2.2}$$

Another formulation of the Dependence Theorem is the:

(2.3) INDUCTION THEOREM. *G is independent over K iff no $1 \neq g$ in G induces the identity in $K/k^{\perp}$, for any $0 \neq k \in K$.*

LOCAL RINGS

The next trivial consequence of the Dependence Theorem generalizes Dedekind's Theorem stating that any group of automorphisms of a field is independent.

3. INDEPENDENCE THEOREM FOR LOCAL RINGS. *If K is a local ring, and if G is a group of automorphisms such that no $g \neq 1$ in G induces the identity in the residue field $\bar{K}$, then G is independent over K.*

Proof. (DR'K) ⇒ that g induces $\bar{g} = 1$ in $\bar{K}$.

Recall that $Q = Q_c(K)$ is a local ring iff the zero divisors of K form an ideal zd(K). In this case, $zd(K) = K \cap J(Q)$, where $J(Q) = \text{rad } Q$, and $J(Q) = zd(K)S^{-1} = zd(K)Q$, where $S = K - zd(K)$. Moreover, every automorphism g of K maps zd(K) onto itself, hence induces an automorphism in the factor ring K/zd(K).

A ring R is Kasch if every simple R-module embeds in R. This holds iff every maximal ideal M has nonzero annihilator (and iff every maximal ideal M is an annihilator ideal).

An annihilator ideal of the form $k^{\perp}$ for some $k \in K$ is said to be a principal annihilator ideal. For example, K and 0 are principal annihilators, and if K is Kasch, so are the maximal ideals.

3.1. REMARK. A commutative ring K has Kasch local quotient ring $Q = Q_c(K)$ iff zd(K) is a principal annihilator.

This holds, since Q Kasch implies that $J(Q) = k^{\perp}$ for some $0 \neq k \in K$, and then $zd(K) = k^{\perp}$. Conversely, $zd(K) = k^{\perp} \Rightarrow J(Q) = k^{\perp}$, since $J(Q) = zd(K)S^{-1}$.

3.2. EXAMPLE. If Q is a local PIR not a field, then Q is Kasch since $J(Q) = tQ$ for some $t \in Q$, and t is a zero divisor, hence $J(Q) = k^{\perp}$ for any $0 \neq k \in t^{\perp}$.

3.3. PROPOSITION. *Assume that* $Q = Q_c(K)$ *is a Kasch local ring. Then an automorphism group* G *is dependent iff some* $1 \neq g \in G$ *induces the identity in* $K/zd(K)$.

Proof. If $g \neq 1$ induces the identity in $K/zd(K)$, then $g(x) - x \in zd(K) \forall x \in K$, so (DR'K) holds for any k such that $k^{\perp} = zd(K)$.

Conversely, if G is dependent, then (DR K) holds for $0 \neq k \in K$ and $1 \neq g \in G$, so $g(x) - x \in zd(K)$ for all $x \in K$, and hence g induces the identity in $K/zd(K)$.

RADICAL EXTENSIONS

When $g \neq 1$ and $k \neq 0$ we set $K^{g,k} = K^g + k^{\perp}$. Another consequence of the Dependence Theorem is:

$$K = K^{g,k} \Rightarrow \text{(DR K)}$$

$$\text{(For } K = K^{g,k} \Rightarrow g(x) - x \in k^{\perp}, \text{ so (DR'K) holds.)}$$

The next theorem shows that (DR K) implies that a "close" relationship exists between K and $K^{g,k}$.

4. THEOREM ON RADICAL-TORSION EXTENSIONS. *If a dependence relation* (DR K) *holds on* K, *and if* g *has finite order* n, *then* K *is* n-*radical and* n-*torsion over* $K^{g,k}$.

Proof. A simple calculation shows that (DR K) $\Rightarrow kg^{i}(x) = kx$ for any i, and hence, $k\beta_x = kx^n$, where $\beta_x = N_g(x)$ is the norm with respect to (g), and $n = |g|$. Thus, $\beta_x \in K^g$, hence $x^n \in K^g + k^{\perp} = K^{g,k}$. Similarly, $nx \in K^{g,k}$.

4.1. COROLLARY. *If* G *is a dependent torsion group over* K, *such that* $|g|^{-1} \in K \forall g \in G$, *then* $K = K^{g,k}$ *for some* $g \in G$, $k \in K$.

4.2. COROLLARY. *If* G *is a dependent torsion group over* K *such that* $|g| \notin zd(K) \forall g \in G$, *then* $Q = Q^{g^{ex},k}$ *for some* $g \in G$, $k \in K$.

Proof. For then G^{ex} is a dependent group over Q, and 4.1 applies.

K is radical over $S \subseteq K$ if $x^n \in S \forall x \in K$, with n depending on x.

4.3. COROLLARY. Assume K is not radical over any proper subring.[6] Then a finite group G of automorphisms of K is dependent iff $K = K^{g,k}$ for some $1 \neq g \in G$ and $0 \neq k \in K$.

4.4. COROLLARY. If K is (a power of n) -radical over a subring $K^{g,k}$, where K has prime characteristic n, and if K is reduced, then (DR K) holds: $k(g-1) = 0$ on K.

Proof. If $x \in K$, then for some $e \geq 1$, $x^{n^e} \in K^g + k^\perp$ so $g(x^{n^e}) - x^{n^e} \in K^\perp$, that is, $k(g(x^{n^e}) - x^{n^e}) = 0$. Then $K(g(x) - x)^{n^e} = 0$, hence $(k(g(x) - x))^{n^e} = 0$. Since K is reduced, then $k(g(x) - x) = 0$, so (Dr K) holds.

4.5. EXAMPLE. Let $K = P[x,y]/(x^2,xy,y^2)$ be the ring of rational functions in two variables x and y over the prime subfield P, modulo the ideal (x^2,xy,y^2). Let $\bar{u}$ denote the image of $u \in P[x,y]$ under the canonical map $P[x,y] \to K$, and let g be the automorphism induced by switching $\bar{x}$ and $\bar{y}$. The fixring F of (g) is the vector subspace over P spanned by $\bar{1}$ and $\bar{x} + \bar{y}$, and the radical is the vector subspece $J = P\bar{x} + P\bar{y}$. The element $\alpha = \bar{x} + \bar{y} \in F$ annihilates J since $\alpha\bar{x} = 0$ and $\alpha\bar{y} = 0$. Since $0 \neq \alpha K = P(\bar{x} + \bar{y}) \subseteq F$, then

$$\alpha g(x) = g(\alpha x) = \alpha x \quad \forall x \in K.$$

KAPLANSKY'S THEOREM REVISITED

We next investigate the situation for a local ring K with residue field $\bar{K} = K/J$ (with radical $K = J$) when (DR K)(g,k) holds for g of finite order $n \neq 1$. For then by the Theorem on Radical-Torsion Extensions, K is radical over $K^g + k^\perp$, hence $\bar{K}$ is radical over $\overline{K^g}$. In the event that $\bar{K} \neq \overline{K^g}$, there is a decisive theorem of Kaplansky [5] on the structure of the radical extension $\bar{K}$ over $\overline{k^g}$, namely, there are just two possibilities

(KAP 1) $\bar{K}$ is a finite field

(KAP 2) $\bar{K}$ is purely inseparable over $\overline{K^g}$.

Now in our situation, KAP 1 is ruled by the following lemma.

5. LEMMA ON $|g|$. If $\bar{K} \neq \overline{K^h}$, for all automorphisms $h \neq 1$, then (DR K)(k,g) implies that $|g| = p^e$, for $e \geq 1$.

Before proving this lemma, we pause to state two theorems.

6. PURELY INSEPARABLE RESIDUE FIELD THEOREM. If $\bar{K} \neq \overline{K^g}$ for a local ring K satisfying (DR K), then $\bar{K}$ is purely inseparable over $\overline{K^g}$.

7. PERFECT RESIDUE FIELD THEOREM. If $\bar{K} \neq \overline{K^g}$ for any $g \neq 1$ in a finite automorphism group G, and if $\overline{K^g}$ is a perfect field $\forall g \in G$, then G is independent over K.

Proofs. The proof of the lemma on $|g|$ follows from Kaplansky's theorem and Theorem 4, since the former implies that $\bar{K}$ has characteristic p, and the latter implies that $n\bar{x} = 0\ \forall\bar{x} \in \bar{K}$, so $p|n$. Write $n = p^e n_0$, with $(n_0,p) = 1$, and set $h = g^{p^e}$. Now $(DR'K)(k,g) \Rightarrow kg(x) = kx$ and thus $kh(x) = kx\ \forall x \in K$, that is, $(DR'K)(k,h)$ holds. By hypothesis, $K \neq K^h + J$ when $h \neq 1$, hence the fact that p does not divide $n_0 = h$ implies by Kaplansky's theorem that $n_0 = 1$, so $g^{p^e} = 1$. Since the Radical Extension Theorem implies that $\bar{x}^{|g|} \in \overline{K^g}$, when $\overline{K^g}$ is perfect (i.e., finite as in KAP 1), then $\bar{x} \in \overline{K^g}\ \forall\bar{x} \in \bar{K}$, contrary to assumption. This proves the Perfect Residue Field Theorem; and the Purely Inseparable Residue Field Theorem is a restatement employing Kaplansky's Theorem.

8. <u>PURELY INSEPARABLE RESIDUE THEOREM FOR</u> $Q = Q_c(K)$. <u>If</u> $Q = Q_c(K)$ <u>is a local ring which is not extended from its radical</u> $J(Q)$ <u>by a Galois proper subring, that is, if</u> $Q \neq J(Q) + Q^H$ <u>for some automorphism group</u> H, <u>and if</u> K <u>has a dependent automorphism torsion group</u> G, <u>then</u> $\bar{Q} = Q/J(Q)$ <u>is purely inseparable over</u> $\bar{Q}^{g^{ex}}$ <u>for some</u> $1 \neq g \in G$. <u>Furthermore, if</u> $x \in K$, <u>then</u> $x^{p^e} \in K^g + k^{\perp} \subseteq K^g + zd(K)$, <u>for some</u> $0 \neq k \in K$, <u>where</u> $|g| = p^e$.

<u>Proof</u>. This follows Theorem 6 (in view of Footnote 3). The last statement is a trivial consequence.

REDUCED RINGS

The next proposition is a partial converse of one part of Corollary 1.3.

8.0. <u>PROPOSITION</u>. <u>If</u> K <u>is a reduced ring, and if</u> g <u>satisfies the dependence relation</u> $(DR\ K)$, <u>then</u> K^g <u>contains a nonzero ideal of</u> K.

<u>Proof</u>. Suppose $(DR\ K)$ holds, so that $k(g-1) = 0$ on K, and by (8), $k(\beta_x - x^n) = 0,\ \forall x \in K$, where $\beta_x = N_g(x) = \Pi_{i=0}^{n-1} g^i(x) \in K^g$. Take $x = k$, $\beta = \beta_k$, so that $k(\beta - k^n) = 0$. If K is reduced, then $k^{n+1} = \beta k \neq 0$, hence $\beta = N_g(k) \neq 0 \in K^g$. Now $(DR'K)$ implies that

$$h(k)hg(x) = h(k)h(x) \qquad \forall h \in (g)$$

Since any $y \in K$ has the form $y = h(x)$, then $hg = gh$ implies that

$$h(k)g(y) = h(k)y \qquad \forall y \in K.$$

Hence

$$N_g(k)g(y) = N_g(k)y$$

or

$$g(\beta y) = \beta g(y) = \beta y \qquad \forall y \in K$$

with $\beta \in K^g$. Since $g(\beta y) = \beta y\ \forall y \in K$, then βK is a nonzero ideal of $K \subseteq K^g$.

8.1. <u>PROPOSITION</u>. <u>If</u> $(DR\ K)$ <u>holds on</u> K, <u>then either</u> K^g <u>contains a nonzero ideal of</u> K, <u>or else</u> $nk^2 = 0$.

<u>Proof</u>. By dualizing the proof of Proposition 8.0 we have

$$\beta g(y) = \beta y \qquad \forall y \in K.$$

where $\beta = T_g(k) = \Sigma_{i=0}^{n-1} g^i(k) \in K^g$. Then, as in the proof of Proposition 8.0, either $\beta = 0$, or else βK is a nonzero ideal of K contained in K^g.

Now suppose $\beta = T_g(k) = 0$. By the proof of Theorem 3.1, $k(T_g(x)-nx) = 0$ $\forall x \in K$; in particular, for $x = k$, we have $k(\beta - nk) = 0$, hence $nk^2 = 0$.

8.2. <u>THEOREM</u>. <u>Let</u> G <u>be a torsion automorphism group</u>. <u>Then</u> G <u>is dependent over</u> K <u>iff for some</u> $1 \neq g \in G$ <u>either the fixring</u> K^g, <u>or the</u> K^g-<u>submodule</u> Y <u>consisting of all</u> $x \in K$ <u>with</u> $T_g(x) = 0$, <u>contains a nonzero ideal of</u> K.

<u>Proof</u>. If G is dependent, then (DR K) holds, and by the proof of 8.1, either K^g contains the nonzero ideal βK, or else $\beta = 0$. Since $kag(x) = kax$ $\forall a \in K$, then if K^g does not contain a nonzero ideal, we must have $ka \in Y$ for all $a \in K$, that is, $kK \subseteq Y$.

Conversely, if K^g contains an ideal $\neq 0$, then G is dependent by Corollary 1.3. Moreover, if Y contains the ideal kK for some $k \neq 0$, then $T_g(kx) = 0$ for all $x \in K$, and this implies a relation

(8.2.1) $$\Sigma_{i=0}^{n} g^i(k) g^i = 0 \text{ on } K \qquad (n = |g|)$$

with nonzero coefficients, so G is dependent.

FOOTNOTES

1. Parts of this paper were presented at the Hudson Symposium, Plattsburgh, April 21-23, 1981, and at the Special Session of the Canadian Mathematical Society held in Vancouver, December 12-14, 1980, and summarized in two abstracts appearing in the 1981 Notices of the American Mathematical Society. (Both had the same title, suffixed by "I" or "II," but the subtitle of "II" was: "Kaplansky's Theorem on Radical Extensions Revisited.") This work in part was supported by a grant by Rutgers University through the auspices of its Faculty Academic Sabbatical Program.

2. Linear independence of automorphisms is assumed in the Galois Theory for commutative rings of Auslander-Goldman [2], and is also the starting point for that in [1]. Usually, we drop the modifier "linear" and speak of independence on dependence of automorphisms over K. Below, K^G denotes the Galois subring corresponding to a group G of a automorphisms. If $G = (g)$, set $K^g = K^G$. The fact that each automorphism g of a commutative ring K has a unique extension g^{ex} to the quotient ring $Q_c(K)$, shows that not only does Dedekind's theorem hold for integral domains, but that a set G of automorphisms of K is independent over K iff the corresponding set G^{ex} is independent over $Q_c(K)$.

3. Theorem 1 is proved using the familiar technique of replacing x in (R'K) by $x = uy$, for a fixed u, and first allowing y to range over K, then u.

4. This proposition has been excised because of editorially suggested space-limitations.

5. This was pointed out to me by F. DeMeyer (Letter of February, 1981).

6. For a study of this hypothesis, see [4].

REFERENCES

1. Artin, E., Galois Theory, Notre Dame University, South Bend, 1955.

2. Auslander, M. and Goldman, O., The Brauer group of a commutative ring, Trans. Amer. Math. Soc. 97 (1960), 367-409.

3. Chase, S. U., Harrison, D. K., and Rosenberg, A., Galois Theory and Galois cohomology of commutative rings, Memoirs of the Amer. Math. Soc. 52 (1965), 15-33.

4. Faith, C., Radical extensions of rings, Proc. Amer. Math. Soc. 12(1961), 274-83.

5. Kaplansky, I., A theorem on division rings, Can. J. Math. 3 (1951), 290-292.

6. Jacobson, N., Structure of Rings, Colloquium Publications of the Amer. Math. Soc. (Revised), Providence, 1964.

7. Faith, C., Algebra I: Rings, Modules, and Categories, Bd. 190, Springer, Berlin-Heidelberg-New York, 1973, 1981.

8. Eakins, P. and Sathaye, A., R-automorphisms of finite order in R[x], J. Algebra 67 (1980), 110-28.

9. Atkins, D. L. and Brewer, J. W., D-automorphisms of power series and the continuous radical of D, J. Algebra 67 (1980), 185-203.

10. DeMeyer, F. and Ingraham, E., Separable Algebras over Commutative Rings, Lecture Notes in Math., Vol. 81, Springer-Verlag, Berlin-Heidelberg-New York, 1971.

Rutgers, The State University
New Brunswick, NJ 08540

Contemporary Mathematics
Volume 13, 1982

ON THE DEFORMATION OF DIAGRAMS OF ALGEBRAS*

Murray Gerstenhaber** and S. D. Schack

By a diagram of k-algebras over a partially ordered set ("poset") I we will mean a family $\mathbb{A} = \{\mathbb{A}^i, i \in I\}$ of unital k-algebras with unital morphisms $\varphi^{ij}:\mathbb{A}^j \to \mathbb{A}^i$ defined whenever $i \leqq j$ such that φ^{ii} is the identity and $\varphi^{ij}\varphi^{jk} = \varphi^{ik}$ whenever $i \leq j \leq k$. That is, $\mathbb{A}$ is a functor from I, considered as a category, to unital k-algebras (.he double use of k should cause no confusion.) For example, let X be a Hausdorff complex space or a separated scheme over k with a covering $\{U_i\}$, let $\mathbb{A}^i = \mathcal{O}_X(U_i)$ (the sections over U_i of the structure sheaf $\mathcal{O}_X$). Set $i \leq j$ if $U_i \subset U_j$, and take φ^{ij} to be the restriction map. The $\mathbb{A}^i$ here are commutative, but that need not always be the case. If intersections of sets in the covering are again in the covering then X may be recaptured from the diagram. When, moreover, the U_i are Stein (in the complex case) or affine (in the scheme case) then we call the covering Stein or affine, respectively. It is not hard to see that morphisms $f:X \to Y$ of complex spaces or schemes, and even commutative diagrams of such, can be represented by commutative diagrams of algebras.

A (two-sided) module $\mathbb{M}$ over $\mathbb{A}$ is a family $\{\mathbb{M}^i, T^{ij}:\mathbb{M}^j \to \mathbb{M}^i\}$ where each $\mathbb{M}^i$ is an $\mathbb{A}^i$ module (and so by virtue of φ^{ij} therefore also a module over any $\mathbb{A}^j$ with $j \geq i$), each T^{ij} is an $\mathbb{A}^j$-module morphism, and where $T^{ii} = \mathrm{id}_{\mathbb{M}^i}$, $T^{ij}T^{jk} = T^{ik}$ whenever $i \leq j \leq k$. For example, a diagram $\mathbb{A}$ is a module over itself. If $\mathbb{A}$ is "commutative", i.e., if each $\mathbb{A}^i$ is commutative, and if also each $\mathbb{M}^i$ is a commutative $\mathbb{A}^i$ module, i.e., if left and right operations coincide, then we call $\mathbb{M}$ commutative. For example, if F is a sheaf on X and $\mathbb{A} = \{\mathcal{O}_X(U_i)\}$, then we may take $\mathbb{M}^i = F(U_i)$.

A module morphism $\lambda:\mathbb{M} \to \mathbb{N}$ is a diagram of the obvious kind. It is "allowable" if each individual $\lambda^i:\mathbb{M}^i \to \mathbb{N}^i$ is (cf. [MacLane]), i.e., if for each i there is a $\mu^i:\mathbb{N}^i \to \mathbb{M}^i$ such that $\lambda^i\mu^i\lambda^i = \lambda^i$. This is always the case when k is a field. A sequence is allowable if each morphism is.

---

* This paper is an abstract of one to appear in the Transactions of the A.M.S. under the title "On the deformation of algebra morphisms and diagrams."

** Supported in part by an N.S.F. grant to the University of Pennsylvania.

0271-4132/82/0526/$02.25

The Hochschild and (in the commutative case) Harrison cohomology theories can be extended to diagrams: An n-cochain $\Gamma \in C^n(\mathbb{A},\mathbb{M})$ is a family $\{\Gamma^i, \text{all } i; \Gamma^{ij}, \text{all } i < j\}$ with

$$\Gamma^i \in C^n(\mathbb{A}^i,\mathbb{M}^i), \; \Gamma^{ij} \in C^{n-1}(\mathbb{A}^j,\mathbb{M}^i)$$

(which is well-defined since $\mathbb{M}^i$ is also an $\mathbb{A}^j$-module) such that $\Gamma^{ik} = T^{ij}\Gamma^{jk} + \Gamma^{ij}\varphi^{jk}$, where the meanings of the terms on the right should be clear. If $\mathbb{A}$ and $\mathbb{M}$ are commutative, let Char^n denote the groups of Harrison n-cochains, i.e., the submodule of C^n consisting of those Γ in which the Γ^i and Γ^{ij} vanish on shuffles (cf. [Harrison]). This will be most important for us when $n = 2$, where the only condition is that $\Gamma^i(a,b) = \Gamma^i(b,a)$ for all $a,b \in \mathbb{A}^i$. Define $\delta: C^n \to C^{n+1}$ by $\delta\{\Gamma^i,\Gamma^{ij}\} = \{\delta\Gamma^i, T^{ij}\Gamma^j - \Gamma^i\varphi^{ij} - \delta\Gamma^{ij}\}$, where on the right δ is the usual Hochschild coboundary. One has $\delta^2 = 0$, and in the commutative case $\delta(\mathrm{Char}^n) \subset \mathrm{Char}^{n+1}$, so one can define $H^n = Z^n/B^n$ and $\mathrm{Har}^n = \mathrm{Zhar}^n/\mathrm{Bhar}^n$, as usual.

A basic problem, when $\mathbb{A}$ comes from a complex space or scheme X and $\mathbb{M}$ from a sheaf F is to determine the relationship between the cohomology groups $H^n(X,F)$ and the $H^n(\mathbb{A},\mathbb{M})$ or $\mathrm{Har}^n(\mathbb{A},\mathbb{M})$. Unfortunately $H^n(\mathbb{A},\mathbb{M})$ need not vanish for $n \geq 1$ when the covering is Stein or affine and F coherent, although we conjecture that $\mathrm{Har}^n(\mathbb{A},\mathbb{M})$ does when X is complex or a scheme over a k containing $\mathbb{Q}$. The situation is even more complicated in characteristic $p > 0$ for there is then another "commutative" cohomology theory due to [André] which need not agree with Harrison's in dimension > 3. Suppose, however, that X is smooth and let T denote its tangent sheaf. Take a Stein or affine covering $\{U_i\}$ and form $\mathbb{A}$ as before. Then there is a natural isomorphism $\mathrm{Har}^2(\mathbb{A},\mathbb{A}) \xrightarrow{\sim} H^1(X,T)$, the classical group of infinitesimal deformations of X. (Sketch of proof: Every γ on the left is represented by a $\Gamma = \{\Gamma^i,\Gamma^{ij}\}$ where the Γ^i in turn represent allowable extensions

$$0 \to \mathcal{O}_X(U_i) \to B \to \mathcal{O}_X(U_i) \to 0$$

with B commutative. In the smooth case these are all split, so we may take $\Gamma^i = 0$ for all i. The Γ^{ij} are then essentially derivations of $\mathcal{O}_X(U_i \cap U_j)$ into itself, i.e., sections of $T(U_i \cap U_j)$, cf. [G4].)

A "deformation" of a diagram of k-algebras $\mathbb{A} = \{\mathbb{A}^i,\varphi^{ij}\}$ will mean a diagram of $k[[t]]$-algebras $\mathbb{A}_t = \{\mathbb{A}^i_t,\varphi^{ij}_t\}$ over the same poset, where i) $\mathbb{A}^i_t$ is a deformation of $\mathbb{A}^i$ in the sense of [G2], i.e., has multiplication given by a power series

$$\alpha^i_t(a,b) = ab + t\alpha^i_1(a,b) + t^2\alpha^i_2(a,b) + \cdots,$$

$a,b \in \mathbb{A}^i$, and ii) φ^{ij}_t is also given by a power series

$$\varphi^{ij} + t\varphi^{ij}_1 + t^2\varphi^{ij}_2 + \cdots \quad .$$

Then $\Gamma = \{\alpha_1^i, \varphi_1^{ij}\}$ is in $Z^2(\mathbb{A},\mathbb{A})$, and when $\mathbb{A}$ is commutative, is even in $\mathrm{Zhar}^2(\mathbb{A},\mathbb{A})$. A second deformation $\{(\mathbb{A}_t^i)', (\varphi_t^{ij})'\}$ is equivalent to the first if there exist isomorphisms $g_t^i : \mathbb{A}_t^i \to (\mathbb{A}_t^i)'$ expressible as power series and reducing to the identity when $t = 0$ such that the g_t^i, φ_t^{ij}, and $(\varphi_t^{ij})'$ form a commutative diagram. The corresponding Γ' is cohomologous to Γ and any cohomologous Γ' can be obtained through an equivalent deformation. We may therefore view the class of Γ in $H^2(\mathbb{A},\mathbb{A})$ (or $\mathrm{Har}^2(\mathbb{A},\mathbb{A})$, in the commutative case) as the "infinitesimal" of the deformation. For smooth X with a covering as before, the group of all infinitesimal deformations of X is thus the same as the group of infinitesimal deformations of the diagram associated with X.

As with a single algebra, a $\Gamma \in Z^2(\mathbb{A},\mathbb{A})$ defines a diagram of $k[t]/t^2$ algebra (reducing, modulo t, to $\mathbb{A}$) which depends, up to isomorphism, only on the cohomology class γ of Γ. Now if $H^*(\mathbb{A},\mathbb{A}) = \oplus H^n(\mathbb{A},\mathbb{A})$ behaved like the cohomology of a single ring, one would have i) a quadratic map $\mathrm{obs}: H^2(\mathbb{A},\mathbb{A}) \to H^3(\mathbb{A},\mathbb{A})$ such that $\mathrm{obs}\,\gamma = 0$ if and only if the diagram of $k[t]/t^2$ - algebras can be extended to one of $k[t]/t^3$ - algebras, ii) a graded Lie product on $H^*(\mathbb{A},\mathbb{A})$, the degree of $\gamma \in H^n$ being counted as $n - 1$, such that if $\gamma \in H^2$ then $[\gamma,\gamma] = 2\,\mathrm{obs}\,\gamma$ and iii) an associative, graded commutative (with the usual degree), cup product on H^*, with Lie multiplication by an element $\gamma \in H^n$ acting as a graded derivation of degree $n - 1$. These are known for a single algebra, [G1]. (In the commutative case one may replace H by Har in i) and ii), but no cup product is defined.) When the diagram is reduced to a single morphism $\varphi: B \to \mathbb{A}$, i) and ii) have been demonstrated by direct computation in [Sch], but a direct approach to iii) or to the case of a general diagram seems hopeless. Nevertheless, all are true for finite diagrams (and some others) because of the following: there is a functor, !, from diagrams $\mathbb{A}$ to (single) algebras $\mathbb{A}!$, and a functor also denoted ! from $\mathbb{A}$ modules $\mathbb{M}$ to $\mathbb{A}!$ modules $\mathbb{M}!$ and natural morphisms $H^n(\mathbb{A},\mathbb{M}) \to H^n(\mathbb{A}!,\mathbb{M}!)$ which are isomorphisms whenever I is finite. We call $\mathbb{A}!$ the "diagram ring" of $\mathbb{A}$. Here are two descriptions.

First set $\mathbb{G}^j = \coprod_{k \geq j} \mathbb{A}^j$, a direct sum of copies of $\mathbb{A}^j$ indexed by all $k \geq j$. To indicate the index, denote the copy indexed by k by $\mathbb{A}^j\varphi^{jk}$, the elements of which we denote by $a^j\varphi^{jk}$, where $a^j \in \mathbb{A}^j$. Then we may write $\mathbb{G}^j = \coprod_{k \geq j} \mathbb{A}^j\varphi^{jk}$. Viewing $\mathbb{A}^j\varphi^{jk}$ as a left $\mathbb{A}^j$-module, for each $i \leq j$ there is a left $\mathbb{A}^j$-module morphism $\mathbb{A}^j\varphi^{jk} \to \mathbb{A}^i\varphi^{ik}$ sending $a\varphi^{jk}$ to $\varphi^{ij}(a)\varphi^{ik}$. These morphisms together give a left $\mathbb{A}^j$ morphism $g^{ij}: \mathbb{G}^j \to \mathbb{G}^i$ and make $\mathbb{G} = \{\mathbb{G}^i, g^{ij}\}$ into a left (not two-sided) $\mathbb{A}$ module which is a projective generator for the category of such modules. (It is small if $\mathbb{A}$ is finite.) Then $\mathbb{A}!$ is $\mathrm{End}_{\mathbb{A}}\mathbb{G}$ in the category of left $\mathbb{A}$ modules. Equivalently, set $\mathbb{A}! = \prod_j \coprod_{k \geq j} \mathbb{A}^j\varphi^{jk}$, with $(a^i\varphi^{ij})(a^k\varphi^{k\ell}) = 0$ if $j \neq k$, while

$(a^i\varphi^{ij})(a^j\varphi^{jk}) = (a^i\varphi^{ij}(a^j))\varphi^{ik}$. For modules, set $\mathbb{M}! = \prod_j \coprod_{k\geq j} M^j\varphi^{jk}$, with

$$(a^i\varphi^{ij})(m^k\varphi^{k\ell}) = 0 \quad \text{if} \quad j \neq k,$$

$$(a^i\varphi^{ij})(m^j\varphi^{jk}) = (a^i(T^{ij}m^j))\varphi^{ik}, \quad \text{and}$$

$$(m^i\varphi^{ij})(a^k\varphi^{k\ell}) = 0 \quad \text{if} \quad j \neq k,$$

$$(m^i\varphi^{ij})(a^j\varphi^{jk}) = (m^i\varphi^{ij}(a^j))\varphi^{ik}.$$

It is easy to prove that an allowable short exact sequence

$$0 \to \mathbb{M}' \to \mathbb{M} \to \mathbb{M}'' \to 0$$

of $\mathbb{A}$-modules induces the usual long exact sequence in cohomology and therefore that

$$H^n(\mathbb{A},\mathbb{M}) \cong \operatorname{Ext}^n_{\mathbb{A}}(\mathbb{A},\mathbb{M}),$$

where for any pair of two-sided $\mathbb{A}$-modules $\mathbb{M}$ and $\mathbb{N}$, $\operatorname{Ext}^n_{\mathbb{A}}(\mathbb{N},\mathbb{M})$ is the group of Yoneda equivalence classes of allowable exact sequence

$$\mathbb{E}: 0 \to \mathbb{M} \to \mathbb{M}_n \to \cdots \to \mathbb{M}_1 \to \mathbb{N} \to 0.$$

There is a natural morphism

$$\omega : \operatorname{Ext}^n_{\mathbb{A}}(\mathbb{N},\mathbb{M}) \to \operatorname{Ext}^n_{\mathbb{A}!}(\mathbb{N}!,\mathbb{M}!)$$

sending the class of $\mathbb{E}$ to that of the sequence $\mathbb{E}!$ obtained by applying ! to every term of $\mathbb{E}$.

MAIN THEOREM: *If $\mathbb{N}$ has finite support, i.e., if $\mathbb{N}^i = 0$ for all but a finite number of $i \in I$, then $\omega: \operatorname{Ext}_{\mathbb{A}}(\mathbb{N},\mathbb{M}) \to \operatorname{Ext}_{\mathbb{A}!}(\mathbb{N}!,\mathbb{M}!)$ is an isomorphism for all $\mathbb{M}$.*

For finite $\mathbb{A}$ one has the composite isomorphism $H^n(\mathbb{A},\mathbb{M}) \xrightarrow{\sim} \operatorname{Ext}^n_{\mathbb{A}}(\mathbb{A},\mathbb{M}) \xrightarrow{\omega} \operatorname{Ext}^n_{\mathbb{A}!}(\mathbb{A}!,\mathbb{M}!) \xrightarrow{\sim} H^n(\mathbb{A}!,\mathbb{M}!)$. Thus $H^*(\mathbb{A},\mathbb{A})$ has the same kind of structure as the cohomology of a single ring. When $\mathbb{A}$ comes from a complex space, sheaf, or diagram of such, it would be interesting to have geometric interpretations (if any exist) for the operations in $H^*(\mathbb{A},\mathbb{A})$.

Unfortunately, simple examples show that $\mathbb{M}$ may be a projective or injective $\mathbb{A}$-module without $\mathbb{M}!$ being such over $\mathbb{A}!$, else the proof of the Main Theorem would be trivial. Our proof proceeds by comparing particular projective resolutions.

As mentioned, the Harrison cohomology of a commutative ring carries a graded Lie structure, so one would expect the same for $\operatorname{Har}^*(\mathbb{A},\mathbb{A})$ when $\mathbb{A}$ is commutative. But $\mathbb{A}!$ is never commutative except in trivial cases so one can not get this from an isomorphism like the foregoing since $\operatorname{Har}^*(\mathbb{A}!,\mathbb{A}!)$ is undefined. There should be an extension of Harrison's theory to algebras like $\mathbb{A}!$. Hopefully we will then find that $\operatorname{Har}^*(\mathbb{A},\mathbb{A})$ does carry a graded Lie structure, and, when $\mathbb{A}$ comes from a Stein covering of a complex manifold X, that $\mathrm{Har}^*(\mathbb{A},\mathbb{A}) \cong H^*(X,\mathcal{T})$ **as graded Lie rings.**

The Main Theorem is the basic tool in proving the following.

THEOREM: <u>If</u> $\mathbb{A}$ <u>is finite then every deformation of</u> $\mathbb{A}!$ <u>is equivalent to one induced by a deformation of</u> $\mathbb{A}$.

This is, so far, purely formal, since deformations have been defined only in terms of formal power series: it says that the formal deformation theories of $\mathbb{A}$ and $\mathbb{A}!$ are functorially equivalent. However, in certain cases it is possible to exhibit deformations of a diagram $\mathbb{A}$ in which the deformed multiplications are given by polynomials rather than power series. One can do this, for example, when $\mathbb{A}$ comes from the standard affine covering of projective 2-space $\mathbb{P}^2$ over a field k of characteristic $p > 0$. The result is a diagram of non-commutative algebras over $k[t]$ each of the sort described in [G3]. One may view this as representing a kind of "non-commutative" $\mathbb{P}^2$, the simplest of a class of "non-commutative varieties" obtained by deforming diagrams which originally defined commutative varieties.

REFERENCES

[André]: M. André, Homologie des algébres commutatives, Springer, N.Y. and Berlin, 1974.

[G]: M. Gerstenhaber, [1] The cohomology structure of an associative ring, Ann. of Math. 78 (1963), 267-288; [2] On the deformation of rings and algebras, ibid., 79 (1964), 59-103; [3] ---, III, ibid., 88(1968), 1-34; [4] On the deformation of sheaves of rings, in Global analysis, papers in honor of K. Kodaira, D. C. Spencer and S. Iyanaga, eds., Tokyo and Princeton U. Presses, 1969.

[Harrison]: D. K. Harrison, Commutative algebras and cohomology, Transactions of the A.M.S., 104 (1962), 191-204.

[MacLane]: S. MacLane, Homology, Academic Press, N.Y., 1963.

[Sch]: S. D. Schack, On the deformation of an algebra morphism, thesis, U. of Pennsylvania, 1980.

University of Pennsylvania
Philadelphia, PA 19104

SUNY
Buffalo, NY 14214

Contemporary Mathematics
Volume 13, 1982

ON THE GELFAND-KIRILLOV DIMENSION OF NOETHERIAN PI-ALGEBRAS

by Martin Lorenz and Lance W. Small

Let $d(.)$ denote Gelfand-Kirillov dimension over a fixed base field k. In this note we prove the following.

THEOREM. *Let* R *be a Noetherian* PI-*algebra over* k *and let* N *denote its nilpotent radical. Then* $d(R) = d(R/N) = \max_P d(R/P)$, *where* P *ranges over the minimal prime ideals of* R.

Here, the last equality is of course clear from [3, Lemma 3.1e]. The corresponding result for (Gabriel-Rentschler) Krull dimension is well-known and is in fact an easy consequence of the so-called partitivity of Krull dimension: For any two modules $M \subseteq N$ over a ring R one has $\text{Kdim } N = \sup\{\text{Kdim } M, \text{Kdim } N/M\}$ whenever these Krull dimensions do exist. In general, this relation does not hold for Gelfand-Kirillov dimensions of modules. Indeed, G. Bergman has recently constructed finitely generated modules $M \subseteq N$ over an affine PI-ring R such that $d(N) > \sup\{d(M), d(N/M)\}$ [2]. Another, non-finitely generated, example, due to McConnell [6], is as follows. Let $R = k\{X,Y\}/\langle Y\rangle^2$, where $k\{X,Y\}$ denotes the free k-algebra generated by X and Y, and set $N = \langle Y\rangle$, the ideal of R generated by Y. Then, as right R-modules, one has $d(R/N) = d(N) = 1$ but $d(R) = 2$. Note that R is an affine PI-algebra, yet the conclusion of the theorem does not hold for R.

Our theorem has the following consequence.

COROLLARY. *Let* R *be a Noetherian* PI-*algebra. Then* $d(R)$ *is either infinite or an integer. If* R *is also affine then* $d(R) = \dim R$, *the prime length of* R.

0271-4132/82/0527/$02.75

Proof. By the theorem, we immediately reduce to the case where R is prime. Let C denote the center of R. Then, using the fact that Q(R) is obtained by localizing at the non-zero elements of C and is a finitely generated module over Q(C), one easily shows that $d(R) = d(C)$ [5]. But d(C) is either infinite or an integer, by [3, Satz 3.2]. In fact, d(C) equals the transcendence degree of C over k, i.e. the maximal number of algebraically independent elements of C over k. By [1. Theorem 7.1], the latter is equal to dim R, and the corollary is proved.

The reader should contrast this with the examples constructed in [3, (2.10 and (2.11)] which show that for any real number $\gamma \geq 2$ there exists an affine (in fact, 2-generated) PI-algebra R with $d(R) = \gamma$.

Throughout this note, k will denote a commutative field. For the basic facts concerning Gelfand-Kirillov dimension we refer to [3].

We begin with two observations that will be needed later on. At least part (i) of the following lemma is well-known, but we include the short argument for the reader's convenience.

LEMMA 1. *Let R be a k-algebra.*

i. If $S \subseteq R$ is a subalgebra such that R is finitely generated as a (right or left) S-module then $d(R) = d(S)$.

ii. Assume that R is right Noetherian and let $e_1, e_2, \ldots, e_t \in R$ be an orthogonal set of idempotents with $\sum_i e_k = 1$. Then $d(R) = \max_i d(e_k Re_i)$.

Proof. (i). Only $d(R) \leq d(S)$ has to be proved. To see this, just note that R is a subalgebra of $\mathrm{End}_S(R)$ which in turn is a subquotient of some matrix algebra $M_n(S)$ over S. Therefore, $d(R) \leq d(M_n(S))$ and the assertion follows, since $d(M_n(S)) = d(S \otimes_k M_n(k)) = d(S)$, by [3, Lemma 3.1a].

(ii). If $e = e^2 \in R$ then Re is a Noetherian right eRe-module. Indeed, if $I \subseteq Re$ is an eRe-submodule then $IR \cap Re = IRe = IeRe = I$. In particular, each Re_i is finitely generated as a right e_iRe_i-module and so R is finitely generated as a right module over $S = \oplus_i e_iRe_i \subseteq R$. Finally, $d(S) = \max_i d(e_iRe_i)$, by [3, Lemma 3.1b]. Hence the assertion follows from part (i).

The next lemma contains the computations needed for the proof of the theorem.

LEMMA 3. *Let R be a k-algebra and let N be a nilpotent ideal of R which is finitely generated as a right ideal of R. Assume that R/N is a finitely generated module over some commutative subalgebra. Then* $d(R) = d(R/N)$.

Proof: We only have to show that $d(R) \leq d(R/N)$ holds and for this we may clearly assume that $d(R/N) = d < \infty$. Let $^- : R \to R/N$ denote the canonical map. Using Lemma 1 (i), we immediately reduce to the case where R is commutative.

Fix a finite generating set Γ for N as a right ideal of R. Since N is nilpotent, we may assume that products of elements of Γ are contained in Γ.

Now let V be a finite dimensional k-subspace of R with $1 \in V$. Then by the Noether normalization theorem, the subalgebra $k[\bar{V}] \subseteq \bar{R}$ is a finitely generated module over some polynomial ring $S = k[x_1, x_2, \ldots, x_r] \subseteq k[\bar{V}]$. Here, $r = d(S) \leq d$. Choose preimages $y_i \in k[V] \subseteq R$ with $\bar{y}_i = x_i$ and set $T = k[y_1, \ldots, y_r] \subseteq k[V]$. Furthermore, let $g_1, g_2, \ldots, g_p \in k[V]$ be preimages for the generators of $k[\bar{V}]$ over S and let $v_1, v_2, \ldots, v_q$ be a k-basis for V. Then there exists elements $t_{hi}, t_{hij} \in T$ and $n_i, n_{ij} \in N$ such that

$$(1) \qquad v_i = w_i + n_i \quad \text{with} \quad w_i = \sum_{h=1}^{p} g_h t_{hi} \quad (i=1,2,\ldots,q)$$

and

$$(2) \qquad g_i g_j = \sum_{h=1}^{p} g_h t_{hij} + n_{ij} \quad (i,j=1,2,\ldots,p).$$

Choose a finite dimensional k-subspace W of R such that

$$(3) \qquad V \subseteq W \quad \text{and} \quad [V,V],\ V\cdot\Gamma,\ \{n_i, n_{ij} \ \text{all}\ i,j\} \subseteq \Gamma\cdot W.$$

Moreover, choose an integer $s \geq 1$ such that $y_i, g_h, t_{hi}, t_{hij}, w_i \in V^s$, the vector space generated by the product of length $\leq s$ with factors $\in V$. From (3) we deduce that

$$(4) \qquad [V^s,V^s] = \sum_{h=0}^{2s-2} V^h[V,V]V^{2s-2-h} \subseteq \sum_{h=0}^{2s-2} V^h\Gamma\cdot W\, W^{2s-2-h} \subseteq \Gamma\cdot W^{2s-1} .$$

Now consider a monimial $v_{i_1}v_{i_2}\cdots v_{i_n} \in V^n$. Successively writing $v_{i_h} = w_{i_h} + n_{i_h}$ $(h=1,2,\ldots n)$ we get

$$v_{i_1}v_{i_2}\cdots v_{i_n} = w_{i_1}w_{i_2}\cdots w_{i_n} + n_{i_1}v_{i_2}v_{i_3}\cdots v_{i_n} + $$
$$+ w_{i_1}n_{i_2}v_{i_3}\cdots v_{i_n} + \cdots + w_{i_1}w_{i_2}\cdots w_{i_{n-1}}n_{i_n} .$$

By (3), all summands but the first one belongs to $\Gamma\cdot W^{sn}$. Thus we have

$$v_{i_1}v_{i_2}\cdots v_{i_n} \in w_{i_1}w_{i_2}\cdots w_{i_n} + \Gamma\cdot W^{sn} .$$

Now consider $w_{i_1}w_{i_2}\cdots w_{i_n} = w$. By (1), w is a sum of monomials of the form $g_{h_1}t_{h_1i_1}g_{h_2}t_{h_2i_2}\cdots g_{h_n}t_{h_ni_n}$. But

$$g_{h_1}t_{h_1i_1}g_{h_2}t_{h_2i_2}\cdots g_{h_n}t_{h_ni_n} = g_{h_1}g_{h_2}\cdots g_{h_n}t_{h_1i_1}\cdots t_{h_ni_n} + c,$$

where c is a sum of monomials of the form $c_1c_2\cdots c_h[c_{h+1},c_{h+2}]c_{h+3}\cdots c_{2n}$ with each $c_h \in \{g_m, t_{hi} \mid m,h,i\}$. Using (3) and (4), we see that $c \in \Gamma.W^{2sn-1}$.

Next, successively using (2), we can express $g_{h_1}\cdots g_{h_n}t_{h_1i_1}\cdots t_{h_ni_n}$ as a sum of monomials of the form

$$g_1\beta t_{h_1i_1}\cdots t_{h_ni_n} ,$$

where β denotes a monomial of length $n-1$ in the t_{hij}'s, plus a sum of monomials of the form

$$\gamma n_{ij}\delta ,$$

where γ denotes a monomial in the g_h's and δ denotes a monomial in the t_{hi}'s such that $\gamma n_{ij}\delta$ has length $2n-1$. By (3), the latter type of monomials again belongs to $\Gamma\cdot W^{2ns-1}$. The first type on monomials, $g_1 t_{h_1i_1}\cdots t_{h_ni_n}$, can be rewritten by expressing each t_{hij} and t_{hi} in terms of the y_i's. So let Y denote the subspace of T generated by $1, y_1, y_2, \ldots, y_r$ and choose $f \geq 1$ with $f_{hij}, t_{hi} \in Y^f$. Then, summarizing the above, we have shown that

$$(5) \quad V^n \subset \sum_{h=1}^{p} g_h Y^{f(2n-1)} + \Gamma \cdot W^{sn} + \Gamma \cdot W^{2sn-1}$$
$$\subseteq \sum_{h=1}^{p} g_h Y^{2fn} + \Gamma \cdot W^{2sn} .$$

Finally, each monomial $y_{i_1} y_{i_2} \cdots y_{i_m} \in Y^m$ can be written as

$$y_{i_1} y_{i_2} \cdots y_{i_m} = y_1^{e_1} y_2^{e_2} \cdots y_r^{e_r} + b ,$$

where b belongs to $\sum_{h=0}^{m} Y^h [Y,Y] Y^{m-(1+2)}$ and $\Sigma e_i = m$. Moreover, by (3) and (4), we have

$$Y^h [Y,Y] Y^{m-(h+2)} \subseteq V^{sh} \Gamma \cdot W^{2s-1} \cdot W^{s(m-h-2)} \subseteq \Gamma \cdot W^{sm-1}.$$

Thus, if Y_m denotes the k-subspace of R generated by the "ordered" monomials $y_1^{e_1} y_2^{e_2} \cdots y_r^{e_r}$ with $\Sigma e_i \leq m$, then we have, for all m,

$$(6) \quad Y^m \subset Y_m + \Gamma\, W^{sm-1} .$$

Using this with $m = 2fn$, (5) yields

$$V^n \subseteq \sum_{h=1}^{p} g_h Y_{2fn} + \sum_{h=1}^{p} g_h \Gamma \cdot W^{2fsn-1} + \Gamma \cdot W^{2sn}$$
$$\subseteq \sum_{h=1}^{p} g_h Y_{2fn} + \Gamma \cdot W^{s+2fsn-1} + \Gamma \cdot W^{2sn}$$
$$\subseteq \sum_{h=1}^{p} g_h Y_{2fn} + \Gamma \cdot W^{h \cdot n}$$

for a suitable constant $h \geq 1$. Set $X = \sum_{h=1}^{p} g_h Y_{2fn}$ and $V_1 = W^h$. Then, since $\dim_K Y_{2fn} \leq \eta\, n^r$ for a suitable constant $\eta > 0$, we get $\dim_K X \leq \xi n^r \leq \xi n^d$ for a suitable $\xi > 0$. Thus we have shown that

(7) $V^n \subseteq X + \Gamma \cdot V_1^{\,n}$, with $V_1 \subseteq R$ finite dimensional and $\dim_K X \leq \xi\, n^d$, $\xi > 0$.

We now iterate (7) by letting V_1 play the role of V above, etc. This gives $V^n \subseteq X + \Gamma\cdot V_1{}^n \subseteq X + \Gamma\cdot(X_1 + \Gamma\cdot V_2{}^n) \subseteq \ldots \subseteq X + \Gamma\cdot(X_1 + \Gamma\cdot(X_2 + \ldots + \Gamma\cdot(X_1 + \Gamma\cdot V_{h+1}{}^n)) \ldots)$, where each V_i is a finite dimensional k-subspace of R and $\dim_k X_i \le \xi_i n^d$ for suitable constants $\xi_i > 0$. Finally, if t denotes the nilpotence of N, then we have

$$X + \Gamma\cdot(X_1 + \Gamma\cdot(X_2 + \ldots + \Gamma\cdot(X_{t-1} + \Gamma\cdot V_t{}^n)) \ldots) =$$
$$X + \Gamma\cdot(X_1 + \Gamma\cdot(X_2 + \ldots + \Gamma\cdot X_{t-1})) \ldots) .$$

Thus, since Γ is closed under products, we deduce that

$$V^n \subseteq X + \Gamma\cdot \sum_{i=1}^{t-1} X_i .$$

Therefore, $\dim_K V^n \le \xi n^d + |\Gamma|\cdot \sum_{i=1}^{t-1} \xi_i n^d = \lambda_V n^d$ for some constant $\lambda_V > 0$. The assertion of the lemma now follows, since V was arbitrary.

We remark that the hypotheses of Lemma 3 do <u>not</u> force R to be a finitely generated module over some commutative subalgebra. (See Sarraille [7].) We are now ready to prove the theorem.

<u>Proof of the theorem</u>. Let R be a Noetherian PI-algebra over k. We have to show that $d(R) = d(R/P)$ for some prime ideal P of R.

Since R is Noetherian, we have $0 = \bigcap_{j=1}^{h} I_j$ for certain meet-irreducible. By Gordon [4], R has an Artinian ring of quotients Q. Let N be the nilpotent radical of R and let $\bar{\ }:R \to R/N$ denote the canonical map. S Since $\bar{\ }$ maps regular elements of R to regular elements of $\bar{R}$, we can extend $\bar{\ }$ to a map $\bar{\ }:Q \to \bar{Q} = Q(\bar{R})$ with kernel NQ, and hence we have an isomorphism $\bar{Q} \cong Q/NQ$.

We claim that $d(\bar{Q}) = d(\bar{R})$. Indeed, $\bar{Q} = \bigoplus_{i=1}^{t} Q(\bar{R}/P_i)$, where $P_i (i=1,2,\ldots,t)$ are the minimal primes of $\bar{R}$. Since each $Q(\bar{R}/P_i)$ is obtained from $\bar{R}/P_i$ by central localization, we have $d(Q(\bar{R}/P_i)) = d(\bar{R}/P_i)$ for all i, by [3, Lemma 31f], and hence $d(\bar{Q}) = \max_i d(Q(\bar{R}/P_i)) = \max_i d(\bar{R}/P_i) = d(\bar{R})$, as we have claimed. Clearly, $d(R) \le d(Q)$ and so it suffices to show that $d(Q) = d(\bar{Q})$.

For this, let $\bar{e}_1, \bar{e}_2, \ldots, \bar{e}_t \in \bar{Q}$ denote the idempotents corresponding to the decomposition $\bar{Q} = \bigoplus_{i=1}^{t} Q(\bar{R}/P_i)$. There are orthogonal idempotents $e_i \in Q(i=1,2,\ldots,t)$ which are preimages for the $\bar{e}_i$'s and satisfy $\sum_{i=1}^{t} e_i = 1$. By Lemma 1(ii), we know that $d(Q) = \max_i d(e_iQe_i)$. Moreover, we have $e_iQe_i/e_iNQe_i \simeq \overline{e_iQe_i} = Q(\bar{R}/P_i) \subseteq \bar{Q}$. Since $Q(\bar{R}/P_i)$ is a finitely generated module over its center, we can apply Lemma 3, with e_iQe_i playing the role of R and e_iNQe_i playing the role of N, to conclude that $d(e_iQe_i) = d(Q(R/P_i))$. Therefore, $d(Q) = \max_i d(Q(\bar{R}/P_i)) = d(\bar{Q})$, which completes the proof of the theorem.

Acknowledgment

This work was being done while the first author was visiting the University of California, San Diego and he would like to thank that institution for its hospitality.

REFERENCES

1. S. A. Amitsur and L. W. Small, Prime ideals in PI-rings, J. Algebra 62 (1980), 358-383.
2. G. Bergman, to appear in Communications in Algebra.
3. W. Borko and H. Kraft, Über die Gelfand-Kirillov-Dimension, Math. Annalen 220 (1976), 1-24.
4. R. Gordon, Artinian quotient rings of FBN rings, J. Algebra 35 (1975), 304-307.
5. M. P. Malliavin, Dimension de Gelfand-Kirillov des algèbres à identités polynomiales, C. R. Acad. Sci. Paris 282 (1976), 679-681.
6. J. C. McConnell, unpublished.
7. J. Sarraille, unpublished.

Fachbereich Mathematik
Universität Essen
D-4300 Essen, West Germany

and

Department of Mathematics
University of California, San Diego
La Jolla, CA 92093

Contemporary Mathematics
Volume 13, 1982

A SIMPLE PROOF OF KOSTANT'S THEOREM, AND AN ANALOGUE FOR THE SYMPLECTIC INVOLUTION

Louis Halle Rowen[1]

ABSTRACT

An elementary proof is given of the theorem of Kostant, that for n even the standard polynomial S_{2n-2} is an identity for skew-symmetric $n \times n$ matrices (with respect to the transpose). The parallel result for the symplectic involution is also proved, that S_{2n-2} is an identity for symmetric $n \times n$ matrices, and this result is sharp. Other identities are also obtained for the symplectic involution.

The object of this paper is threefold - to give an easy proof of an interesting theorem of Kostant [2], that for n even the standard polynomial S_{2n-2} vanishes identically for all $n \times n$ matrices which are skew-symmetric with respect to the transpose, to give an analogue for the symplectic involution, and to show that these results follow from the same trace identity (arising from the generic minimal polynomial for a symmetric element with respect to the symplectic involution). I would like to take the opportunity of giving heartfelt thanks to Professor Jacobson for the patience and encouragement he showed during my student years under his guidance, and for acquainting me to Kostant's theorem at that time.

§1. PRELIMINARIES.

An involution is an antiautomorphism of degree 2. The thrust of this paper will be to check certain identities of $n \times n$ matrix rings with involution, and standard methods (given in detail in [6]) show that any multilinear identity of all matrix rings with or without involution can be verified via the special case in which the base field is the field of complex numbers $\mathbb{C}$. So we work in $M_n(\mathbb{C})$, the $n \times n$ matrices over $\mathbb{C}$; now, up to isomorphism, there are two types of involution - the transpose, denoted (t), and the canonical symplectic involution (s), defined (only when n is even) by the formula

[1] Currently on sabbatical at Yale University.

0271-4132/82/0528/$03.25

$$\begin{pmatrix} A & B \\ C & D \end{pmatrix}^s = \begin{pmatrix} D^t & -B^t \\ -C^t & A^t \end{pmatrix}$$

where the original matrix is partitioned into $(n/2) \times (n/2)$ blocks.

Define the standard polynomial $S_k = \Sigma_\pi (\text{sg } \pi) X_{\pi 1} \cdots X_{\pi k}$. By the Amitsur-Levitzki theorem S_{2n} is an identity of $M_n(\mathbb{C})$, and an easy matrix unit argument show there is no identity of smaller degree. However, the question remained open whether the Amitsur-Levitzki method could be improved if only certain kinds of matrices were considered. In other words, writing $L(n,d,k, (*))$ for the sentence

$$S_d(B_1, \ldots, B_k, A_{k+1}, \ldots, A_d) = 0 \text{ for all symmetric } n \times n$$

matrices $B_1, \ldots, B_k$ and all akew-symmetric $n \times n$ matrices $A_{k+1}, \ldots, A_d$ with respect to the involution $(*)$, one had $L(n,d,k, (*))$ by Amitsur-Levitzki whenever $d \geq 2n$. Kostant [2] reproved the Amitsur-Levitzki theorem by showing it is equivalent to a theorem in Lie cohomology, and demonstrated the power of his method by proving $L(n,2n-2,0,(t))$ for all even n. An "elementary" proof of Kostant's theorem plus the additional results $L(n,2n-2,0,(t))$, $L(n,2n-1,1,(t))$, $L(n,2n-1,0,(t))$, for all n and $L(n,2n-2,1,(t))$ for odd n, was found in [5], and reproved in a different way in [1], but both arguments involved careful and complicated graph-theoretical approaches. Moreover the analogous question of finding results for (s) remained open. When it was determined in [3], [4] that the Amitsur-Levitzki theorem followed formally from the Hamilton-Cayley equation, with Razmyslov in fact providing a rather straightforward proof, renewed interest was aroused to prove Kostant's results by this method. This is the point of view presented here, and we shall rely principally on the following well-known analogue to the Hamilton-Cayley theorem and Newton's identities:

Suppose $n = 2m$ and $x^s = x \in M_n(\mathbb{C})$. Then x satisfies a polynomial of the form $p(\lambda) = \Sigma_{k=0}^m (-1)^k \mu_k \lambda^{m-k}$, where $\mu_0 = 1$ and, inductively $2k\mu_k = \Sigma_{i=1}^k (-1)^{i-1} \mu_{k-i} \text{tr}(x^i)$ for $1 \leq k \leq m$. (This is obtained by taking the Pfaffian instead of the determinant in order to halve the degree of the "characteristic polynomial", and then obtaining the above formulae in a manner analogous to the proof of Newton's formulae.) We shall call this the "generic minimal equation of x". In order to carry out the proof, we shall also need the following easily verified remarks; many are due to Kostant and have already been given in detail in [5].

1. Suppose $g(X_1, \ldots, X_k) = \Sigma\, \alpha_\pi X_{\pi 1} \cdots X_{\pi k}$, where π runs over all

permutations of $(1, \ldots, k)$; then $\sum_\tau (\mathrm{sg}\ \tau)\, g(X_{\tau 1}, \ldots, X_{\tau k}) = (\sum_\tau (\mathrm{sg}\ \tau)\alpha_\tau) S_k$, summed over all permutations τ of $(1, \ldots, k)$ (c.f. [6, remark 1.2.15]).

2. Given an involution $(*)$ of $M_n(\mathbb{C})$, write $M_n(\mathbb{C})^+$ (resp. $M_n(\mathbb{C})^-$) for the symmetric (res. skew-symmetric) elements of $M_n(\mathbb{C})$ with respect to $(*)$. For any $A \in M_n(\mathbb{C})^-$ and $B \in M_n(\mathbb{C})^+$ we have $\mathrm{tr}(AB) = 0$ since $\mathrm{tr}(AB) = \mathrm{tr}(AB)^* = \mathrm{tr}(B^*A^*) = \mathrm{tr}(A^*B^*) = -\mathrm{tr}(AB)$. Consequently, the trace is a nondegenerate bilinear form both on $M_n(\mathbb{C})^+$ and on $M_n(\mathbb{C})^-$ (i.e., if $A_0 \in M_n(\mathbb{C})^-$ and $\mathrm{tr}(A_0A) = 0$ for all A in $M_n(\mathbb{C})^-$ then $A_0 = 0$; likewise for $M_n(\mathbb{C})^+$.)

3. Suppose $(*)$ is a given involution with $B_1, \ldots, B_k \in M_n(\mathbb{C})^+$ and $A_{k+1}, \ldots, A_d \in M_n(\mathbb{C})^-$; then $S_d(B_1, \ldots, B_k, A_{k+1}, \ldots, A_d)^* = (-1)^{[d/2]+ d-k}\, S_d(B_1, \ldots, B_k, A_{k+1}, \ldots, A_d)$.

4. For all d and all matrices $x_1, \ldots, x_{2d}$ in $M_n(\mathbb{C})$, $\mathrm{tr}\, S_{2d}(x_1, \ldots, x_{2d}) = 0$ (c.f. [6, remark 1.3.4]). Also $\mathrm{tr}\, S_{2d-1}(x_1, \ldots, x_{2d-1}) = (2d-1)\mathrm{tr}(S_{2d-2}(x_1, \ldots, x_{2d-2})x_{2d-1})$.

§2. KOSTANT'S THEOREM

In this section we deal with the transpose involution (t) and prove Kostant's identity (Theorem 1: $L(n,2n-2,0,(t))$ holds for n even). (An easy argument given in [5] then yields $L(n,2n-1,0,(t))$ for all n, but I do not see how to recover all of the positive results of [5] without some work.)

Suppose $n = 2m$, and $A, A' \in M_n(\mathbb{C})^-$, with A' invertible. The map $x \to A'x^t(A')^{-1}$ is well-known to be a symplectic-type involution (seen by counting the dimension of the space of symmetric elements) and in particular $A'A$ is symmetric with respect to this new involution; as stated in §1, $A'A$ satisfies a generic minimal equation

$$(1) \qquad 0 = (A'A)^m + \sum_{k=1}^m (-1)^k \mu_k (A'A)^{m-k} \quad \text{where inductively}$$

$$2k\mu_k = \sum_{i=1}^k (-1)^i \mu_{k-i}\, \mathrm{tr}((A'A)^i) \qquad \text{with} \quad \mu_0 = 1.$$

(Thus μ_k has the form $(-1)^k \sum_u q_u \mathrm{tr}((A'A)^{u_1}) \ldots \mathrm{tr}((A'A)^{u_j})$ for suitable $q_u \in \mathbb{Q}$ and j-tuples $u = (u_1, \ldots, u_j)$ with $\sum_{i=1}^j u_i = k$. Either a generic matrix argument or a Zariski topology argument (depending on one's taste) enables one to remove the assumption that A' is invertible. Then we may apply the usual multilinearization procedure to equation (1) to obtain

$$0 = \sum_{\pi,\sigma} A_{\pi 1}A_{m+\sigma 1}A_{\pi 2}A_{m+\sigma 2} \cdots A_{\pi m}A_{m+\sigma m}$$

$$+ \sum_{k,u,\pi,\sigma} q_u \mathrm{tr}(A_{\pi 1}A_{m+\sigma 1}\cdots A_{\pi u_1}A_{m+\sigma u_1})\cdots A_{\pi(m-k+1)}A_{m+\sigma(m-k+1)} \cdots A_{\pi m}A_{m+\sigma m}$$

where the sum at the right is over all tuples $u = (u_1, \ldots, u_j)$ such that $u_1 + \ldots + u_j = k$, all k, and all permutations π,σ of $(1, \ldots, m)$. We have mimicked the first step of Razmyslov's proof of the Amitsur-Levitzki theorem (c.f. [6, p. 21]) and now modify the second step a bit. Namely, leaving A_1 and A_2 as is, substitute $[A_3,A_4]$ for A_3, $[A_5,A_6]$ for A_4, etc. Applying remark 1 (of the series of remarks above) we get

$$0 = S_{2m-2}(A_1, \ldots, A_{2m-2}) + \text{sum of terms}$$

each term being a product of traces of standard polynomials times at most one standard polynomial. But the remarks show $\mathrm{tr}\, S_j(A_{i_1}, \ldots, A_{i_j}) = 0$ unless j is odd; also if $j \equiv 1 \mod 4$ then $S_j(A_{i_1}, \ldots, A_{i_j}) \in M_n(\mathbb{C})^-$ and has trace 0. Thus the only ways to get possibly nonzero terms in the sum is to have either terms of the form $\mathrm{tr}\, S_{u_1}(\quad)\ \mathrm{tr}\, S_{u_2}(\quad)\ S_{u_3}(\quad)$, where $u_1, u_2 \equiv 3$ modulo 4 and $u_1+u_2+u_3 - 2m-2$, or else of the form $\mathrm{tr}\, S_{u_1}(\quad)\ S_{u_2}(\quad)$, where $u_1 \equiv 3 \mod 4$ and $u_2 = 2m-2-u_1$ (so also $u_2 \equiv 3$ modulo 4). So terms of the latter form are in $M_n(\mathbb{C})^+$; considering only the skew-symmetric part leaves us then with

$$0 = S_{2m-2}(A_1, \ldots, A_{2m-2})$$

$$+ \sum \text{ terms of the form } \mathrm{tr}\, S_{u_1}(\quad)\ \mathrm{tr}\, S_{u_2}(\quad)\ S_{u_3}(\quad),$$

where u_1 and u_2 are odd. But then we have $\mathrm{tr}\, S_{u_2}(\quad)\mathrm{tr}\, S_{u_1}(\quad)S_{u_3}(\quad)$ occurring with the opposite sign, so the terms in the sum occur in opposite pairs, yielding $0 = S_{2m-2}(A_1, \ldots, A_{2m-2})$, as desired. Q.E.D.

This proof was hard to write down because of the cumbersome notation, but perhaps some special cases would cast some light. For $n = 2$ we have

$$(A'A) - \frac{1}{2}\mathrm{tr}(A'A) = 0 \text{ yielding}$$

$$0 = [A', A] - \frac{1}{2}\mathrm{tr}[A', A] = [A', A], \text{ i.e. } [A_1A_2] = 0;$$

For $n = 4$ we have

$$0 = (A'A)^2 - \frac{1}{2}\mathrm{tr}(A'A)A'A + \frac{1}{8}(\mathrm{tr}(A'A))^2 - \frac{1}{4}\mathrm{tr}(A'A)^2)$$

yielding

$$\begin{aligned}
0 = {} & A_1A_2[A_3,A_4][A_5,A_6] + [A_3,A_4]A_2A_1[A_5,A_6] \\
& \quad + A_1[A_5,A_6][A_3,A_4]A_2 + [A_3,A_4][A_5,A_6]A_1A_2 \\
& - \tfrac{1}{2}\,([A_3,A_4][A_5,A_6]\mathrm{tr}(A_1A_2) \\
& \quad + A_1[A_5,A_6]\mathrm{tr}([A_3,A_4]A_2) + [A_3,A_4]A_2\mathrm{tr}(A_1[A_5,A_6]) \\
& \quad + A_1A_2\mathrm{tr}([A_3,A_4][A_5,A_6])) \\
& + \tfrac{1}{8}\,(\mathrm{tr}A_1A_2\mathrm{tr}([A_3,A_4][A_5,A_6]) \\
& \quad + \mathrm{tr}([A_3,A_4]A_2)\mathrm{tr}(A_1[A_5,A_6])+\mathrm{tr}(A_1[A_5,A_6])\mathrm{tr}([A_3,A_4]A_2) \\
& \quad + \mathrm{tr}([A_3,A_4][A_5,A_6])\mathrm{tr}(A_1A_2)) \\
& - \tfrac{1}{4}\,\mathrm{tr}(A_1A_2[A_3,A_4][A_5,A_6] + [A_3,A_4]A_2A_1[A_5,A_6] + A_1[A_5,A_6][A_3,A_4]A_2 \\
& \quad + [A_3,A_4][A_5,A_6]A_1A_2).
\end{aligned}$$

Writing $\psi(A_1, \ldots, A_6)$ for the right-hand side we now take $\sum_{\tau} (\mathrm{sg}\,\tau)\, \psi(A_{\tau 1}, \ldots, A_{\tau 6})$ to get (by remark 1) 0 equals a sum of terms which we treat line by line. The first line yields

$$\begin{aligned}
& 4S_6(A_1, \ldots, A_6) - 4S_6(A_1, \ldots, A_6) + 4S_6(A_1, \ldots, A_6) \\
& \quad + 4S_6(A_1, \ldots, A_6) = 8S_6(A_1, \ldots, A_6)
\end{aligned}$$

The second line divided by 2 yields four sums, which we treat one by one:

$$\begin{aligned}
& S_4(A_3,A_4,A_5,A_6)\mathrm{tr}[A_1,A_2] - S_4(A_2,A_4,A_5,A_6)\mathrm{tr}[A_1,A_3] \\
& \quad + S_4(A_2,A_3,A_5,A_6)\mathrm{tr}[A_1,A_4] \ \text{ etc.}
\end{aligned}$$

equals 0 term by term, by remark 4;

$S_3(A_1,A_5,A_6)\mathrm{tr}S_3(A_2,A_3,A_4) - S_3(A_1,A_4,A_6)\mathrm{tr}S_3(A_2,A_3,A_5)$ etc. is symmetric;

$S_3(A_2,A_3,A_4)\mathrm{tr}\ S_3(A_1,A_5,A_6)$ etc. is also symmetric;

$[A_1,A_2]\mathrm{tr}\ S_4(A_3,A_4,A_5,A_6) - [A_1,A_3]S_4(A_2,A_4,A_5,A_6)$ etc. equals 0 by remark 4.

Thus the second line equals a symmetric matrix. (In fact the second and third terms cancel, so the second line equals 0, but we do not need that). The terms of the third line yield

$\frac{1}{2}$ (tr$[A_1,A_2]$tr $S_4(A_3,A_4,A_5,A_6)$ - tr$[A_1,A_3]$tr $S_4(A_2,A_4,A_5,A_6)$ etc.) —— equals 0 by remark 4, plus

$\frac{1}{2}$ (tr $S_3(A_2,A_3,A_4)$tr $S_3(A_1,A_5,A_6)$ - tr $S_3(A_1,A_3,A_4)$tr $S_3(A_2,A_5,A_6)$ etc.)

This sum has $\binom{6}{3} = 20$ summands, which partition into 10 canceling pairs; for example $-$tr $S_3(A_1,A_5,A_6)$tr $S_3(A_2,A_3,A_4)$ cancels tr $S_3(A_2,A_3,A_4)$tr $S_3(A_1,A_5,A_6)$. Thus this term is 0; the other two terms in the third line are also 0, analogously.

The fourth line is tr$(S_4(A_1, \ldots, A_6) - S_4(A_1, \ldots, A_6) + S_4(A_1,\ldots, A_6) + S_4(A_1, \ldots, A_6)) = 2$ tr $S_4(A_1, \ldots, A_6) = 0$ by remark 4.

Thus we are left with $0 = 8S_6(A_1, \ldots, A_6) + B$ for some $B \in M_4(\mathbb{C})^+$; since $S_6(A_1, \ldots, A_6) \in M_4(\mathbb{C})^-$ and $M_4(\mathbb{C})^+ \cap M_4(\mathbb{C})^- = 0$ we get $S_6(A_1, \ldots, A_6) = 0$.

§3. THE SYMPLECTIC INVOLUTION (s).

In §2 we used the symplectic involution in a roundabout way to derive Kostant's theorem, so one may hope to derive an identity for (s) with the generic minimal polynomial. In fact, one may proceed as follows. Here we deal exclusively with the symplectic involution (s). For $B \in M_n(\mathbb{C})^+$, and $m = n/2$,

$$0 = B^m + \Sigma_{k=1}^m (-1)^k \mu_k B^{m-k} \text{ where inductively } 2k\mu_k = \Sigma_{i=1}^k (-1)^i \mu_{k-i} \mathrm{tr}(B^i)$$

$$\text{with } \mu_0 = 1,$$

so μ_k has the form $(-1)^k \Sigma_u q_u \mathrm{tr}(B^{u_1})\ldots\mathrm{tr}(B^{u_j})$ for $q_u \in Q$ and j-tuples $u = (u_1, \ldots, u_j)$ with $\Sigma_{i=1}^j u_i = k$. Multilinearizing now leaves

$$(2) \quad 0 = \Sigma_\pi B_{\pi 1}\cdots B_{\pi m} + \Sigma_{k,u,\pi} q_u \mathrm{tr}(B_{\pi 1}\cdots B_{\pi u_1})\mathrm{tr}(B_{\pi(u_1+1)}\cdots B_{\pi u_2}) \ldots B_{\pi(m-k+1)}\cdots B_{\pi m}$$

Until now, this case has followed Razmyslov's argument very closely. At this point, we want a substitution which permits us to apply remark 1 without destroying the right-hand side (and hopefully which yields more than the Amitsur-Levitzki theorem). Write $[A_1,B_1]$ in place of B_1, where $A_1 \in M_n(\mathbb{C})^-$ (noting that $[A_1,B_1] \in M_n(\mathbb{C})^+$). There are two useful substitutions for the other B_i:

<u>Case I</u>. For $i \geq 2$, write $B_{4i-6}B_{4i-5}B_{4i-4}B_{4i-3} + B_{4i-3}B_{4i-4}B_{4i-5}B_{4i-6}$ in

place of B_i. These elements are still in $M_n(\mathbb{C})^+$, yielding a new equation from (2) of degree 1 in A_1 and degree $2n-3$ in the B_i. Let us write the right side as $\psi(A_1, B_1, \ldots, B_{2n-3})$ where ψ is a function linear in $(2n-2)$ indeterminates. Note that, in any term, A_1 can appear only in the $(4j+1)$ or $(4j+2)$ position for suitable j, so that we cannot apply remark 1. Nevertheless, if we antisymmetrize only so far as to form

$$\sum_\pi (\mathrm{sg}\ \pi)\ \psi(A_1, B_{\pi 1}, \ldots, B_{\pi(2n-3)})$$

we see that the analogue of remark 4 holds, namely that all of the terms involving "tr" are 0. For example, for $i \equiv 1 \mod 4$,
$\mathrm{tr}(B_{\pi 1}\cdots B_{\pi(i-1)}A_1B_{\pi 1}\cdots B_{\pi(2n-3)})$ may occur in
$\psi(A_1, B_{\pi i}, B_{\pi 1}, B_{\pi 2}, B_{\pi 3}, B_{\pi 4}, \ldots)$, but equals
$\mathrm{tr}(B_{\pi(2n-3)}B_{\pi 1}\ B_{\pi 2}\cdots B_{\pi(i-1)}A_1B_{\pi i}\cdots B_{\pi(2n-4)})$ which appears in
$\psi(A_1, B_{\pi(i-1)}, B_{\pi(2n-3)}, B_{\pi 1}, B_{\pi 2}, B_{\pi 3}, \ldots)$ with opposite sign (since $B_{\pi(i-1)}A_1$ occurs negatively in $[A_1, B_{\pi(i-1)}]$). Thus all terms involving "tr" cancel, and $\sum_\pi (\mathrm{sg}\ \pi)\psi(A_1, B_{\pi 1}, \ldots, B_{\pi(2n-3)})$ has the same value as

$$(3)\quad \sum_\pi (\mathrm{sg}\ \pi) f([A_1, B_{\pi 1}], B_{\pi 2}B_{\pi 3}B_{\pi 4}B_{\pi 5} + B_{\pi 5}B_{\pi 4}B_{\pi 3}B_{\pi 2}, B_{\pi 6}B_{\pi 7}B_{\pi 8}B_{\pi 9} + B_{\pi 9}B_{\pi 8}B_{\pi 7}B_{\pi 6}, \ldots)\ ,$$

where f is the symmetric polynomial on m letters.

Hence (3) vanishes identically for $A_1 \in M_n(\mathbb{C})^-$ and $B_1, \ldots, B_{2n-3} \in M_n(\mathbb{C})^+$.

Case II. For $i \geq 2$, write $A_{4i-6}A_{4i-5}A_{4i-4}A_{4i-3}+A_{4i-3}A_{4i-4}A_{4i-4}A_{4i-5}A_{4i-6}$. Then the analogous argument to case I shows that (4) below vanishes identically:

$$(4)\quad \sum_\pi (\mathrm{sg}\ \pi) f([A_{\pi 1}, B_1], A_{\pi 2}A_{\pi 3}A_{\pi 4}A_{\pi 5} + A_{\pi 5}A_{\pi 4}A_{\pi 3}A_{\pi 2}, A_{\pi 6}A_{\pi 7}\ A_{\pi 8}A_{\pi 9} + A_{\pi 9}A_{\pi 8}A_{\pi 7}A_{\pi 6}, \ldots)$$

To recapitulate, we have

THEOREM 2. *Suppose* n *is even and* f *is the symmetric function* $\sum_\sigma X_{\sigma 1}X_{\sigma 2}\cdots X_{\sigma(n/2)}$, *summed over all permutations* σ *of* $1, \ldots, n/2$. *Then* (3), (4) *vanish identically for all* $A_i \in M_n(\mathbb{C})^-$ *and* $B_i \in M_n(\mathbb{C})^+$ *(with respect to* (s)).

We would rather obtain a standard polynomial, to obtain an analogue to Kostant's theorem. To do this, we see that theorem 2 certainly implies for all B_{2n-2} in $M_n(\mathbb{C})^+$

$$\operatorname{tr}(\sum_{\pi}(\operatorname{sg}\pi) f([A_1,B_{\pi 1}],B_{\pi 2}B_{\pi 3}B_{\pi 4}B_{\pi 5}+B_{\pi 5}B_{\pi 4}B_{\pi 3}B_{\pi 2},B_{\pi 6}B_{\pi 7}B_{\pi 8}B_{\pi 9}$$
$$+B_{\pi 9}B_{\pi 8}B_{\pi 7}B_{\pi 6}, \ldots)\ B_{\pi(2n-2)}) = \operatorname{tr}(0) = 0$$

Using the fact that the trace of a cyclic permutation of a monomial equals the trace of that monomial, we see easily (via remark 1) that $0 = \operatorname{tr} S_{2n-1}(A_1,B_1, \ldots, B_{2n-2})$. By remark 4 we have

$\operatorname{tr}(S_{2n-2}(A_1,B_1, \ldots, B_{2n-3})B_{2n-2}) = 0$ and $\operatorname{tr}(S_{2n-2}(B_1, \ldots, B_{2n-2})A_1) = 0$.

But the A_i,B_i are arbitrary in $M_n(C)^-$ (resp. $M_n(C)^+$), and by remark 3, $S_{2n-2}(A_1,B_1, \ldots, B_{2n-3}) \in M_n(C)^+$ and $S_{2n-2}(B_1, \ldots, B_{2n-2}) \in M_n(C)^-$. Thus by remark 2, $S_{2n-2}(A_1,B_1, \ldots, B_{2n-3}) = 0 = S_{2n-2}(B_1, \ldots, B_{2n-2})$; a quick application of the formula $S_k = \Sigma_{i=1}^{k} (-1)^{i-1}X_kS_{k-1}(X_1, \ldots,X_{i-1},X_{i+1},\ldots,X_k)$ yields

<u>THEOREM 3.</u> $L(n,2n-2,2n-2,(s))$, $L(n,2n-2,2n-3,(s))$, $L(n,2n-1,2n-1,(s))$ <u>and</u> $L(n,2n-1,2n-2,(s))$ <u>all hold (for</u> n <u>even)</u>.

Note these are precisely the analogues to the results of [5]. We now test the sharpness of theorem 3.

<u>Example 1</u>. Suppose $n = 2$. Then all (s)-symmetric matrices are scalar, so theorem 3 is immediate (and is clearly sharp).

<u>Example 2</u>. Suppose $n = 4$. Then, letting e_{ij} denote the usual matric units, we see the (s)-symmetric matrices have the following base: $B_1 = e_{12}+ e_{43}$, $B_2 = e_{21}+ e_{34}$, $B_3 = e_{11}+ e_{33}$, $B_4 = e_{22}+ e_{44}$, $B_5 = e_{14}- e_{23}$, and $B_6 = e_{41}-e_{32}$. The coefficient of e_{14} in $S_5(B_1, \ldots, B_5)$ has the following contributions:

Positive: $B_1B_4B_2B_3B_5$, $B_5B_4B_1B_3B_2$, $-B_1B_4B_5B_3B_2$

Negative: $-B_1B_2B_3B_5B_4$, $-B_3B_1B_2B_5B_4$, $-B_3B_1B_4B_2B_5$, $-B_5B_1B_3B_2B_4$, $-B_3B_5B_1B_2B_4$, $-B_3B_5B_4B_1B_2$, $B_3B_1B_4B_5B_2$, $B_3B_1B_5B_2B_4$, $B_1B_5B_3B_2B_4$.

Thus $e_{11}S_5(B1, \ldots, B_5)e_{14} = -6e_{14}$, proving $S_5(B_1, \ldots, B_5) \neq 0$. Similarly, $S_5(B_1,B_2,B_3,B_4,e_{14} + e_{32}) \neq 0$. In view of example 2, we see that the seemingly inefficient substitution used to prove theorem 2 (4 matrices for all but one of the original symmetric matrices) in fact captures as much as possible vis-a-vis the standard polynomial.

REFERENCES

[1] Hutchinson, J., Eulerian graphs and polynomial identities for skew-symmetric marrices, Canad. Math. Bull. 13(1975), 590-609

[2] Kostant, B., A theorem of Frobenius, a theorem of Amitsur-Levitzki, and cohomology theory, Indiana J. Math. (and Mech.) 7(1958), 237-264.

[3] Procesi, C., The invariant theory of $n \times n$ matrices, Advances in Math. 19 (1976), 306-381.

[4] Razmyslov, Yu. P., Trace identities of full matrix algebras over a field of characteristic zero, Math. USSR-Izv. 8(1974), 727-760.

[5] Rowen, L. H., Standard polynomials in matrix algebras, Trans. Amer. Math. Soc. 19(1974), 253-284.

[6] __________, Polynomial Identities in Ring Theory, Academic Press, New York, 1980.

Department of Mathematics
Bar Ilan University
Ramat gan, ISRAEL

Contemporary Mathematics
Volume 13, 1982

AFFINE RINGS SATISFYING A POLYNOMIAL IDENTITY

by
William F. Schelter

By an affine pi ring we mean a ring, finitely generated as an algebra, over an algebraically closed field k. In addition, we assume it satisfies all the identities of $M_n(k)$, and so is a homomorphic image of the ring of generic matrices ([2], p. 63).

Simple homomorphic images of affine rings are just isomorphic to $M_q(k)$, for some q between 1 and n. Spec R denotes the set of maximal two-sided ideals.

$$\text{Spec } R = \text{Spec}_1 R \cup \ldots \cup \text{Spec}_n R$$

where $\text{Spec}_q R$ denotes the maximal ideals with factor ring equal to $M_q(k)$.

The topology of Spec R is the Jacobson-Zariski topology. Thus a subset is closed iff it consists of all maximal ideals containing the intersection of the subset. Spec R is noetherian, since we have DCC on closed subsets, or alternately ACC on semiprime ideals ([2] p. 106). Because of the Nullstellensatz, we are able to restrict attention to maximal ideals, since a general prime ideal just corresponds to the irreducible closed set of maximal ideals containing it.

The dimension of R, or of Spec R, is the length of the longest chain of closed irreducible subsets, not counting the bottom one. Thus a finite set of points has dimension 0. Spec R being 0-dimensional is equivalent to R being artinian ([2], p. 122). If R is prime and of dimension 1, then R is a finite module over its centre and hence is noetherian. A closed irreducible subset of Spec R of dimension 1 is called a curve. We can alternately think of a curve as being given by a map

$$\text{Spec } A \longrightarrow \text{Spec } R$$

0271-4132/82/0529/$02.25

which is induced by a central homomorphism from R to A, where A is a prime order over a Dedekind domain D. This is more general in that the closure of the image will be 1-dimensional irreducible; the image itself may have a few points left out. In the following $\overline{x}$ denotes x^{-1}.

EXAMPLE 1:

$$k\{X\ ,\ Y\} \longrightarrow k\left\{\begin{pmatrix} 1 & \\ & -1 \end{pmatrix},\ \begin{pmatrix} & 1 \\ v & \end{pmatrix},\ \begin{pmatrix} \overline{v-1} & \\ & \overline{v-1} \end{pmatrix}\right\}$$

is a curve in the spectrum of the generic matrix ring. Schematically it looks like

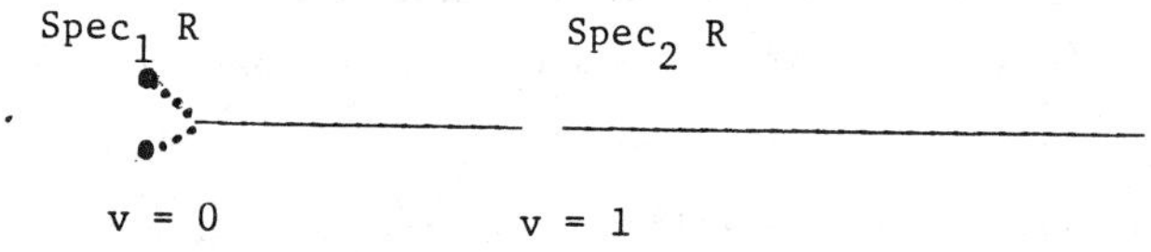

The following theorem is joint work with M. Artin.

THEOREM 1: A subset Y of Spec R is closed iff

a) $Y \cap Spec_i$ R is closed for each i; and

b) for all curves $C \xrightarrow{f} Spec\ R$, $f^{-1}(Y)$ is closed.

For a proof of this see [1].

As an example let us look at the ring of two 2×2 generic matrices. $Spec_1$ is just k^2. $Spec_2$ is k^5 minus the locus determined by $\det(XY-YX) = 0$. The plane k^2 fits together with $Spec_2$, so that most curves in $Spec_2$ hit $Spec_1$ in two points. We saw this happen in Example 1. Even though one can easily construct (see Example 2) a commutative ring with a closed subset equal to k^2 and whose complement is just $Spec_2$, the topology is different, on the commutative ring, from the natural one on the generic matrix ring.

EXAMPLE 2: Let C be the commutative affine domain generated by

$$trX,\ trY,\ tr(XY),\ detX,\ detY.$$

It is the trace ring of the generic matrix ring in X and Y. The five elements are algebraically independent and every trace polynomial can be expressed in terms of them. Let $f = \det(XY-YX)$. Let $B = k[trX, trY] + fC$.

$$\begin{array}{ccc} \text{Spec } B = & \text{Spec } C[f-1] \cup & \text{Spec } k[\text{tr}X, \text{tr}Y] \\ & \wr\wr & \wr\wr \\ & \text{Spec}_2 R & \text{Spec}_1 R \end{array}$$

Where R is the ring generated by X and Y. But the image of the curve in Example 1 in $\text{Spec}_2 R$ has two extra points in its closure in $\text{Spec}_1 R$ whereas in the corresponding part of Spec B there would only be one.

DEFINITION: A variety Y is complete if the map from Y to Spec k is proper.

The reader should refer to [1] for the basic facts concerning proper maps. The idea is simply that curves in Y can be extended.

We shall construct a complete variety Y containing Spec of the generic matrix ring in two variables in pideg 2.

Let M_i be a set of m elements of a domain R. Let R_j be the ring generated by the terms $M_i\overline{M_j}$, where $\overline{}$ denotes inverse. We thus obtain m rings.

LEMMA 1: *The rings R_i and R_j taken together may be generated by R_j and $(M_i\overline{M}_j)^{-1}$.*

LEMMA 2: *If we started with the set $L_i = M_i\overline{M}_j$ instead of the M_j, the set of rings constructed would be the same.*

The proofs of the above are straightforward.

If we take for the M_i, the elements $1, X, Y, X + Y, X + 1$, and $Y + 1$ of the ring of two 2×2 generic matrices, then the six rings above are generated by

R_0:	X, Y	R_3:	$X\overline{X+Y}$, $\overline{X+Y}$
R_1:	$Y\overline{X}$, $\overline{X}$	R_4:	$X\overline{Y+1}$, $Y\overline{X+1}$
R_2:	$X\overline{Y}$, $\overline{Y}$	R_5:	$X\overline{Y+1}$, $Y\overline{Y+1}$

Thus each of the R_i can be generated by only 2 elements instead of 6, and so will be isomorphic to the ring of generic matrices. By Lemma 1, the maps from R_i to R_{ij} are localisations and so geometric. We identify Spec R_i and Spec R_j on Spec R_{ij} to form a scheme Y.

THEOREM 2: *The variety* Y *constructed above is complete, that is proper over* k.

PROOF: To check properness over k, we suppose we are given a prime order A over a Dedekind domain D, t an element of D, and a map $\operatorname{Spec} A[\bar{t}] \longrightarrow Y$. Without loss of generality the image meets $\operatorname{Spec} R_0$ non-trivially, and t is not constant. Thus we have a map of rings $R_0 \longrightarrow A[\bar{t}]$.

If A is commutative, then we are in the pideg 1 part of Y, but Y_1 is just isomorphic to P_k^2. Hence we can find one of the other R_i mapping to A. Thus we have found something in $\operatorname{Spec} R_i$ to map $t = 0$ to.

If pideg A is 2, extend the map to $R_0[T]$, that is to R_0 with all the coefficients of the characteristic polynomial adjoined. The coefficients are generated by

$$\det X,\ \det Y,\ \det(X+Y),\ \det(X+1),\ \det(Y+1). \qquad (1)$$

If, when these are mapped to $D[\bar{t}]$, none of the above has a pole in t, the image of R_0 is actually in A. This is because the image will actually be integral over D.

Now suppose one has a pole, say $u = \det(X+1)$, and suppose it has maximal order among the poles of the images of the five. Then we claim R_4 is mapped to $A[\bar{u}]$. The reason for this is that the trace ring of R_4 is generated by elements which will map to elements where the pole has been cleared: $\det(X)\det(\bar{u}), \ldots \det(X+Y+1)\det(\bar{u})$, (set Table I for lists of generators). Since $\det X, \ldots, \det(X+Y+1)$ are linear combinations of the terms in (1), the poles are cleared. The image will be integral over $D[\bar{u}]$ and the inverse of t is not in this ring. Thus t generates a proper ideal in the image of R_4 with D adjoined, and so Y is proper over k. Q.E.D.

As we remarked earlier $Y_1 = P_k^2$. We claim $Y_2 = P_k^5$ minus the closure of $(XY - YX)^2 = 0$. Indeed if we look at the generators for the trace rings of the six rings, we obtain (where dX stands for detX in this table)

Table I:

R_0:	dX, dY, $d(X+Y)$, $d(X+1)$, $d(Y+1)$
R_1:	$d\bar{X}dY$, $d\bar{X}$, $d(Y+1)d\bar{X}$, $d(Y+X)d\bar{X}$, $d(X+1)d\bar{X}$
R_2:	$dXd\bar{Y}$, $d\bar{Y}$, $d(X+1)d\bar{Y}$, $d(X+Y)d\bar{Y}$, $d(1+Y)d\bar{Y}$
R_3:	$dXd\overline{X+Y}$, $d\overline{X+Y}$, $d(X+1)d\overline{X+Y}$, $d(2X+Y)d\overline{X+Y}$, $d(X+Y+1)d\overline{X+Y}$
R_4:	$dXd\overline{X+1}$, $dYd\overline{X+1}$, $d(X+Y)d\overline{X+1}$, $d(2X+1)d\overline{X+1}$, $d(X+Y+1)\overline{dX+1}$
R_5:	$dXd\overline{Y+1}$, $dYd\overline{Y+1}$, $d(X+Y)d\overline{Y+1}$, $d(X+Y+1)d\overline{Y+1}$, $d(2Y+1)d\overline{Y+1}$

Since $d(X+Y+1) = 1 - tr(X+Y) + d(X+Y)$, it is expressible as a linear combination of the generators of R_0. Since $d(aX+Y) = a^2 dX + a(d(X+Y) - dX - dY) + dY$ we see the same is true for $d(2X+Y)$. ETC. Thus the six trace rings whose generators are given above in Table I, form the standard affine open covering of P_k^5.

We verify that the closed sets of $\text{Spec}_2\ R_0$ and $\text{Spec}_2\ R_1$ defined by $\det(XY-YX)$ and $\det(Y\overline{XX}-\bar{X}Y\bar{X})$ coincide on Spec R_{0_1}, the other verifications being similar. Simply observe:

$$\det(Y\overline{XX}-\bar{X}Y\bar{X}) = \det\bar{X}\det(XY-YX)\det\bar{X}^2 .$$

Question: One can construct a Y as in Lemma 1 and subsequent remarks, by using a set of small monomials (small relative to the pideg) of an arbitrary generic matrix ring. For a suitably chosen set, is Y complete? If not, can one find some other complete scheme containing Spec of the generic matrix ring?

BIBLIOGRAPHY

[1] M. Artin and W. Schelter, Integral Ring Homomorphisms, Advances in Mathematics, vol. 39 (1981), 289-329.

[2] C. Procesi, Rings with Polynomial Identities, Marce Dekker, New York, 1973.

Department of Mathematics
University of Texas at Austin
Austin, Texas 78712

Contemporary Mathematics
Volume 13, 1982

ALGEBRAIC STRUCTURE OF POWER SERIES RINGS

by M. Artin

This talk describes some results and open problems about regular local rings. Most of the results were obtained in joint work with Popescu. For simplicity, we restrict ourselves to (commutative) algebras over a ground field of characteristic zero.

Our point of view is that the algebraic structure of algebra consists of the set of algebraic relations among its elements. Therefore its structure is determined by the inductive system of finitely generated sub-algebras of A, or more precisely, by the ind-object represented by this inductive system. Many properties of A can be described directly on subalgebras. For example, A is a normal domain if and only if there is a cofinal subsystem of $\{S_i\}$ of normal domains. Here we consider how smoothness of the algebras forming an inductive system is related to the limit algebra. (A domain C of Krull dimension d is _smooth_ if it can be presented in the form $C \approx k[y_1,\ldots,y_n]/(f_1,\ldots,f_r)$ in such a way that the Jacobian matrix $\partial f_i/\partial y_j$ has rank $n - d$ on Spec C.) It is not too difficult to prove

THEOREM 1: _Let $A = \varinjlim C_i$ where $\{C_i\}$ is an inductive system of smooth k-algebras. If A is noetherian, then A is a regular ring._

This being so, the main interest is in establishing the converse:

Conjecture 1: Every excellent regular local k-algebra A is a direct limit of smooth algebras.

Our first result says that the conjecture is true for the ring $k[[x]]$ of formal series in $x = x_1,\ldots,x_m$.

THEOREM 2: _The power series ring $k[[x]]$ is the direct limit of smooth k-algebras:_

$$k[[x]] \approx \varinjlim C_i,$$

where the C_i are smooth, finitely generated integral domains over k.

The same result holds if $k = \mathbb{C}$ and if $\mathbb{C}[[x]]$ is replaced by the ring of convergent series $\mathbb{C}\{x\}$ in $x_1,\dots,x_m$.

The intuitive meaning of theorem 2 is that there are "essentially" no non-trivial relations among formal series. It is slightly easier to grasp in the following analytic analogue. In this analogue, all series are assumed to be with constant term zero.

THEOREM 3: *Let $\bar{y}_i \in \mathbb{C}[[x]]$, $i = 1,\dots,n$, be a formal power series, and let $f(Y) = f_1(Y),\dots,f_r(Y)$ be convergent series in $Y = Y_1,\dots,Y_n$ which are analytic relations among the $\bar{y}_i$, i.e., $f(\bar{y}) = 0$. There exist formal series $\bar{z} = \bar{z}_1,\dots,\bar{z}_N \in \mathbb{C}[[x]]$, and convergent series $g(Z) = g_1(Z),\dots,g_n(Z)$ in $Z_1,\dots,Z_N$, such that*

(i) $\bar{y} = g(\bar{z})$, *and*

(ii) $f(g(Z)) = 0$ *identically.*

In other words, the relations $f(\bar{y}) = 0$ become trivial when the set $\bar{y}$ is replaced by $\bar{z}$. Equivalently, the formal power series ring $\mathbb{C}[[x]]$ is a direct limit of convergent series rings $\mathbb{C}\{Z\}$. Theorem 3 is a restatement of a result of Płoski [7], and theorem 2 was proved independently by us (see also [6]). Both proofs are analyses of the arguments used in [1,2].

Two open problems arise naturally when one tries to refine these results. First of all, theorem 2 does not make any assertion of injectivity of the maps $C_i \to k[[x]]$.

Problem 1: Is $k[[x]]$ the direct limit of smooth subalgebras? Or, in theorem 3, can one choose $\bar{z}_1,\dots,\bar{z}_N$ to be analytically independent?

The second problem concerns the nature of the resolution of singularities implied by theorem 2. Let A denote either the ground field k or the polynomial ring $k[x]$, and let $\bar{A} = k[[x]]$, where as always $x = x_1,\dots,x_m$. Let B be any finitely generated A-algebra, say $B = A[y]/(f(y))$, and suppose given a map $B \to \bar{A}$ of A-algebras, or a solution $\bar{y} \in \bar{A}$ of the system of equations $f(y) = 0$. By theorem 2, it factors through some smooth A-algebra C. The data are summed up in the diagrams below, in which the one on the right represents the spectra of the rings on the left:

(*) 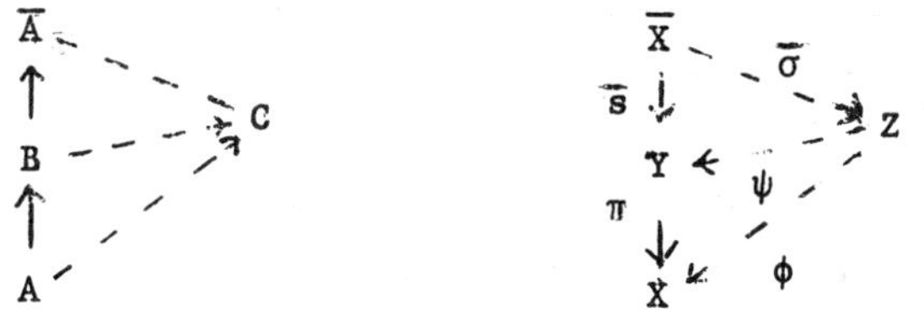

Theorem 1 tells us that a diagram of dotted arrows exists with φ smooth, but one can also put conditions on the map ψ.

<u>Problem 2</u>: Given $\bar{s}$, π, does there exist a diagram (*) such that

(i) φ is smooth

(ii) Let $\bar{p} \in \bar{X}$. If π is smooth at $\bar{s}(\bar{p})$, then ψ is smooth at $\bar{\sigma}(\bar{p})$.

This problem can be posed for any pair A, $\bar{A}$ of rings such that the map $A \to \bar{A}$ is regular (which means that the map is flat, and its fibres are regular and remain so after finite field extension). It seems plausible that these conditions can be satisfied, for most such maps.

<u>Conjecture 2</u>: Let $A \to B \to \bar{A}$ be ring homomorphisms such that B is finitely generated over A and that the map $A \to \bar{A}$ is regular. Suppose that A, $\bar{A}$ are excellent local rings. Then a diagram (*) exists which satisfies conditions (i,ii) of problem 2.

A proof of this conjecture would solve conjecture 1 and most of the open problems about approximation in henselian local rings [2,3,5,6]. At this time, the only case that is completely solved is that $\bar{A}$ is a Dedekind domain. Problem 1 has an affirmative answer, and the conjecture is true in that case, because of Néron's p-desingularization theorem [2,6]. In fact, a diagram (*) exists so that in condition (ii) of problem 2, ψ is <u>étale</u> at $\bar{\sigma}(\bar{p})$. Conjecture 2 can be viewed as a possible extension of Néron's result to higher dimension. In trying to generalize his results, one has to take into account the following complication, which necessitates the change from étale to smooth in condition (ii): Let $\Delta \subset Y$ be the singular locus of the map π, and let $\bar{U} = \bar{X} - s^{-1}(\Delta)$. The normal bundle $\bar{N}$ to $\bar{U}$ in $\bar{X} \times_X Y$ is a vector bundle which must be trivial if $\bar{X}$ is local and ψ is étale on $\bar{U}$. When $\bar{N}$ is not trivial, extra parameters are needed in Z.

There is some interest even in the special case $A = \bar{A}$, and we can prove that case:

THEOREM 3: *Conjecture* 2 *is true if* A *is a normal noetherian domain, not necessarily local, and* $\bar{A} = A$.

An analytic version states that the solutions of a system of equations can be parametrized: Let $\{f_\nu(x,y)\}$ be convergent series in $x = x_1,\ldots,x_m$; $y = y_1,\ldots,y_n$ such that $f(x,0) = 0$. Let $\delta \subset \mathbb{C}\{x\}$ be an ideal defining the singular locus of the system of equations $f(x,y) = 0$ along the solution $y = 0$.

THEOREM 4: *There is an integer* N, *a set of variables* $z = z_1,\ldots,z_r$, *and a convergent series solution* $y = Y(x,z)$ *of the system of equations* $f(x,y) = 0$ *with* $Y(x,0) = 0$, *so that every solution* $y = y(x)$ *with* $y(x) \equiv 0$ (modulo δ^N) *is obtained by substitution*: $y(x) = Y(x,z(x))$ *for some convergent series* $z(x)$.

Theorem 4 is a fairly elementary fact, and related results are known, but so far as we know it has not appeared in this form before.

Finally, we have the following result which extends a theorem of Pfister [4,6,8]:

THEOREM 5: *Conjecture* 1, 2 *are true if* $\bar{A}$ *is a regular local ring of dimension* 2.

REFERENCES

[1] A. Artin, On the solutions of analytic equations, Invent. Math. 5 (1968) 277-291.

[2] M. Artin, Algebraic approximation of structures over complete local rings, Pub. Math. Inst. Hautes Études Sci. 36 (1969), 23-58.

[3] J. Becker, J. Denef, L. Lipschitz and L. van den Dries, Ultraproducts and approximation in local rings I, Invent. Math. 51 (1979), 183-203.

[4] M. Brown, Artin's approximation property, (to appear).

[5] R. Elkik, Solutions d'équations à coefficients dans un anneau hensélien, Ann. Sci. École Normale Sup. 4^{e}sér. 6 (1973), 553-601.

[6] H. Kurke, T. Mostowski, G. Pfister, D. Popescu, and M. Roczen, Die Approximationseigenschaft lokaler Ringe, Lec. Notes in Math. 634, Springer Verlag Berlin 1978.

[7] A Płoski, Note on a theorem of M. Artin, Bull. Acad. Polonaise Sci., ser. Math. 22 (1974), 1107-1109.

[8] D. Popescu, A remark on two-dimensional local rings with the property of approximation, Math. Zeitschrift 173 (1980), 235-240.

[9] D. Popescu, Global form of Néron's p-desingularization and approximation, (to appear).

Massachusetts Institute of Technology
Cambridge, MA 02139

Contemporary Mathematics
Volume 13, 1982

LINKAGE AND THE SYMMETRIC ALGEBRA OF IDEALS

Craig Huneke*

Dedicated to Nathan Jacobson

INTRODUCTION.

Let R be a Noetherian domain and let M be a finitely generated torsion-free R-module. When is the symmetric algebra of M, Sym(M), a domain? This question dates back to papers of Samuel [14] and Micali [9], and has been studied by many researchers since the appearance of these papers, particularly in the case when M is an ideal of R. ([1], [2], [4], [5], [6], [15], [16].)

In this paper we study this question for height two unmixed ideals I in Cohen-Macaulay local rings R. Recall an ideal I is said to be unmixed if every associated prime ideal of I has the same height. Our analysis of this question considers the equivalence relation of linkage, defined by Peskine and Szpiro [12]. They define linkage only for regular local rings; we define linkage for Cohen-Macaulay local rings.

Definition 0.1. Let R be a Cohen-Macaulay local ring. Two unmixed ideals I and J of the same height are said to be linked (written $I \sim J$) if there is an ideal $A \subseteq I \cap J$ such that A is generated by a regular sequence in R, ht A = ht I = ht J, and

$$I/A = \mathrm{Hom}_R(R/J, R/A)$$
$$J/A = \mathrm{Hom}_R(R/I, R/A).$$

This definition is what Peskine and Szpiro call algebraic linkage; however, in this paper we simply refer to this as linkage. If I and J are unmixed ideals, we say that J is in the linkage class of I if there exist unmixed ideals $K_1,\ldots,K_n$ such that $I \sim K_1 \sim \ldots \sim K_n \sim J$. In [12], Peskine and Szpiro prove the following theorem.

THEOREM. *Let R be a regular local ring. If I is a height two ideal such that R/I is Cohen-Macaulay, then I is in the linkage class of a complete intersection.*

If I is generated by a regular sequence and R is a domain, it has long been known that Sym(I) is also a domain. This suggests an intriguing

*Partially supported by the National Science Foundation.

question: can the property of Sym( ) being a domain be passed along a linkage class? If Sym(I) is a domain, then one must take into account some necessary conditions on I.

If M is an R-module and R is local, let $v(M)$ denote the least number of generators of M. If Sym(I) is a domain, then I must satisfy

$$v(I_P) \leq \text{ht } P \qquad (\#)$$

for all non-zero prime ideals P. In [6] it was shown that (#) was also a sufficient condition for certain ideals.

THEOREM [6]. *Let R be a regular local ring. If I is a height two ideal such that R/I is Cohen-Macaulay, then Sym(I) is a domain if and only if I satisfies (#).*

These two results suggest the following question. Suppose R is a regular local ring (or R is a Cohen-Macaulay local ring) and I and J are two unmixed ideals such that $I \sim J$. If Sym(I) is a domain, and J satisfies condition (#), is Sym(J) a domain? If R is not regular, it is easy to produce examples where the answer is no. If R is regular, this author knows no counterexample to the above question, however, it is probably not true. It is known if R is a regular local ring and I is an ideal such that R/I is Gorenstein, then if $J \sim I$, Sym(J) is a domain [5]. However it is very unlikely that Sym(I) is a domain for every Gorenstein ideal I in R which satisfies (#). This author does not know if Sym(I) is a domain in the case where $R = K[x_{ij}]_{1 \leq i \leq n}^{1 \leq j \leq n}$ with K a field, and $I = I_{n-1}(X)$, the ideal generated by the submaximal minors of the matrix $X = (x_{ij})$. In this case I is a height four Gorenstein ideal. This author will show in a later paper that if I is a height three Gorenstein ideal which satisfies (#) in a regular local ring R, then Sym(I) is a domain.

Although it appears that linkage is too weak a bond to study the symmetric algebra, much can be said if one instead considers unmixed ideals I and J which are *doubly linked*, that is there is an unmixed ideal K such that $I \sim K \sim J$. The main purpose of this paper is to obtain the following result.

THEOREM 2.1. *Let R be a Cohen-Macaulay local domain and let I and J be two unmixed height two ideals such that $I \sim K \sim J$. Suppose Sym(I) is a Cohen-Macaulay domain. Then Sym(J) is a Cohen-Macaulay domain if and only if J satisfies (#).*

Suppose there exists a sequence of unmixed height two ideals $K_1, \ldots, K_n$ such that $I \sim K_1 \sim \ldots \sim K_n \sim J$ and n is odd. We refer to all such J as the <u>even linkage class</u> of I. Then if $K_2,\ldots,K_{n-1}$, and J satisfy (#) and $\mathrm{Sym}(I)$ is a Cohen-Macaulay domain, then $\mathrm{Sym}(J)$ are Cohen-Macaulay domains by repeated application of Theorem 2.1.

The proof of Theorem 2.1 involves considering exact sequences

$$0 \to F \to M \to I \to 0$$

where F is a free R-module. We compare the properties "$\mathrm{Sym}(I)$ is a domain" and "$\mathrm{Sym}(M)$ is a domain". This analysis is in Section 1. All rings considered will be commutative Noetherian with identity.

§1.

Before we begin the proof of the main results we prove a few simple remarks concerning the symmetric algebra of a module. Throughout this discussion we fix a Noetherian domain R and a finitely generated torsion-free R-module M. Since M is torsion-free there is an embedding of M into a free module F of rank $d = \operatorname{rank} M$. This embedding induces a map $f\colon \mathrm{Sym}(M) \to \mathrm{Sym}(F)$.

<u>LEMMA 1.1</u>. (see [9]). <u>Let</u> R,M,F,f <u>be as above</u>. <u>Then the following are equivalent</u>.

1) $\mathrm{Sym}(M)$ <u>is torsion-free over</u> R.
2) <u>The map</u> f <u>is injective</u>.
3) $\mathrm{Sym}(M)$ <u>is a domain</u>.

<u>Proof</u>. Clearly 2) implies 3) implies 1). Suppose 1) and set K = fraction field of R. Then $\mathrm{Sym}(M)$ embeds in $\mathrm{Sym}(M) \otimes_R K$. Consequently there is a commutative diagram,

$$\begin{array}{ccccc} 0 \to & \mathrm{Sym}(M) & \to & \mathrm{Sym}(M) \otimes_R K \\ & \downarrow f & & \downarrow f \otimes 1 \\ & \mathrm{Sym}(F) & \to & \mathrm{Sym}(F) \otimes_R K \end{array}.$$

If we show $f \otimes 1$ is injective then clearly f must also be injective. However, $\mathrm{Sym}(M) \otimes_R K = \mathrm{Sym}(M \otimes_R K) = \mathrm{Sym}(F \otimes_R K) = \mathrm{Sym}(F) \otimes_R K$, and so $f \otimes 1$ is an isomorphism.

<u>LEMMA 1.2</u>. <u>Let</u> (R,P) <u>be a universally catenarian local domain, and</u> M <u>a finitely generated torsion-free</u> R-<u>module</u>. <u>Let</u> J = <u>the unique maximal ideal of</u> $\mathrm{Sym}(M)$ <u>containing</u> P <u>and all elements of positive degree</u>. <u>If</u> $\mathrm{Sym}(M)$ <u>is a domain, then</u> $\dim \mathrm{Sym}(M)_J = \dim R + \operatorname{rank} M$.

Proof. Set $I = \mathrm{Sym}_+(M)$ = ideal of elements of positive degree. Then $\mathrm{Sym}(M)/I \simeq R$, and I is a prime ideal. R is universally catenarian, and thus $\mathrm{Sym}(M)$ is catenarian. As $\mathrm{Sym}(M)$ is by assumption a domain,

$$\dim \mathrm{Sym}(M)_J = \dim(\mathrm{Sym}(M)/I)_J + \mathrm{ht}\, I = \dim R + \mathrm{ht}\, I.$$

It remains to prove $\mathrm{ht}\, I = \mathrm{rank}\, M$. However since $I \cap R = (0)$, it is enough to compute $\mathrm{ht}\, I \otimes_R K$, where K is the fraction field of R. However, $\mathrm{Sym}(M) \otimes_R K = \mathrm{Sym}(M \otimes_R K) \simeq K[T_1,\ldots,T_d]$ where $d = \mathrm{rank}\, M$. It is clear $d = \mathrm{ht}\, J$.

THEOREM 1.1. *Let (R,P) be a Cohen-Macaulay local domain and let I be an unmixed ideal of height two. Suppose there is an exact sequence,*

$$0 \to R^d \to M \to I \to 0. \tag{1}$$

Then the following hold:

(1) *If* $\mathrm{Sym}(I)$ *is a domain, then* $\mathrm{Sym}(M)$ *is a domain.*

(2) *If* $\mathrm{Sym}(I)$ *is a Cohen-Macaulay domain, then* $\mathrm{Sym}(M)$ *is a Cohen-Macaulay domain.*

(3) *If* $\mathrm{Sym}(M)$ *is a Cohen-Macaulay domain, then the following statements are equivalent:*

(a) $\mathrm{Sym}(I)$ *is a domain.*

(b) $\mathrm{Sym}(I)$ *is a Cohen-Macaulay domain.*

(c) *For every non-zero prime ideal* Q, $v(I_Q) \le \mathrm{ht}\, Q$.

Proof. (1) It is clear that $\mathrm{Sym}(I) \simeq \mathrm{Sym}(M)/(L_1,\ldots,L_d)$ where the L_i are linear forms in $\mathrm{Sym}(M)$ corresponding to the images of basis elements under the map $R^d \to M$. By assumption $(L_1,\ldots,L_d)$ is a prime ideal Q. We claim $\mathrm{ht}\, Q = d$. Since $Q \cap R = (0)$, it is enough to show $\mathrm{ht}\, Q \otimes_R K = d$, where as always, K is the fraction field of R. However, after we tensor with K, the map

$$0 \to R^d \to M$$

becomes

$$0 \to K^d \to K^{d+1}, \quad \text{and}$$

the induced map $K^d \to \mathrm{Sym}(K^{d+1}) = K[T_1,\ldots,T_{d+1}]$ has an image of height d. Hence $\mathrm{ht}\, Q \otimes_R K = d$. Thus Q is a prime ideal of height d generated by d elements. We use a lemma due to Davis:

LEMMA 1.3. *If R is a local ring and P is a prime ideal of height d generated by d elements $a_1,\ldots,a_d$, then*

(1) R <u>is a domain, and</u>

(2) $a_1,\dots,a_d$ <u>is an</u> R-<u>sequence</u>.

Let J be the ideal $P + \mathrm{Sym}_+(M)$. To show Sym(M) is a domain it is enough to prove $(\mathrm{Sym}(M))_J$ is a domain. Clearly $Q \subseteq P$. We apply Lemma 1.3 to conclude Sym(M) is a domain.

To prove (2) it is enough to show $(\mathrm{Sym}(M))_J$ is Cohen-Macaulay [8]. By assumption $(\mathrm{Sym}(M)/(L_1,\dots,L_d))_J = (\mathrm{Sym}\ (I))_J$ is Cohen-Macaulay. By Lemma 1.3, $L_1,\dots,L_d$ are an $\mathrm{Sym}(M)_J$-sequence. Thus $\mathrm{Sym}(M)_J$ is Cohen-Macaulay and consequently Sym(M) is also Cohen-Macaulay.

We now prove (3). Clearly (3b) implies (3a). Conversely, if Sym(I) is a domain then the proof of part (1) shows $L_1,\dots,L_d$ are a $\mathrm{Sym}(M)_J$-sequence. Since Sym(M) is Cohen-Macaulay by assumption, we conclude Sym(I) is Cohen-Macaulay. Hence (3a) implies (3b).

Assume (3a). We will prove (3c). We may suppose R is local with maximal ideal Q and Sym(I) is a domain. We need to prove $v(I) \le \dim R$. Clearly $\mathrm{Sym}(I)/Q\ \mathrm{Sym}(I) \simeq (R/Q)[T_1,\dots,T_n]$, where $n = v(I)$. Hence

$$\begin{aligned}\dim \mathrm{Sym}(I)_J &= \dim(\mathrm{Sym}(I)/Q\ \mathrm{Sym}(I))_J + \mathrm{ht}\ Q\ \mathrm{Sym}(I)_J \\ &\ge v(I) + 1.\end{aligned}$$

By Lemma 1.2, $\dim \mathrm{Sym}(I)_J = \dim R + 1$. Combining these inequalities gives the desired result.

Now assume (3c) and we will prove (3a). Let A be the kernel of the composite maps, $\mathrm{Sym}(M) \to \mathrm{Sym}(I) \to R[t] = \mathrm{Sym}(R)$, where the last map is induced by the inclusion of I into R. We will show $A = (L_1,\dots,L_d)$ by induction upon dim R. Set $N = (L_1,\dots,L_d)$. If $\dim R = 0$, R is a field and it is clear $A = N$. We may therefore assume (R,P) is a local ring of dimension > 1 and $A_Q = N_Q$ for all prime ideals Q of R which are not maximal. We may also assume $v(M) = v(I)$. Together these statements imply there is an $m >> 0$ such that $P^m A \subseteq N$.

It is a simple exercise to compute $\dim \mathrm{Sym}(M)/A = \dim R + 1$. By Lemma 1.2, $\dim \mathrm{Sym}(M)_J = \dim R + \mathrm{rank}\ M$. Hence $\mathrm{ht}\ A = \mathrm{rank}\ M - 1 = d$. On the other hand, $\dim \mathrm{Sym}(M)/P = v(M) = v(I)$, so that $\mathrm{ht}\ P\ \mathrm{Sym}(M) = \dim R + \mathrm{rank}\ M - v(I) = \mathrm{rank}\ M$ by assumption (3c). Since $P^m A \subseteq N$, either $\mathrm{ht}\ N = \mathrm{ht}\ A$ or $\mathrm{ht}\ N = \mathrm{ht}\ P\ \mathrm{Sym}(M)$. Since N is generated by d elements, the latter statement cannot be true, and so $\mathrm{ht}\ N = \mathrm{ht}\ A = d$. Since Sym(M) is assumed to be Cohen-Macaulay, N is generated by a regular sequence; in particular P is not associated to N. This last statement forces $N = A$.

§2

We will now apply Theorem 1.1 to the linkage classes of unmixed height 2 ideals. Our starting point is a Proposition found in Miller's thesis [11], which in turn was inspired by the paper of Rao [13].

PROPOSITION 2.1. [11]. *Let R be a Cohen-Macaulay local ring and K an unmixed ideal of height two with presentation*

$$0 \to M \to R^d \to K \to 0.$$

If I is linked to K then there is an exact sequence,

$$0 \to R^d \to M^* \oplus R^2 \to I \to 0,$$

where $M^* = \mathrm{Hom}(M,R)$.

THEOREM 2.1. *Let R be a Cohen-Macaulay local domain and let I and J be unmixed height two ideals such that* $I \sim K \sim J$. *Suppose* Sym(I) *is a Cohen-Macaulay domain. Then* Sym(J) *is a Cohen-Macaulay domain if and only if* $v(J_P) \leq \mathrm{ht}\ P$ *for all non-zero prime ideals* P.

Proof. By *Proposition 2.1*, there exist exact sequences

$$0 \to R^{d+2} \to N \to I \to 0$$
$$0 \to R^{d+2} \to N \to J \to 0$$

where $N = M^* \oplus R^2$. We are assuming Sym(I) is a Cohen-Macaulay domain; by Theorem 1.1.2, Sym(N) is a Cohen-Macaulay domain. By Theorem 1.1.3, Sym(J) is then a Cohen-Macaulay domain if and only if

$$v(J_P) \leq \mathrm{ht}\ P$$

for all non-zero prime ideals P.

COROLLARY 2.1. [6]. *Suppose R is a Cohen-Macaulay local domain and I is a height two ideal with* pd I = 1. *In addition, assume I is generically a complete intersection, i.e.* I_P *is generated by a regular sequence whenever P is in* Ass(R/I). *Then* Sym(I) *is a Cohen-Macaulay domain if and only if*

$$v(I_P) \leq \mathrm{ht}\ P$$

for all non-zero prime ideals P.

Proof. Using the paper of Peskine-Szpiro [12], we see there is an ideal

J, evenly linked to I having the following properties: pd J = 1, ht J = 2, J is unmixed, J is generically a complete intersection, and either v(J) = 2 or v(J) = 3. By Theorem 2.1 it is enough to prove Sym(J) is a Cohen-Macaulay domain. If v(J) = 2, J is generated by a regular sequence and the result follows [9]. If v(J) = 3, J is generated by a d-sequence [5], and again the result follows [5].

BIBLIOGRAPHY

1. L. Avramov, Complete intersections and symmetric algebras, preprint.
2. M. Brodmann, Rees rings and form rings of almost complete intersections, preprint.
3. M. Fiorentini, On relative regular sequences, J. Algebra, 18 (1971), 384-389.
4. M. Hochster, Properties of Noetherian rings stable under general grade reduction, Archiv. Math. 24 (1973).
5. C. Huneke, On the symmetric and Rees algebra of an ideal generated by a d-sequence, J. Alg. 62 (1980), 268-275.
6. C. Huneke, On the symmetric algebra of a module, to appear, J. Alg.
7. C. Huneke, Symbolic powers of prime ideals and special graded algebras, to appear, Comm. in Alg.
8. J. Matijevic, Three local conditions on a graded ring. TAMS 205 (1975), 275-284.
9. A. Micali, Sure les algebras symetrique et de Rees d'un ideal, Ann. Inst. Fourier.
10. A. Micali, Sur les algebres de Rees, Bull. Soc. Math. Belgique, 20 (1968), 215-235.
11. M. Miller, Thesis, University of Illinois, 1979.
12. C. Peskine and L. Szpiro, Liaison des variétes algébriques, Inven. Math. 26 (1974), 271-302.
13. A. P. Rao, Liaison among curves in $\mathbb{P}^3$, Inven. Math. 50 (1979), 205-217.
14. P. Samuel, Anneaux gradues factoriels et modules reflexifs. Bull. Soc. Math. France, 92 (1964), 237-249.
15. A. Simis and W. Vasconcelos, The dimension and integrality of the symmetric algebra, preprint.
16. G. Valla, On the symmetric and Rees algebras of an ideal, mans. Math. 30 (1980), 239-255.

UNIVERSITY OF MICHIGAN
ANN ARBOR, MICHIGAN 48109

Contemporary Mathematics
Volume 13, 1982

POLYNOMIAL CONSTRAINTS IN LATTICE-ORDERED RINGS

by

Stuart A. Steinberg

A lattice-ordered ring is ℓ-prime if the product of two nonzero ℓ-ideals is nonzero, and an ℓ-domain if the product of two nonzero positive elements is nonzero. In 1968 John Diem asked if an ℓ-prime ℓ-ring R in which the square of every element is positive must be an ℓ-domain. We show that if R is unital, then it must be a domain. More generally, we get the same conclusion if the condition $x^2 \geq 0$ is replaced by: for each $a \in R$ there is a nonzero polynomial $p(x)$ in $Z[x]$ with zero constant term and positive coefficients such that $p(a) \geq 0$.

Details are expected to appear in the A.M.S. <u>Proceedings</u>.

The University of Toledo
2801 W. Bancroft Street
Toledo, Ohio 43606

0271-4132/82/0532/$01.25

Contemporary Mathematics
Volume 13, 1982

REALITY PROPERTIES OF CONJUGACY CLASSES IN SPIN GROUPS AND SYMPLECTIC GROUPS

BY

Walter Feit*

and

Gregg J. Zuckerman**

To Nathan Jacobson

§1. INTRODUCTION

Let Γ be a (connected) Dynkin diagram. Let α be an automorphism of order 2 of Γ if such an automorphism exists; otherwise let $\alpha = 1$.

If F is an algebraically closed field and let $G = G(F)$ be a semi-simple group over F of type Γ such that $G^\alpha = G$. Define $H = H(F) = H\langle\alpha\rangle$.

Suppose that F is the algebraic closure of a finite field. Let σ be an endomorphism of $G = G(F)$ whose fixed point set G_σ is finite. Extend σ to H by $\alpha^\sigma = \alpha$. Let $H_\sigma = G_\sigma\langle\alpha\rangle$. Thus $|H_\sigma : G_\sigma| \leq 2$.

An element x of a group M is _real_ in M if x is conjugate to x^{-1} in M. We will first prove:

THEOREM A (i) _If F is algebraically closed, then every semi-simple element s in $G(F)$ is real in $H(F)$._

(ii) _If F is the algebraic closure of a finite field and σ is as above, then every semi-simple element s in $G(F)_\sigma$ is real in $H(F)_\sigma$._

In general the conclusion of Theorem A does not hold if s is replaced by an arbitrary element of G_σ. A counterexample is provided if G is of type G_2. See [2]. However the stronger result is true if G is a suitable form of a classical group. If G is the orthogonal group this follows from results of Wall [5]. If G is the reductive group $GL_n(F)$, then $x^\alpha = x'^{-1}$, where $'$ denotes transpose. It is easily seen that any element of $GL_n(q)$ is conjugate to its inverse in $GL_n(q)\langle\alpha\rangle$. This implies that any element of $U_n(q)$ is

*Supported by NSF Grant MCS 79-04473

**Supported by NSF Grant MCS 80-05151, and the A. P. Sloan Foundation.

conjugate to its inverse in $U_n(q)\langle\alpha\rangle$. See [5]. The main object of this paper is to prove an analogous result for Spin groups. We will also sketch a proof of a similar (and simpler) result for symplectic groups.

If F has characteristic 2 then $\mathrm{Spin}_n(F) = SO_n(F) = O_n(F)$ and the result is known. Throughout the rest of this section F is assumed to be a field of characteristic not 2.

The following results can for instance be found in [1] pp. 186-196 and [3] Section 6.8.

Let V be an n-dimensional vector space over F with $n \geq 2$. Let f be a nondegenerate quadratic form on V which has nonzero isotropic vectors. Let $O(f)$ be the orthogonal group corresponding to f. Define

$$SO(f) = O(f) \cap SL_n(F),$$
$$\Omega(f) = O(f)', \text{ the commutator subgroup of } O(f).$$

Let $\mathcal{Z}(M)$ denote the center of the group M.

Then $\Omega(f)/\mathcal{Z}(\Omega(f))$ is simple unless $n = 4$, or $n = 3$ and $F = \mathbb{F}_3$. In all cases $SO(f)/\Omega(f) \simeq F^\times/(F^\times)^2$, where $(F^\times)^2$ denotes the group of squares in $F^\times$. Thus in particular $\Omega(f) = SO(f)$ if F is algebraically closed and $|SO(f):\Omega(f)| = 2$ if F is finite. Observe also that if F is algebraically closed and f is an arbitrary quadratic form then f has nonzero isotropic vectors.

There exists a group $R(f)$ such that $F^\times \subseteq \mathcal{Z}(R(f))$ and $R(f)/F^\times \simeq O(f)$. Then $R(f)$ contains a subgroup $SR(f)$ of index 2 (denoted by D in [1]) with $F^\times \subseteq SR(f)$ and $SR(f)/F^\times \simeq SO(f)$. Furthermore $\mathrm{Spin}(f) \subseteq SR(f)$ and $|F^\times \cap \mathrm{Spin}(f)| = 2$. See [1] p. 190 for the definition of $R = R(f)$. There exist exact sequences

$$1 \to F^\times \to R(f) \to O(f) \to 1$$
$$1 \to F^\times \to SR(f) \to SO(f) \to 1$$
$$1 \to F^\times \cap \mathrm{Spin}(f) \to \mathrm{Spin}(f) \to \Omega(f) \to 1.$$

If F is algebraically closed then $SR(f)$ is a connected algebraic group. If $n = 2m$ is even, and so $O(f)$ is of type D_m, let α be defined as above. If $G(F) = \mathrm{Spin}(f)$ then $H(f) \subseteq R(f)$; in fact $R(f) = F^\times H(f) = \mathcal{Z}(R(f))H(f)$. Thus the following result is the desired analogue.

THEOREM B (i) *If F is algebraically closed, then every element in $\mathrm{Spin}(f)$ is real in $R(f)$.*

(ii) *If F is finite, then every element in $\mathrm{Spin}(f)$ is real in $R(f)$.*

If n is odd then $R(f) = \mathbb{Z}(R(f))SR(f)$, See e.g. [1] p. 190. Thus one gets the following Consequence of Theorem B.

Corollary C. *Suppose that* n *is odd. If* F *is either algebraically closed or finite then every element in* $\mathrm{Spin}(f)$ *is real in* $SR(f)$.

If F is algebraically closed then $SR(f) = \mathbb{Z}(SR(f))\mathrm{Spin}(f)$ and so Corollary C implies

Corollary D. *Suppose that* n *is odd and* F *is algebraically closed. Then every element in* $\mathrm{Spin}(f)$ *is real in* $\mathrm{Spin}(f)$.

The conclusion of Corollary D is false if F is finite. This can be seen from the following example.

Let $F = \mathbb{F}_q$ with $q \equiv 3 \pmod 4$. Then no unipotent element $n \neq 1$ in $\mathrm{Spin}_3(q) \simeq SL_2(q)$ is real in $\mathrm{Spin}_3(q)$.

It is known that if F is an algebraically closed field then every element of $Sp_{2m}(F)$ is real. The same conclusion holds if F is finite of characteristic 2. See [5].

Suppose that F is a finite field of odd characteristic p with $|F| = q$. Let V be a 2m-dimensional vector space over F and let f be a nondegenerate alternating form on V. If $x \in GL(V)$ define $f^x(v,w) = f(xv,xw)$. Let

$$Sp_{2m}^{\pm}(q) = \{x \mid x \in GL(V),\ f^x = \pm f\} .$$

In Section 5 a proof of the following result will be sketched.

THEOREM E. (i) *If* $q \equiv 1 \pmod 4$ *then every element of* $Sp_{2m}(q)$ *is real.*

(ii) *If* $q \equiv 3 \pmod 4$ *then every element of* $Sp_{2m}(q)$ *is real in* $Sp_{2m}^{\pm}(q)$.

Throughout this paper we use standard notation. In particular if M is a group, S is a subset of M and $x \in M$ then

$$\mathbb{C}_M(S) = \{y \mid y \in M,\ y^{-1}xy = x \text{ for all } x \in S\}.$$
$$\mathbb{N}_M(S) = \{y \mid y \in M,\ y^{-1}Sy = S\}.$$
$$\mathbb{C}_M^*(x) = \{y \mid y \in M,\ y^{-1}xy = x^{\pm 1}\}.$$

If q is a power of a prime then $\mathbb{F}_q$ denotes the field of q elements.

§2. PROOF OF THEOREM A

Proof of Theorem A. It suffices to prove the result in case G is simply connected.

Let T be a maximal torus in G = G(F). Then $\mathbb{N}_G(T)/T = W$, and $\mathbb{N}_H(T)/T = \widetilde{W}$, where W is the Weyl group of G, and $W \subseteq \widetilde{W}$ with $|\widetilde{W}:W| = |H:G|$. Furthermore, $-1 \in \widetilde{W}$. Thus, every element of T is conjugate to its inverse in H. Because F is algebraically closed, every semi-simple element of G is conjugate to an element of T. This proves (i).

Suppose that F is the algebraic closure of a finite field. Thus, if s is a semi-simple element in G_σ, there exists $y \in G$ such that either $y^{-1}sy = s^{-1}$ or $\alpha^{-1}y^{-1}sy\alpha = s^{-1}$. In the latter case, $y^{-1}sy = \alpha s^{-1}\alpha \in G_\sigma$. Because G is simply connected, there exists $x \in G_\sigma$ with either $x^{-1}sx = s^{-1}$ or $x^{-1}sx = \alpha s^{-1}\alpha$. See [4] (12.5). Thus, either $x^{-1}sx = s^{-1}$ or $\alpha^{-1}x^{-1}sx\alpha = s^{-1}$.

Corollary 2.1. Let F be a field of characteristic not 2. Let f be an anisotropic form on a 2-dimensional vector space over F. Then every element in Spin(f) is real in R(f).

Proof. In this case every element of Spin(f) is semi-simple. See [1] p. 199. Hence the result follows from Theorem A.

The next result will be needed in the sequel.

THEOREM 2.2. Let F be a field of characteristic not 2. Let f be a a nondegenerate quadratic form on a vector space V over F of dimension n = 2m which has nonzero isotropic vectors. Let d = d(f) denote the discriminant of f. Then the following hold.

(i) There exists $t \in SR(f)$ with $t^4 = 1$ such that t maps onto -1 in SO(f).

(ii) $\mathbb{C}_{R(f)}(t) = SR(f)$ for any such t.

(iii) $-1 \in \Omega(f)$ if and only if d is a square in F. In that case there exists $t \in Spin(f)$ which maps onto -1. Furthermore if $z \neq 1$ is in Spin(f) and z maps onto 1 then t and tz are the only elements of Spin(f) which map onto -1.

(iv) If $-1 \in \Omega(f)$ and $t \in Spin(f)$ then $t^2 = 1$ if and only if $n \equiv 0 \pmod 4$.

(v) If $t^2 \neq 1$ and $x \in R(f)$, $x \notin SR(f)$ then $x^{-1}tx = t^{-1}$.

Proof. (iii) follows from [1] Theorem 5.19; (i) and (ii) follows from [13, p. 189] and (iii).

iv) By (iii) there exists an orthonormal basis $v_1, \ldots, v_n$ of V. Let J denote the anti-automorphism of the Clifford algebra defined by

$$(v_{i_1} \cdots v_{i_s})^J = v_{i_s} \cdots v_{i_1} .$$

Thus J defines an anti-automorphism of $SR(f)$ and $\Omega(f)$ consists of all elements x with $xx^J = 1$. Hence $v_1 \cdots v_n \in \mathrm{Spin}(f)$ and maps onto -1. Thus $t = v_1 \cdots v_n$ by (iii). By [1], p. 189 $t^2 = (-1)^{n/2}$ which implies the result.

(v) As $t^4 = 1$, t is semi-simple. By Theorem A there exists $x_0 \in R(f)$ with $x_0^{-1}tx_0 = t^{-1}$. Thus x_0 does not centralize t and so $x_0 \notin SR(f)$ by (ii). Therefore $x = yx_0$ for some $y \in SR(f)$. We finally have $x^{-1}tx = x_0^{-1}tx_0 = t^{-1}$.

§3. SOME PROPERTIES OF ORTHOGONAL GROUPS

Throughout this section K is a field of characteristic not 2 and q is a power of an odd prime.

If $n = 2m$ is even, define the group

$$C_n(K) = \{ \begin{pmatrix} x & 0 \\ 0 & x'^{-1} \end{pmatrix} \mid x \in GL_m(K) \}.$$

Thus, $C_{2m}(K) \cong GL_m(K)$. Since

$$\begin{pmatrix} x & 0 \\ 0 & x'^{-1} \end{pmatrix} \begin{pmatrix} 0 & 1 \\ 1 & 0 \end{pmatrix} \begin{pmatrix} x' & 0 \\ 0 & x^{-1} \end{pmatrix} = \begin{pmatrix} 0 & 1 \\ 1 & 0 \end{pmatrix} \tag{3.1}$$

it follows that $C_{2m}(K) \subseteq O_{2m}(K)$, the orthogonal group corresponding to the form of maximal Witt index. Thus, if $K = \mathbb{F}_q$, we have $C_{2m}(K) \subseteq O^+_{2m}(q)$.

Now, $\alpha = \begin{pmatrix} 0 & 1 \\ 1 & 0 \end{pmatrix}$ is in $O_{2m}(K)$, and by (3.1), $\alpha^{-1}y\alpha = y'^{-1}$ for $y \in C_{2m}(K)$. Let $C^*_{2m}(K) = C_{2m}(K)\langle\alpha\rangle$. Then $C^*_{2m}(K) \cong GL_m(K)\langle\alpha\rangle$, where the latter α is the graph automorphism of $GL_m(K)$.

Let λ be in K with $\lambda \neq 0, \pm 1$. Let λ also denote the $m \times m$ scalar matrix λI. If $\begin{pmatrix} \lambda & 0 \\ 0 & \lambda^{-1} \end{pmatrix}$ is in $O_{2m}(K)$, then

$$C_{2m}(K) = \mathbb{C}_M \begin{pmatrix} \lambda & 0 \\ 0 & \lambda^{-1} \end{pmatrix},$$

(3.2)

$$C^*_{2m}(K) = \mathbb{C}^*_M \begin{pmatrix} \lambda & 0 \\ 0 & \lambda^{-1} \end{pmatrix},$$

where $M = 0_{2m}(K)$.

For a natural number m let J_m denote the Jordan block of size m, i.e. J_m is the $m \times m$ matrix (a_{ij}) with

$$a_{ii} = a_{i,i+1} = 1 \quad \text{and}$$
$$a_{ij} = 0 \quad \text{otherwise.}$$

A unipotent linear transformation is of type $\Sigma a_m J_m$ if J_m occurs with multiplicity a_m is the Jordan form of the transformation.

We will now state some results without proof. Theorem 3.3 is proved in [5]. See especially the Corollary to 1.4.6, Theorem 2.3.1 and Theorem 2.4.2.

THEOREM 3.3. *Let f be a nondegenerate quadratic form on the vector space V over F. Let u be a unipotent element in $0(f)$ of type $\Sigma a_m J_m$.*

(i) $V = \oplus\Sigma V_m^{(u)}$, *where the sum is an orthogonal direct sum of u-invariant subspaces $V_m^{(u)}$ and the restriction u_m of u to $V_m^{(u)}$ is of type $a_m J_m$.*

(ii) *Let v be another unipotent element of $0(f)$ of type $\Sigma a_m J_m$. Let $V_m^{(v)}$ be defined as in (i). Let $f_m^{(u)}, f_m^{(v)}$ be the quadratic form on $V_m^{(u)}, V_m^{(v)}$ respectively induced by f. Then u is conjugate to v in $0(f)$ if and only if $f_m^{(u)}$ is equivalent to $f_m^{(v)}$ for all m.*

(iii) *If m is even then a_m is even and $f_m^{(u)}$ is a form of maximal Witt index $\frac{1}{2}(a_m m)$.*

THEOREM 3.4. *Let $K = \mathbb{F}_q$ and let g be a nondegenerate quadratic form on the vector space V over K where V has odd dimension m. Let c be a nonsquare in K. Then ag and $(a-1)g \oplus c\cdot g$ are inequivalent and any nondegenerate quadratic form on an am-dimensional vector space over K is equivalent to exactly one of these.*

Theorem 3.4 is well known. See e.g. [1] Chapter III.

For the remainder of this section assume that either K is algebraically closed or $K = \mathbb{F}_q$ is finite.

If K is algebraically closed then up to equivalence there is only one nondegenerate quadratic form in each dimension. This necessarily has maximum

Witt index. In any case let f be a nondegenerate quadratic form on an n-dimensional vector space V over K.

If n is odd then $O(f)$ contains unipotent elements of type J_n. See [5] Theorem 2.3.1. Call such an element <u>indecomposable in $O(f)$</u> or <u>indecomposable on V</u>.

Suppose that $n \equiv 0 \pmod 4$. Let $n = 2m$. An element of $O(f)$ is <u>indecomposable in $O(f)$</u> or <u>indecomposable on V</u> if it is conjugate in $O(f)$ to $\begin{pmatrix} J_m & 0 \\ 0 & J_m'^{-1} \end{pmatrix}$ in $C_m(K)$. Clearly an indecomposable element in $O(f)$ is of type $2J_m$.

Observe that if $n \equiv 2 \pmod 4$ we have not defined an indecomposable element in $O(f)$. This omission is justified by the next result.

<u>Lemma 3.5</u>. <u>Let u be a unipotent element in $O(f)$. Then $V = \oplus\Sigma V_i$, where the sum is an orthogonal direct sum of u-invariant spaces and the restriction of u to each V_i is indecomposable on V_i.</u>

<u>Proof</u>. By Theorem 3.3(i) it may be assumed that u is of type aJ_m for some natural numbers a and m.

If m is even then $a = 2b$ is even and f has maximum Witt index by Theorem 3.3(iii). As any two forms of maximum Witt index are equivalent the result follows from Theorem 3.3(ii) and the existence of $C_m(K)$.

If m is odd then $O_m(K)$ contains a unipotent element of type J_m. If K is algebraically closed then $f = af_m$, where f_m is a nondegenerate quadratic form on an m-dimensional vector space over K. By Theorem 3.4 the same is true if K is finite. Then there exists $v \in O(f)$ such that v is of type aJ_m and $V = \oplus\Sigma V_i$, where the sum is an orthogonal direct sum of v-invariant subspaces and for each i the restriction of v to V_i is of type J_m on V_i. By Theorem 3.3(ii) and Theorem 3.4 v is conjugate to u in $O(f)$.

<u>Lemma 3.6</u>. <u>Let u be a unipotent element in $O(f)$. Let u be of type $\Sigma a_m J_m$.</u>

(i) <u>If $a_m \neq 0$ for some odd m, then there exists $y \in O(f)$ with $y \notin SO(f)$ such that $uy = yu$.</u>

(ii) <u>There exists $x \in SO(f)$ with $x^{-1}ux = u^{-1}$.</u>

<u>Proof</u>. By Lemma 3.5 it may be assumed that u is indecomposable in $O(f)$.

Suppose that n is odd. Let $y = -1 \notin SO(f)$. Then $uy = yu$. By [5] Theorem 2.4.2 there exists $x_0 \in O(f)$ with $x_0^{-1}ux_0 = u^{-1}$. Thus $(yx_0)^{-1}u(yx_0) = u^{-1}$ and one of x_0, yx_0 is in $SO(f)$.

Suppose that $n = 2m$ is even. Then $n \equiv 0 \pmod 4$. By (3.1)

$$\begin{pmatrix} 0 & 1 \\ 1 & 0 \end{pmatrix}^{-1} \begin{pmatrix} J_m & 0 \\ 0 & J_m'^{-1} \end{pmatrix} \begin{pmatrix} 0 & 1 \\ 1 & 0 \end{pmatrix} = \begin{pmatrix} J_m'^{-1} & 0 \\ 0 & J_m \end{pmatrix} .$$

There exists $w \in GL_m(K)$ with $w^{-1}J_m w = J_m'$. Define

$$x = \begin{pmatrix} 0 & 1 \\ 1 & 0 \end{pmatrix} \begin{pmatrix} w & 0 \\ 0 & w'^{-1} \end{pmatrix} = \begin{pmatrix} 0 & w'^{-1} \\ w & 0 \end{pmatrix}$$

Then $x^{-1}ux = u^{-1}$. Furthermore $x \in O(f)$ and even in $SO(f)$ as $n \equiv 0 \pmod 4)$.

§4. PROOF OF THEOREM B.

In this section F will always denote an algebraically closed field of characteristic not 2. As above q is a power of an odd prime and K is either F or $\mathbb{F}_q$.

V is an n-dimensional vector space over K and f is a nondegenerate quadratic form on V.

If $K = F$ then f is unique up to equivalence and we will write $R = R_n = R(f)$, $O = O_n = O(f)$ and similarly for all the groups related to f.

If $x \in R(f)$ let $\bar{x}$ denote its image in $O(f)$.

Let s be a semi-simple element of $Spin(f)$. We will describe $C^*_{R(f)}(s)$.

Consider first the following situation: $n = 2m$, $\lambda \in F$, $\lambda \neq 0, \pm 1$, and $s \in Spin_{2m}$ is one of the two semi-simple elements such that $\bar{s} = \begin{pmatrix} \lambda & 0 \\ 0 & \lambda^{-1} \end{pmatrix}$. We know by 3.2 that $\mathbb{C}_{O_{2m}}(\bar{s}) \cong GL_m(F)$. Let μ be the covering homomorphism from $Spin_{2m}$ to SO_{2m}; extend μ to a homomorphism μ from H_{2m} to O_{2m}. Clearly, $\mathbb{C}_H(s)$ is a subgroup of $\mu^{-1}(\mathbb{C}_{O_{2m}}(\bar{s}))$ of index at most two.

It is well known from the theory of spinors that $\mu^{-1}(\mathbb{C}_{O_{2m}}(\bar{s}))$ is a connected subgroup of H_{2m}. This subgroup is sometimes referred to as the metalinear group, $ML_m(F)$, a double cover of $GL_m(F)$. Thus $\mathbb{C}_{H_{2m}}(s)$ is a subgroup of a connected group and has index at most two. It follows that

$$\mathbb{C}_{H_{2m}}(s) = \mu^{-1}(\mathbb{C}_{O_{2m}}(\bar{s})) \cong ML_m(F). \tag{4.1}$$

Let $\tilde{\mu}$ denote the homomorphism from R_{2m} to O_{2m} whose restriction to H_{2m} is μ. Since $R_{2m} = F^{\times}H_{2m}$ with $F^{\times} \subseteq \mathcal{Z}(R_{2m})$ it follows that

$$\tilde{\mu}^{-1}(GL_m(F)) = F^{\times}\mu^{-1}(GL_m(F)) \approx F^{\times}ML_m(F),$$

and $\tilde{\mu}^{-1}(GL_m(F)) \subseteq SR_{2m}$. Thus (4.1) implies

$$\mathbb{C}_{R_{2m}}(s) = \tilde{\mu}^{-1}(\mathbb{C}_{O_{2m}}(\bar{s})) \simeq F^{\times}ML_m(F). \tag{4.2}$$

<u>Lemma 4.3.</u> $\mathbb{C}^*_{H_{2m}}(s) = \mathbb{C}_{H_{2m}}(s)\langle\alpha\rangle$, <u>which as an abstract group is isomorphic to</u> $ML_m(F)\langle\alpha\rangle$ <u>where now</u> α <u>is the lift to</u> ML <u>of transpose inverse</u>.

<u>Proof</u>. It is enough to show $\alpha s\alpha^{-1} = s^{-1}$; for then $\mathbb{C}_H(s)\langle\alpha\rangle$ will be a subgroup of $\mathbb{C}^*_H(s)$, and both $\mathbb{C}_H(s)\langle\alpha\rangle$ and $\mathbb{C}^*_H(s)$ contain $\mathbb{C}_H(s)$ as a subgroup of index two.

We know that $\bar{\alpha}\bar{s}\bar{\alpha}^{-1} = \bar{s}^{-1}$ from the form of $\bar{s}$. We know by Theorem A that there exists $\gamma \in H_{2m}$ such that $\gamma s\gamma^{-1} = s^{-1}$. Hence, $\overline{\gamma s\gamma}^{-1} = \bar{s}^{-1}$, and $\bar{\alpha}\bar{\gamma}^{-1}$ is in $\mathbb{C}_{O_{2m}}(\bar{s})$. By (4.1), $\alpha\gamma^{-1} \in \mathbb{C}_{H_{2m}}(s)$; hence $\alpha s\alpha^{-1} = s^{-1}$.

<u>Lemma 4.4.</u> <u>Let</u> s <u>be a semi-simple element of</u> $Spin_n$. <u>Then there exists an exact sequence</u>

$$\{ML_{n_1}(F) \times\ldots\times ML_{n_k}(F) \times H_{m(1)} \times Spin_{m(-1)}\}\langle\beta\rangle \to \mathbb{C}^*_H(s) \to 1,$$

<u>with</u> $m(-1)$ <u>even, (where possibly some of the numbers</u> k, $m(1)$, $m(-1)$ <u>are zero). Moreover</u> β <u>acts on each</u> $ML_{n_i}(F)$ <u>as the lift of transpose inverse, on</u> $H_{m(1)}$ <u>as the identity and on</u> $Spin_{m(-1)}$ <u>as</u> 1 <u>or an element of</u> $H_{m(-1)}$ <u>according to whether</u> $m(-1) \equiv 0 \pmod 4$ <u>or</u> $2 \pmod 4$.

<u>There exist elements</u> $s_1, \ldots, s_k$ <u>with</u> $s_i \in \mathcal{Z}(ML_{n_i}(F))$ <u>and</u> t_1, t_{-1}, <u>where</u> $t_1 \in H_{m(1)}$ <u>and maps onto</u> 1 <u>in</u> $O_{m(1)}(F)$, <u>while</u> $t_{-1} \in Spin_{m(-1)}$ <u>and maps onto</u> -1 <u>in</u> $O_{m(-1)}(F)$, <u>such that</u> $(\Pi s_i) \times t_1 \times t_{-1}$ <u>maps to</u> s. <u>Finally</u> t_{-1} <u>has order</u> 2 <u>or</u> 4 <u>according to whether</u> $m(-1) \equiv 0$ <u>or</u> $2 \pmod 4$ <u>and</u> $\beta^{-1}t_{-1}\beta = t_{-1}^{-1}$.

<u>Proof</u>. Let V be the underlying vector space for O. For each $\lambda \in F$, let $W_\lambda = \{v \in V \mid \bar{s}v = \lambda v\}$. If v and w are in V with $\bar{s}v = \lambda v$ and $\bar{s}w = \mu w$, then

$$v\cdot w = \bar{s}v \cdot \bar{s}w = \lambda\mu\, v\cdot w$$

an so either $v\cdot w = 0$ or $\lambda\mu = 1$. Thus, if $V_\lambda = V_{\lambda^{-1}} = W_\lambda + W_{\lambda^{-1}}$ then

$V = \oplus\Sigma\, V_\lambda$, where the sum is an orthogonal direct sum and one element from each pair $\{\lambda,\lambda^{-1}\}$ occurs. Let $\bar{s}_\lambda$ denote the restriction of $\bar{s}$ to V_λ. By (3.2),

$$\mathbb{C}_0(\bar{s}) = \Pi\, \mathbb{C}_{0(V_\lambda)}(\bar{s}_\lambda) \cong \prod_{\lambda \neq \pm 1} GL_{m(\lambda)}(F) \times 0_{m(1)} \times 0_{m(-1)}$$

Each $\bar{s}_\lambda$ is in the center of $GL_{m(\lambda)}(F)$. Moreover, if $\bar{t}_j$ is the coordinate of $\bar{s}$ in $0_{m(j)}$, then $\bar{t}_1 = 1$, and $\bar{t}_{-1} = -1$. Since $\bar{s} \in SO_n$, it follows that $m(-1)$ is even and $\bar{t}_{-1} \in SO_{m(-1)}$. Clearly, $\bar{t}_j \in Z(0_{m(j)})$. Therefore

$$C_0^*(\bar{s}) = \{\prod_{\lambda \neq \pm 1} GL_{m(\lambda)}(F)\} \rtimes \langle\alpha\rangle \times 0_{m(1)} \times 0_{m(-1)} ,$$

where $x^\alpha = x'^{-1}$ for $x \in GL_{m(\lambda)}(F)$ for any $\lambda \neq \pm 1$.

Define 0_s to be the subgroup of 0 given by

$$0_s = \Pi\, 0(V_\lambda) \cong \{\prod_{\lambda \neq \pm 1} 0_{2m(\lambda)}\} \times 0_{m(1)} \times 0_{m(-1)} .$$

Clearly, $\bar{s} \in 0_s$, and

$$\mathbb{C}_0^*(s) = \mathbb{C}_{0_s}^*(s) .$$

Let H_s be the inverse image of 0_s in H. Then, $s \in H_s$, and we have

$$\mathbb{C}_H^*(s) = \mathbb{C}_{H_s}^*(s) .$$

Let $s_\lambda \in Spin_{2m(\lambda)}$, $\lambda \neq \pm 1$, and $t_j \in Spin_{m(j)}$, $j = \pm 1$, be chosen so that $(\prod_{\lambda \neq \pm 1} s_\lambda) \times t_1 \times t_{-1}$ maps to s, with $\bar{t}_1 = 1$ and $\bar{t}_{-1} = -1$. We then have, by Lemma 2.2, an exact sequence

$$(4.5) \qquad [\{\prod_{\lambda \neq \pm 1} \mathbb{C}_{H_{2m(\lambda)}}(s)\} \times H_{m(1)} \times Spin_{m(-1)}] \rtimes \langle\beta\rangle \to \mathbb{C}_{H_s}^*(s) \to 1,$$

where β acts as the identity on an element of $H_{m(1)}$, and either the identity or an element of $H_{m(-1)}$ on $Spin_{m(-1)}$ according to whether $m(-1) \equiv 0$ or 2 (mod 4). Furthermore, by Lemma 4.2, the component of β_λ of β in $H_{2m(\lambda)}(F)$ has the property that $\beta_\lambda^{-1} s_\lambda \beta_\lambda = s_\lambda^{-1}$.

By (4.1) we know that

$$\mathbb{C}_{H_{2m(\lambda)}}(s_\lambda) = \mathbb{C}_{Spin_{2m(\lambda)}}(s_\lambda) \cong ML_{m(\lambda)} .$$

So the first statement of Lemma 4.3 follows from (4.5) after some relabeling. Note that β_λ operates on $ML_{m(\lambda)}$ by the lift of transpose inverse. The second statement follows because we know that each s_λ is in the center of the corresponding ML group. By Theorem 2.2(iv) t_{-1} has the required order. By Theorem 2.2(v) $\beta^{-1}t_{-1}\beta = t_{-1}^{-1}$.

Corollary 4.6. *Let f be a nondegenerate quadratic form on an n-dimensional vector space over* $\mathbb{F}_q$. *Let s be a semi-simple element of* Spin(f). *Then there exists an exact sequence*

$$\{F_1^{\times}A_1 \times\ldots\times F_\ell^{\times}A_\ell \times R(f_1) \times SR(f_{-1})\}\rtimes\langle\beta\rangle \to \mathbb{C}^*_{R(f)}(s) \to 1,$$

where each F_i *is a finite extension of* $\mathbb{F}_q$, *each* $A_i \simeq ML_{n_i}(q^{r_i})$ *or* $MU_{n_i}(q^{r_i})$, *the twisted version of* $ML_{n_i}(q^{v_i})$, *and* β *is as in Lemma* 4.4. *For* $j = \pm 1$, f_j *is a nondegenerate quadratic form on an* $m(j)$ *dimensional vector space over a finite extension of* $\mathbb{F}_q$.

Furthermore $s = (\prod s_i) \times t_1 \times t_{-1}$ *with* s_i *in the center of* A_i *where* $\bar{t}_j = j$ *for* $j = \pm 1$ *and* t_{-1} *has order* 2 *or* 4 *according to whether* $m(-1) \equiv 0$ *or* 2 (mod 4). *In any case* $\beta^{-1}t_{-1}\beta = t_{-1}^{-1}$.

Proof. Let F be the algebraic closure of $\mathbb{F}_q$ and apply Lemma 4.4. In view of (4.2) there exists an exact sequence

$$(4.7)\quad \{F^{\times}ML_{n_1}(F) \times\ldots\times F^{\times}ML_{n_k}(F) \times R_{m(1)} \times SR_{m(-1)}\}\rtimes\langle\beta\rangle \to \mathbb{C}^*_{R_m}(s) \to 1.$$

There exists an endomorphism σ of the algebraic group R_n with $(R_n)_\sigma = R(f)$, and hence $(\mathrm{Spin}_n)_\sigma = \mathrm{Spin}(f)$. Thus $s^\sigma = s$, $\beta^\sigma = \beta$ and so $\mathbb{C}^*_{R(f)}(s) = (\mathbb{C}^*_{R_n}(s))_\sigma$.

Apply σ to the exact sequence (4.7) and take fixed points. The endomorphism will amalgamate the $F^{\times}ML_{n_i}$ corresponding to conjugate eigenvalues of s acting on V. The last two factors cannot be amalgamated since they correspond to eigenvalues ± 1 in $\mathbb{F}_q$. Thus $t_j^\sigma = t_j$ for $j = \pm 1$. By Lemma 4.4 t_{-1} has the required order and $\beta^{-1}t_{-1}\beta = t_{-1}^{-1}$.

Apply the known classification of $\mathbb{F}_q$-forms of groups $F^{\times}A_{n_i} \times \ldots \times F^{\times}A_{n_i}$, $R_{m(1)}$ and $SR_{m(-1)}$ to get the required exact sequence.

Lemma 4.8. *Let* $B(K) = ML_n(K)$ *or* $MU_n(K)$. *Let* β *be the lift of the transpose inverse automorphism on* $B(K)$. *Let* $s \in \mathbb{Z}(B(K))$ *and let u be a unipotent element in* $B(K)$. *Then there exists* $x \in B(K)\langle\beta\rangle$ *with* $x \notin B(K)$ *such that* $x^{-1}(us)x = (us)^{-1}$.

Proof. The map from $B(K)$ to $GL_n(K)$ or $U_n(K)$ defines a bijection from the unipotent elements in B to the unipotent elements in $GL_n(K)$ or $U_n(K)$

respectively. Every unipotent element in $GL_n(K)$ or $U_n(K)$ is conjugate to its transpose. Thus there exists $x_0 \in B(K)$ with $x_0^{-1}ux_0 = u'$. Since s is in the center of $B(K)$ $\beta^{-1}s\beta = s^{-1}$. Thus $(x_0\beta)^{-1}(us)(x_0\beta) = (us)^{-1}$.

Proof of Theorem B. Let $y \in \mathrm{Spin}(f)$. Then $y = su = us$, where s is semi-simple and u is unipotent with $s,u \in \mathrm{Spin}(f)$. Hence $su \in \mathbb{C}_{R(f)}(s)$. By Lemma 4.4 or Corollary 4.6, $su = (\Pi\, s_iu_i) \times t_1v_1 \times t_{-1}v_{-1}$, where $s_iu_i \in B_i(K_i)$ is defined as in Lemma 4.8 with K_i a finite extension of K, $t_1v_1 \in R(f_1)$ and $t_{-1}v_{-1} \in SR(f_{-1})$. By Lemma 4.8 every element s_iu_i is conjugate to its inverse in $B_i(K_i)\langle\beta\rangle$ by an element which is not in $B_i(K_i)$.

The map from $R(f_1)$ to $O(f_1)$ defines a bijection from the unipotent elements in $R(f_1)$ to those in $O(f_1)$. Thus Lemma 3.6 implies the existence of $x \in R(f_1)$ with $x_1^{-1}t_1v_1x_1 = (t_1v_1)^{-1}$.

The map from $SR(f_{-1})$ to $SO(f_{-1})$ defines a bijection from the unipotent elements in $SR(f_{-1})$ to those in $O(f_{-1})$.

If $m(-1) \equiv 0 \pmod 4$ then t_{-1} has order 2 by either Lemma 4.4 or Corollary 4.6. Thus Lemma 3.6(ii) implies the existence of $x \in SR(f_{-1})$ with $x^{-1}(t_{-1}v_{-1})x = (t_{-1}v_{-1})^{-1}$.

If $m(-1) \equiv 2 \pmod 4$ then $t = t_{-1}$ has order 4 by either Lemma 4.4 or Corollary 4.6. By Lemma 3.5 $v = v_{-1}$ is of type $\Sigma\, a_mJ_m$ with $a_m \neq 0$ for some odd m. Hence Lemma 3.6 implies the existence of $x_0 \in SR(f_{-1})$ with $x_0^{-1}vx_0 = v^{-1}$ and $y \in R(f_{-1})$ with $vy = yv$. By Theorem 2.2(v) $y^{-1}ty = t^{-1}$. Thus if $x = yx_0$ then $x^{-1}(v_{-1}t_{-1})x = (v_{-1}t_{-1})^{-1}$. Furthermore $x \in R(f_{-1})$ and $x \notin SR(f_{-1})$.

It now follows that in any case there exists $w \in \mathbb{C}_{R(f)}(su)$ such that $(w\beta)^{-1}(su)(w\beta) = (su)^{-1}$.

§5. SYMPLECTIC GROUPS

Throughout this section $F = F_q$ with q a power of the odd prime p. Let $C_{2m}(q) = C_{2m}(F) \simeq GL_m(F)$ be defined as in Section 3. Then

$$\begin{pmatrix} x & 0 \\ 0 & x'^{-1} \end{pmatrix}\begin{pmatrix} 0 & 1 \\ -1 & 0 \end{pmatrix}\begin{pmatrix} x' & 0 \\ 0 & x'^{-1} \end{pmatrix} = \begin{pmatrix} 0 & 1 \\ -1 & 0 \end{pmatrix}, \tag{5.1}$$

and so $C_{2m}(q) \subseteq Sp_{2m}(q)$. Let $\alpha = \begin{pmatrix} 0 & 1 \\ -1 & 0 \end{pmatrix} \in Sp_{2m}(q)$. By (5.1) $\alpha^{-1}y\alpha = y'^{-1}$ for $y \in C_{2m}(q) \simeq GL_m(q)$. Let $C^*_{2m}(q) = C_{2m}(q)\langle\alpha\rangle \simeq GL_m(q)\langle\alpha\rangle$. Thus $C^*_{2m}(q) \subseteq Sp_{2m}(q)$.

An analogue of Theorem 3.3(i) and (ii) is true for alternating forms.

See [5]. We will also state the following result without proof. See [5] p.36.

Lemma 5.2. Let a,m be natural numbers with am even.

(i) If m is odd then $Sp_{am}(q)$ contains exactly one conjugate class of unipotent elements of type aJ_m.

(ii) If m is even then $Sp_{am}(q)$ contains exactly two conjugate classes of unipotent elements of type aJ_m.

Lemma 5.3. Let n be an even integer and let u be a unipotent element of type J_n in $Sp_n(q)$. Then the following are equivalent.

(i) u is real in $Sp_n(q)$.

(ii) $q \equiv 1 \pmod 4$.

Proof. Let V be the underlying vector space and let f be the nondegenerate alternating form. Then f is u-invariant. As u is of type J_n there exists a unique u-invariant subspace V_i with $\dim V_i = i$ for each $i = 0, \ldots, n$. Thus $V_0 \subseteq \ldots \subseteq V_n = V$.

Let x be a semi-simple element in $GL(V)$ with $x^{-1}ux = u^{-1}$. Thus x^2 is a scalar. Each V_i is x-invariant. Thus there exist eigenvectors $v_1, \ldots, v_n$ of x so that $\{v_1, \ldots, v_i\}$ is a basis of V_i for each i. In particular, the eigenvalues of x lie in F. If $xv_1 = \lambda v_1$ then it follows that $xv_i = (-1)^{i-1} \lambda v_i$ for each i.

$V_1^{\perp}$ is a u-invariant subspace with $\dim V_1^{\perp} = n-1$. Hence $V_1^{\perp} = V_{n-1}$. Similarly $V_i^{\perp} = V_{n-i}$ for each i. This implies that

$$V = \oplus \sum_{i=1}^{n/2} \langle v_i, v_{n+1-i} \rangle , \tag{5.4}$$

where the sum is an orthogonal direct sum and each $\langle v_i, v_{n-i} \rangle$ is a non-degenerate plane.

(i) $\Rightarrow$ (ii). As u is real it may be assumed that x above is an element in $Sp_n(q)$ whose order is a power of 2 and so x is semi-simple. Thus

$$f(v_1,v_n) = f(xv_1,xv_n) = (-1)^{n-1} \lambda^2 f(v_1,v_n).$$

As $f(v_1,v_n) \neq 0$ this implies that $\lambda^2 = (-1)^{n-1} = -1$. As $\lambda \in F$, $q \equiv 1 \pmod 4$.

(ii) $\Rightarrow$ (i). Let λ be a primitive 4-th root of 1 in F. Then

$$f(xv_i, xv_{n+1-i}) = (-1)^{n-1}\lambda^2 f(v_i,v_{n+1-i}) = f(v_i, v_{n+1-i}) .$$

Thus $x \in Sp_n(q)$ by (5.4).

Lemma 5.4. Let u be a unipotent element in $Sp_n(q)$.

(i) If $q \equiv 1 \pmod 4$ then u is real in $Sp_n(q)$.

(ii) If $q \equiv 3 \pmod 4$ then there exists $x \in Sp_n^{\pm}(q)$, $x \notin Sp_n(q)$ with $x^{-1}ux = u^{-1}$.

Proof. It may be assumed that u is of type aJ_m with $am = n$. By Lemma 5.2 there are at most two conjugate classes of unipotent elements of type aJ_m in $Sp_n(q)$. Thus it suffices to show that the conclusions of the Lemma holds for some unipotent element u of type aJ_m in $Sp_n(q)$.

Suppose that a is even. Let $u = \begin{pmatrix} w & 0 \\ 0 & w'^{-1} \end{pmatrix} \in C_n(q)$ with w a unipotent element of type $(\frac{1}{2}a)J_m$. There exists $y \in C_n^*(q) \subseteq Sp_n(q)$ with $y^{-1}uy = u^{-1}$ and so u is real in $Sp_n(q)$. Let $z = \begin{pmatrix} -1 & 0 \\ 0 & 1 \end{pmatrix}$. Then $(zy)^{-1}u(zy) = u^{-1}$, $zy \in Sp_n^{\pm}(q)$ and $zy \notin Sp_n(q)$.

Suppose that $a = 2b+1$ is odd. Let $V = V_0 \oplus V_1$ where the sum is an orthogonal direct sum and $u = u_0 + u_1$, where u_0 acts on V_0 and is of type $2bJ_m$, while u_1 acts on V_1 and is of type J_m. It is enough to prove the result for each of u_0 and u_1. By the previous paragraph it holds for u_0. Hence it may be assumed that $u = u_1$ is of type J_n.

If $q \equiv 1 \pmod 4$ then u is real in $Sp_n(q)$ by Lemma 5.3.

Suppose that $q \equiv 3 \pmod 4$. Choose $x_0 \in Sp_n^{\pm}(q)$ with $x_0 \notin Sp_n(q)$ and x_0 semi-simple. Since u is not real in $Sp_n(q)$ by Lemma 5.3 it follows from Lemma 5.2 (ii) that $x_0^{-1}ux_0$ is conjugate to u or u^{-1} in $Sp_n(q)$. Hence there exists $y \in Sp_n(q)$ so that $(x_0y)^{-1}uxy = u$ or u^{-1}. Let $x = x_0y$. Replacing x by an odd power it may be assumed that x is semi-simple.

Suppose that $x^{-1}ux = u$. Then $x = c$ is a scalar as u is of type J_n. Hence $f^x = c^2f$. As $x^2 \in Sp_n(q)$ it follows that $c^4 = 1$. Thus $c^2 = 1$ since $q \equiv 3 \pmod 4$. Hence $x \in Sp_n(q)$, contrary to the definition of x. Therefore $x^{-1}ux = u^{-1}$ as required.

The proof of Theorem E now proceeds exactly as in Section 4. Actually it is somewhat simpler as it is not necessary to go to covering groups.

REFERENCES

[1] E. Artin, "Geometric Algebra" Interscience, N. Y. 1957

[2] B. Chang and R. Ree, "The character table of $G_2(q)$", Symposia Mathematica XIII, 395-413, Academic Press. London 1974.

[3] N. Jacobson, "Basic Algebra I" Freeman, San Francisco, 1974.

[4] R. Steinberg, "Endomorphisms of linear algebraic groups", Memoirs of the A.M.S. 80, 1968, Providence, R.I.

[5] G. E. Wall, "On the conjugacy classes in the unitary, symplectic and orthogonal groups", J. Austral. Math. Soc. 3, (1963), 1-63.

YALE UNIVERSITY
BOX 2155 YALE STATION
NEW HAVEN, CT. 06520

Contemporary Mathematics
Volume 13, 1982

VARIATIONS ON EVEN CLIFFORD ALGEBRAS

Dedicated to my teacher, Nathan Jacobson

George B. Seligman*

1. INTRODUCTION.

Much of the interest in Clifford algebras has been due to the fact that they play an essential part in completing the list of finite-dimensional representations for the Lie algebras (or for the simply-connected covering groups) of special orthogonal groups. In other classical cases, tensor representations suffice.

When one considers algebraic groups over non-algebraically closed fields, it is natural to try to realize forms of the spin groups. Then one has to construct forms for these non-tensor representations. In [3], Jacobson dealt with aspects of this problem by methods of Galois descent. For vector spaces over quaternion algebras, Seip-Hornix [5] constructed, by presentation, an algebra whose minimal one-sided ideals afford the missing representations. Both approaches were refined by Tits [8] and extended to characteristic two (see also van Drooge [2]). In particular, Tits gave a presentation generalizing that of Seip-Hornix and applicable to general central simple algebras with involution "of orthogonal type". The resulting associative algebras generalize (and are forms of) the even Clifford algebra rather than the full Clifford algebra. This is not surprising, since it is the even Clifford algebra for a quadratic form whose minimal one-sided ideals afford the spin and half-spin modules.

This paper offers a new presentation for forms of even Clifford algebras, but in a broader context. Namely, given a division algebra D with involution $a \mapsto a^*$, finite-dimensional over its center Z (say with $[D:Z] = d^2$), a finite-dimensional left D-module V carrying a non-degenerate form b (either hermitian or anti-hermitian), and a non-negative integer k, we give a presentation

*Research supported by National Science Foundation grant MCS 79-04473.

for an algebra $A_k = A_k(D,b)$. The associative algebra A_k is always finite-dimensional, and is either zero or semisimple. We have $A_0 = F$, the *-fixed elements of Z.

When $\mathrm{End}_D(V)$ is of orthogonal type, one has $A_k = 0$ for $1 \leq k < d$, while A_d is the even Clifford algebra of Jacobson and Tits. In all other cases, $A_k = 0$ for $1 \leq k < 2d$, with A_{2d} affording a complete set of "fundamental" modules for the Lie algebra of the corresponding special unitary group. Our generators and relations, as well as the role of the space V and the form b, are somewhat more explicit than those of Tits in the overlapping case; the price paid here is that they are very awkward to work with directly.

In the cases distinguished above we have $A_k = 0$ except when k is divisible by d (resp. $2d$), and the irreducible modules for A_{jd} (resp. $2jd$) are realized as submodules of j-fold tensor products of those for A_d (resp. A_{2d}). Each finite-dimensional irreducible module for the Lie algebra of the last paragraph is a minimal right ideal in one of the corresponding A_k. In their heavy reliance on the representation theory of semisimple Lie algebras <u>the results only apply in characteristic zero</u>, a setting we assume throughout. Whether the algebras so presented, or some close variants, play an analogous part in prime characteristic remains to be seen.

2. DEFINITION OF THE ALGEBRAS A_k.

With the conventions above, let $b(u,v)$ be the value of b at the pair $(u,v) \in V \times V$. If b is hermitian, let $\varepsilon = 1$, with $\varepsilon = -1$ if b is anti-hermitian. Thus we assume $b(au,v) = ab(u,v)$ for $a \in D$, and $b(v,u) = \varepsilon\, b(u,v)^*$. For $u, v, w \in V$, define

(1) $$wS_{u,v} = b(w,u)v - \varepsilon\, b(w,v)u,$$

(2) $$[w,u,v] = b(w,u)v - \varepsilon\, b(w,v)u - \varepsilon\, b(u,v)w.$$

For $a \in D$, let $\tau(a) \in Z$ be $^1/_d$ times the reduced trace of a; that is, $\tau(a)$ is the unique element of Z such that $a = \tau(a) + a_0$, where $a_0 \in [DD]$.

If F is the field of *-fixed elements of Z and k is a fixed non-negative integer, we define A_k to be the quotient of the tensor algebra $T(V \otimes_F V)$ (where $V \otimes_F V$ is regarded as vector space over F) by the ideal generated by the following elements:

i) $au \otimes v - u \otimes a^{*}v$;

ii) $(u \otimes v)(w \otimes x) - (w \otimes x)(u \otimes v) - uS_{w,x} \otimes v - u \otimes vS_{w,x}$;

iii) $u \otimes v + \varepsilon\, v \otimes u - \frac{k}{2}\,(\tau(b(v,u)) + \tau(b(v,u))^{*})1_{T}$;

iv) $f_{k+1}(u_1,\ldots,u_{k+1};\ v_1,\ldots,v_{k+1})$, where $f_t(u_1,\ldots,u_t;\ v_1,\ldots,v_t)$ is defined inductively by $f_t(u_1,\ldots,u_t;\ v_1,\ldots,v_t)$

$$= \sum_{i=1}^{t} f_1(u_i;v_1)f_{t-1}(u_1,\ldots,\hat{u}_i,\ \ldots,\ u_t;\ v_2,\ldots,v_t)$$

$$+ \sum_{i<j} f_{t-1}(u_1,\ldots,\ [u_i,\overset{i}{u_j},v_1],\ldots,\ \hat{u}_j,\ \ldots,\ u_t;\ v_2,\ldots,v_t)$$

for $t > 1$, with $f_1(u_1;v_1) = u_1 \otimes v_1$.

The relations resulting from (i) evidently mean that the canonical mapping $T(V \otimes_F V) \to A_k$ factors through the mapping

$$T(V \otimes_F V) \to T(V \otimes_D V)\ ,$$

where the F-vector space $V \otimes_D V$ is the tensor product when the left-hand factor "V" is regarded as <u>right</u> D-module with $va = a^{*}v$.

3. <u>EXAMPLES. STRUCTURE OF</u> A_k.

When $D = F$, $*$=identity, $\varepsilon = 1$ and $k = 1$, the relators iii) and iv) become, respectively,

(iii)' $$u \otimes v + v \otimes u - b(u,v)1_T \quad \text{and}$$

(iv)' $$(u_1 \otimes v_1)(u_2 \otimes v_2) + (u_2 \otimes v_1)(u_1 \otimes v_2) + b(u_1,u_2)v_1 \otimes v_2$$
$$- b(u_1,v_1)u_2 \otimes v_2 - b(u_2,v_1)u_1 \otimes v_2\ .$$

In this case, if the Clifford algebra $C\ell(b)$ is defined as the quotient of $T(V)$ by the ideal generated by all

$$uv + vu - 2b(u,v)1\ ,$$

and if $\bar{u}$ is the image in $C\ell(b)$ of $u \in V \subset T(V)$, then the homomorphism of $T(V \otimes_F V)$ into $C\ell(b)$ sending $u \otimes v$ to $\frac{1}{2}\bar{u}\,\bar{v}$ annihilates (iii)' and (iv)', as well as (i) and (ii). We therefore have a homomorphism of F-algebras with unit: $A_1 \to C\ell(b)$, whose image is the even Clifford algebra $C\ell^{+}(b)$.

With a little care, one sees that this is an <u>isomorphism</u> $A_1 \to C\ell^+(b)$.

When $D = F$, $*$=identity and $\varepsilon = -1$ (the symplectic case), we find $A_1 = 0$. The algebra A_2 is the quotient of $T(V \otimes V)$ by the relations (ii) and the appropriate versions of (iii) and (iv):

(iii)'' $$u \otimes v - v \otimes u - 2b(v,u)1,$$

and a rather lengthy (iv)''. The complexity of the relators already makes a direct assault unpromising. I know of no approach except by way of Lie theory that yields the following conclusions for this case:

<u>If</u> $[V:F) = 2r$, <u>then</u> A_2 <u>is a semisimple algebra over</u> F, <u>the direct sum of</u> $r + 1$ <u>minimal ideals. These minimal ideals are matrix algebras of degrees</u>

$$1,\ 2r,\ \binom{2r}{2} - \binom{2r}{0},\ \binom{2r}{3} - \binom{2r}{1},\ \dots,\ \binom{2r}{r} - \binom{2r}{r-2}.$$

Thus $[A_2:F] = \binom{4r+1}{2r} \cdot \frac{1}{r+1}$.

More generally: <u>If</u> $[V:D] = n$ <u>and if the involutorial algebra</u> $\mathrm{End}_D(V)$ <u>is of</u> (first kind and) <u>symplectic type, then</u> A_{2d} <u>is semisimple of dimension</u>

$$\binom{2nd + 1}{nd} \frac{2}{nd + 2}.$$

<u>Upon extension to a splitting field,</u> A_{2d} <u>has</u> $\frac{nd}{2} + 1$ <u>minimal ideals, matrix algebras of degrees</u>

$$1,\ nd,\ \binom{nd}{2} - \binom{nd}{0},\ \dots,\ \binom{nd}{\frac{nd}{2}} - \binom{nd}{\frac{nd-4}{2}}.$$

There is a corresponding conclusion when $D = Z$, $[Z:F] = 2$: Here $A_1 = 0$; A_2 is a semisimple algebra over F, becoming upon suitable extension of the base field a direct product of matrix algebras of degrees

$$\binom{2n}{j},\ 0 \le j \le 2n,\quad \text{so of } F\text{-dimension } \binom{4n}{2n}.$$

If n is <u>odd</u>, each minimal ideal in A_2 is a Z-algebra, and this also holds for <u>all but one</u> of the minimal ideals when n is even. There is a corresponding generalization to the case of general D, with $[Z:F] = 2$ (involution of <u>second</u> kind). Here A_{2d} is the first non-zero A_k for positive k, and "dn" must replace "n" in the formulas above.

In the case treated by Tits and Jacobson, that were $Z = F$ and the involution in $\mathrm{End}_D(V)$ is of orthogonal type, the center of A_d is either $F \oplus F$ or a quadratic extension field of F, and A_d is, correspondingly, either a simple F-algebra or the sum of two (central simple) minimal ideals, each of dimension 2^{dn-2}. In either case, A_d has F-dimension 2^{dn-1}.

4. CONNECTION WITH LIE THEORY. INDICATIONS OF PROOFS.

Now let me give a few words of explanation about the connection with Lie theory that is the basis for the assertions of the last section. First of all, the set of T in $\mathrm{End}_D(V)$ such that $b(uT,v) + b(u,vT) = 0$ for all u and v is a Lie algebra $\mathfrak{g}$ over F. It is generally simple (not when $Z \neq F$, but close to simple then, too), indeed central simple. (As Jacobson and Landherr showed in the 1930's - see [4], Chapter 10 - these and the derived algebras $\mathfrak{sl}(n,D)$ for central division algebras D account for all central simple Lie algebras over F except for algebras of dimensions 14, 28, 52, 78, 133, 248.) Moreover, $\mathfrak{g}$ is spanned over F by the endomorphisms $S_{u,v}$ of V, and for each k there is a homomorphism of F-Lie algebras $\mathfrak{g} \to A_k$ sending $S_{u,v}$ to

$$\frac{k}{4}\,\tau(b(u,v) + b(u,v)^*)1_A$$

plus the image of $u \otimes v \in \mathcal{T}$. The image of this mapping generates A_k, so the theory of A_k-modules is equivalent to that of certain $\mathfrak{g}$-modules. The $\mathfrak{g}$-modules concerned are (finite-dimensional and) completely reducible, so A_k is semisimple (including possibly $A_k = 0$, when the associated set of $\mathfrak{g}$-modules is empty).

Lie theory determines this set of $\mathfrak{g}$-modules, too, as follows: Let $m > 1$, and enlarge V by adding a 2m-dimensional hermitian resp. anti-hermitian space U, orthogonal to V and of maximal Witt index m. Consider $\tilde{V} = U \oplus V$, with $\tilde{\mathfrak{g}}$ defined relative to $\tilde{V}$ as $\mathfrak{g}$ was relative to V. Let $(e_1,\ldots,e_m)$ and $(e_{m+1},\ldots,e_{2m})$ be ordered bases for dual totally isotropic subspaces of U, with $b(e_i, e_{2m+1-j}) = \delta_{ij}$ for $1 \le i, j \le m$. We consider finite-dimensional irreducible $\tilde{\mathfrak{g}}$-modules M of highest weight relative to a split toral subalgebra of $\tilde{\mathfrak{g}}$ and an ordering of roots relative to this subalgebra determined by V and the ordered basis above. Explicitly, these are irreducible $\tilde{\mathfrak{g}}$-modules generated by a subspace M^+ with the following property:

M^+ is annihilated by all of these:

$$S_{e_j,v},\ m+1 \le j \le 2m,\ v \in V;$$

(3) $$S_{ae_i,e_{2m+1-j}},\ 1 \le j < i \le m,\ a \in D;$$

$$S_{ae_i,e_j},\ m+1 \le i,\ j \le 2m,\ a \in D;$$

$$S_{e_i,e_{2m+1-i}} - S_{e_j,e_{2m+1-j}},\ 1 \le i,\ j \le m;$$

$$S_{e_m,e_{m+1}} + \frac{k}{2}\cdot 1.$$

Then M^+ is a $\mathcal{g}$-submodule of M, necessarily irreducible for $\mathcal{g}$, and the $\mathcal{g}$-module structure of M^+ determines the $\tilde{\mathcal{g}}$-module structure of M. A necessary condition for a given irreducible $\mathcal{g}$-module to be an M^+ as above is that, within M, it is annihilated by all iterated operations of length $k+1$ of the form

$$S_{e_m,u_1} \cdots S_{e_m,u_{k+1}},\ u_i \in V.$$

It is therefore annihilated by all iterations

(4) $$S_{e_m,u_1} \cdots S_{e_m,u_{k+1}} S_{e_{m+1},v_1} \cdots S_{e_{m+1},v_{k+1}},$$

where the u_i, $v_i \in V$. This follows from a relativized highest weight theory, as in [7] - see also [6].

The condition of annihilation by the elements (4) may be translated into a necessary condition, which is given in terms of the annihilation of M^+ by certain elements of the universal associative algebra $\mathcal{U}(\mathcal{g})$ of $\mathcal{g}$, each such element being associated with a double sequence $(u_1,\dots,u_{k+1};\ v_1,\dots,v_{k+1})$ of elements of V. The other necessary conditions entail that each

$$S_{e_m,ae_{m+1}},\ a \in D,\ a^* = a,$$

should act in M^+ as scalar multiplication by $-\frac{k}{2}\tau(a)$, where $\tau(a) \in Z$ is as in §2. (With $a^* = a$, $\tau(a) \in F$.) For each t and for each double sequence $(u_1,\dots,u_t;\ v_1,\dots,v_t)$, we obtain an F-endomorphism of M^+,

$$g_t(u_1,\dots,u_t;\ v_1,\dots,v_t),$$

identified with the action of an element of $\mathcal{U}(\mathcal{g})$, and realized within the action of $\tilde{\mathcal{g}}$ on M as the action of

$$S_{e_m,u_1} S_{e_m,u_2} \quad \cdots \quad S_{e_m,u_t} S_{e_{m+1},v_1} \quad \cdots \quad S_{e_{m+1},v_t} .$$

The mapping $u \otimes v \to g_1(u;v)$ of $V \otimes_F V$ into $\mathrm{End}_F(M^+)$ then extends to a homomorphism φ of $T(V \otimes_F V)$ into $\mathrm{End}_F(M^+)$, whose kernel contains the sets of relators (i) - (iii) of §2. The recursive definition of the f_t in (iv) was chosen so that

$$\varphi(f_t(u_1,\ldots,u_t;\ v_1,\ldots,v_t)) = g_t(u_1,\ldots,u_t;\ v_1,\ldots,v_t).$$

It follows that φ induces a homomorphism of A_k into $\mathrm{End}_F(M^+)$, so that M^+ is an irreducible (right) A_k-module.

Conversely, the Lie morphism $\mathcal{g} \to A_k$ makes each irreducible A_k-module into an irreducible $\mathcal{g}$-module, annihilated by elements of $U(\mathcal{g})$ that correspond to (a canonical projection on $U(\mathcal{g})$ of) the elements (4) of $U(\tilde{\mathcal{g}})$. If Q is such a module, we specify that Q is to be annihilated by the elements (3) and by suitable others to make Q into a module for a suitable (parabolic) subalgebra p of $\tilde{\mathcal{g}}$ containing $\mathcal{g}$, so the $\tilde{\mathcal{g}}$-module

$$U(\tilde{\mathcal{g}}) \otimes_{U(p)} Q$$

has a unique irreducible quotient. This quotient is finite-dimensonal, and Q is its $\mathcal{g}$-submodule annihilated by the elements (3). This procedure gives a one-one correspondence between irreducible A_k-modules and irreducible $\tilde{\mathcal{g}}$-modules generated by irreducible $\mathcal{g}$-submodules satisfying the conditions imposed on M^+.

All the above is multilinear, so is preserved under extension to splitting fields, where the $\tilde{\mathcal{g}}$-modules are determined in terms of highest weights. When $\tilde{\mathcal{g}}$ is split, it is possible to associate with a given k an explicitly determined set of highest weights for the corresponding $\tilde{\mathcal{g}}$-modules M, and a set of irreducible $\mathcal{g}$-modules M^+. These are the irreducible A_k-modules, and provide the basis for our assertions.

REFERENCES

[1] Allen, H. P., Hermitian forms II. Jour. of Algebra 10 (1968), 503-515.

[2] van Drooge, D. C., Spinor theory of quadratic quaternion forms. Proc. Kon. Ned. Akad. Wet. A. 70 (1967), 487-523.

[3] Jacobson, N., Clifford algebras for algebras with involution of type D. Jour. of Algebra 1 (1964), 288-300.

[4] ______________, Lie Algebras. Interscience-Wiley, New York, 1962.

[5] Seip-Hornix, E. A. M., Clifford algebras of quadratic quaternion forms. Proc. Kon. Ned. Akad. Wet. A. 68 (1965), 326-363.

[6] Seligman, G. B., Representations of isotropic simple Lie algebras over general non-modular fields. Queen's Papers in Math. 48 (1978), 528-574.

[7] _____________, Rational constructions of modules for simple Lie algebras. Contemporary Math.

[8] Tits, J., Formes quadratiques, groupes orthogonaux et algèbres de Clifford. Invent. Math. 5 (1968), 19-41.

Department of Mathematics
Yale University
2155 Yale Station
New Haven, Connecticut 06520

Contemporary Mathematics
Volume 13, 1982

ON THE DETERMINATION OF RANK ONE LIE ALGEBRAS OF PRIME CHARACTERISTIC

Georgia M. Benkart*
and
J. Marshall Osborn*

Let L be a finite dimensional Lie algebra over an algebraically closed field K of characteristic $p > 3$. Such an algebra L will be termed a *rank one Lie algebra* provided L has a one-dimensional Cartan subalgebra. Aside from the algebra $s\ell(2)$ of traceless 2×2 matrices, every known simple Lie algebra of rank one is an Albert-Zassenhaus Lie algebra. By an Albert-Zassenhaus algebra ([1, p. 138] or [2, p. 311]), we mean a K-algebra with basis $\{g_\alpha \mid \alpha \in G\}$, where G is any finite additive group of K, and with multiplication given by

$$[g_\alpha, g_\beta] = \{\alpha\psi(\beta) - \beta\psi(\alpha) + \beta - \alpha\}g_{\alpha+\beta}$$

where ψ is any additive mapping of G into K. Each of these algebras is a simple Lie algebra in which g_0 spans a one-dimensional Cartan subalgebra.

The purpose of this present note is to announce the following result.

THEOREM 1. *Let L be a finite dimensional rank one simple Lie algebra over an algebraically closed field K of characteristic $p > 3$. Then L is $s\ell(2)$ or an Albert-Zassenhaus Lie algebra.*

The study of rank one algebras was begun over 20 years ago by Kaplansky [5]. Kaplansky investigated rank one Lie algebras having a one-dimensional Cartan subalgebra spanned by an element h with the roots of h lying in the prime field P and showed that such algebras were $s\ell(2)$ or the Witt algebra. (The Witt algebra is the particular Albert-Zassenhaus algebra obtained by taking $G = P$ and $\psi = 0$).

*Partially supported by NSF grant #MCS-800 2765.

Suppose now that $L = Kh \oplus \sum_{\alpha\neq 0} L_\alpha$ is the Cartan decomposition of a finite dimensional Lie algebra L relative to a one-dimensional Cartan subalgebra spanned by h, and for $\alpha \neq 0$ define $M_\alpha = \{x \in L_\alpha \mid [x, L_-] = 0\}$. Kaplansky determined that in any rank one algebra $\dim L_\alpha / M_\alpha \leq 1$. Kaplansky's investigations have been extended in two different directions. In [4] Block proved that if L is a finite dimensional simple Lie algebra over an algebraically closed field K of characteristic $p > 3$ which is rank one relative to Kh, and if $M_\alpha = 0$ for each root space corresponding to Kh, then L is $s\ell(2)$ or Albert-Zassenhaus. In [6] Yermolaev studied arbitrary finite dimensional rank one Lie algebras L such that the roots of h lie in P and such that $h \in [L, L]$ and showed that these algebras consist of an abelian radical extended by $s\ell(2)$ or the Witt algebra. (The extension is not necessarily split, but Yermolaev is able to provide detailed information concerning the structure of these algebras). Both these results play an important role in our investigations.

To describe our work we let L be a simple rank one Lie algebra relative to Kh, and $L = \sum_\alpha L_\alpha$ be the corresponding Cartan decomposition. If α is a root such that $h \in [L_\alpha, L_{-\alpha}]$, or equivalently, such that $L_\alpha \neq M_\alpha$, then the subalgebra $Y(\alpha) = \sum_{j\in P} L_{j\alpha}$ can be seen to be of the type studied by Yermolaev if one replaces h with $\alpha^{-1}h$. In addition, for each root β and for each $k \in P$, the space $A(k) = \sum_{j\in P} L_{k\beta+j\alpha}$ is a $Y(\alpha)$-module. Thus, our chief technique in proving Theorem 1 is to use the representation theory of the algebras $Y(\alpha)$ developed in [3] to study the modules A(k). We illustrate the type of results obtained by his approach below.

Since the subalgebras $Y(\alpha)$ play a distinguished role in these investigations we define

$$\Omega = \{\alpha \in K \mid L_\alpha \neq M_\alpha\}.$$

Furthermore we denote the radical of $Y(\alpha)$ by $Y'(\alpha)$.

Lemma 2. If $\alpha, \beta \in \Omega$ with $\beta \neq j\alpha$ for any $j \in P$, then $\dim L_{i\alpha+j\alpha} = \dim L_{k\alpha+j\beta}$ for all nonzero $i\alpha + j\beta$, $k\alpha + \ell\beta$.

THEOREM 3. For $\alpha,\beta \in \Omega$, the radical $Y'(\alpha)$ acts nilpotently on each A(k).

This theorem implies that any irreducible $Y(\alpha)$-submodule or irreducible quotient of $Y(\alpha)$-submodules of L can be regarded as an irreducible module for $Y(\alpha) / Y'(\alpha)$ which is a Witt algebra or is $s\ell(2)$. Since each Witt algebra contains a copy of $s\ell(2)$, one can use the theory of $s\ell(2)$-modules [3] to further study the modules $A(k)$. This module approach combines with certain filtration arguments to give

THEOREM 4. Let $\alpha, \beta \in \Omega$ with $\beta \neq j\alpha$ for any $j \in P$. Then there exist roots $\zeta = i\alpha + j\beta$ and $\eta = k\alpha + \ell\beta$ with $\zeta \neq j\eta$ for any $j \in P$ such that one of the following holds for $q, r \in P$ not both zero,

(i) $q\zeta + r\eta \in \Omega$ if and only if $q = \pm 1$, or 0;

(ii) $q\zeta + r\eta \in \Omega$ for all q and r.

Using this theorem it is easy to show that the sum of all root spaces L_γ such that $j\gamma \notin \Omega$ for any $j \in P$ generates a proper ideal, which must then be zero. Thus, each subalgebra generated by a single root string has its structure described by Yermolaev, and each subalgebra generated by two root strings is described by Theorem 4. The remainder of the investigation is devoted to proving that $\sum_{i,j} M_{i\alpha+j\beta}$ is an ideal in $\sum_{i,j} L_{i\alpha+j\beta}$. Once this is accomplished it follows that the root spaces are one-dimensional. If case (i) occurs for any pair of roots, then we can construct a one-dimensional Cartan subalgebra such that case (ii) holds for all pairs of roots. Therefore $M_\alpha \neq 0$ for all roots α relative to that Cartan subalgebra, and by Block's result L is $s\ell(2)$ or an Albert-Zassenhaus algebra.

BIBLIOGRAPHY

[1] A. A. Albert and M. S. Frank, Simple Lie algebras of characteristic p, Univ. e. Politec. Torino Rend Sem. Mat. 14 (1954)-1955), 117-139.

[2] G. M. Benkart I. M. Isaacs, and J. M. Osborn, Albert-Zassenhaus Lie algebras and isomorphisms, J. of Alg. 57 (1979), 310-338.

[3] G. M. Benkart and J. M. Osborn, Representations of rank one Lie algebras of characteristic p, to appear Proc. Rutgers Conference on Lie algebras, Rutgers University, May 1981.

[4] R. E. Block, On Lie algebras of rank one, Trans. A.M.S. 112 (1964),19-31, MR 28, #4013

[5] I. Kaplansky, Lie algebras of characteristic p, Trans. A.M.S. 89 (1958), 149-183, MR 20, #5799.

[6] J. B. Yermolaev, Lie algebras of rank 1 with root systems in the prime field, (Russian) Izv. Vyšs Učebn, Zaved. Mat 5 (120) (1972) 38-50.

Contemporary Mathematics
Volume 13, 1982

RESTRICTED SIMPLE LIE ALGEBRAS OF RANK 2 OR OF TORAL RANK 2

Richard E. Block

Dedicated to Nathan Jacobson

The concept of restricted Lie algebra was introduced by Nathan Jacobson in the late 1930's; he also did much of the early work on restricted simple Lie algebras. Recall that a Lie algebra L (herein always assumed to be finite-dimensional over an algebraically closed field F of prime characteristic $p > 7$) is called restricted (or a p-algebra) if the pth power of an inner derivation is always inner. This gives an additional operation (uniquely determined if L is centerless), denoted $x \mapsto x^p$ and satisfying $(\text{ad } x)^p = \text{ad } x^p$, which generalizes the associative p^{th} power in the full linear Lie algebra. For some of the properties of restricted Lie algebras, see e.g. [J], [S] or [H]. The standard conjecture (that of Kostrikin and Šafarevič) on restricted simple Lie algebras is that they are all either of classical type (i.e. the simple analogues over F of the finite-dimensional simple Lie algebras $A_n, \ldots, F_4, G_2$ over $\mathbb{C}$) or (restricted) of Cartan type (i.e. the p-truncated analogues over F of the infinite-dimensional simple Lie algebras over $\mathbb{C}$ used in classifying Lie pseudogroups). The restricted simple algebras of Cartan type are the algebras W_n $(n \geq 1)$, $S_n^{(1)}$ $(n \geq 3)$, $H_n^{(2)}$ (n even, ≥ 2), and $K_n^{(1)}$ (n odd, ≥ 3). The algebra W_n is the np^n-dimensional Jacobson-Witt algebra $\text{Der } F[x_1,\ldots,x_n]/(x_1^p,\ldots,x_n^p)$ and S_n, H_n and K_n are certain subalgebras of W_n; for more details see [KS,W1].

In this note we shall describe recent work determining the restricted simple (and indeed the restricted semisimple) Lie algebras of rank 2 (i.e., with a 2-dimensional Cartan subalgebra (C.s.a.)); the rank 1 case, in which the algebras are sl(2) and W_1, is a classical theorem of Kaplansky [Kp]. We shall also indicate some current work on a conjectured extension of the rank 2 case to the case of toral rank 2 (i.e., having a 2-dimensional maximal torus). Both the work on the rank 2 case, which is to appear with full details in [BW], and the current work on the toral rank 2 case are joint work with Robert L. Wilson. The results for rank 2 simple algebras are as follows.

THEOREM 1. *Let L be a restricted simple Lie algebra (finite-dimensional over an algebraically closed field F of characteristic $p > 7$) of rank 2. Then L is classical* (i.e., $\cong A_2$, C_2 *or* G_2) *or* $L \cong W_2$.

*Research supported in part by National Science Foundation grant MCA79-03161.

0271-4132/82/0536/$02.25

THEOREM 2. *There are, up to isomorphism, precisely* 19 *restricted non-simple semisimple Lie algebras of rank* 2.

The precise list of these algebras is given in [BW]. In addition to $sl(2) \oplus sl(2)$, $sl(2) \oplus W_1$ and $W_1 \oplus W_1$, they consist of certain algebras between $S \otimes (F[x]/(x^p))$ and $S \otimes (F[x]/(x^p)) + 1 \otimes W_1$ where $S = sl(2)$ or W_1 (8 algebras), the p-closure $\bar{S}$ of 3 nonrestricted simple algebras S of toral rank one, and 5 algebras between $H_2^{(2)}$ and Der $H_2^{(2)}$.

Suppose L is restricted. Recall that a (necessarily abelian) subalgebra T is called a torus if it contains no nonzero element x which is nilpotent (i.e., such that $x^{p^e} = 0$ for some e). A theorem of Jacobson says that T has a basis $\{x_1,\ldots,x_r\}$ consisting of toral elements, i.e., $x_i^p = x_i$. The C.s.a.'s are the centralizers of the maximal tori. We call a C.s.a. H *standard* if H has the form $T + I$ where T is the (unique) maximal torus of H and I is a nil ideal of H. If L is simple, then [W2] any C.s.a. is standard. If L has a maximal torus of dimension n, then L is said to have *toral rank* n.

Conjecture 1. If L *is restricted simple (over* F *as above) of toral rank* 2, *then* L *is classical or* $L \cong W_2$, $S_3^{(1)}$, $H_4^{(1)}$, *or* $K_3^{(1)}$.

The proof of Theorem 1 uses Theorem 2 in an inductive argument. The proof of Theorem 2 is based on the structure theorem of [B1] for semisimple Lie algebras, but also uses further delicate arguments, some involving nonrestricted simple algebras (of toral rank 1), to obtain the precise list of algebras. This classification of restricted semisimple algebras of rank 2 in Theorems 2 and 1 has already been used as a crucial tool in Wilson's classification (announced in [W3]) of the restricted simple Lie algebras of arbitrary rank with a C.s.a. which is a torus.

Similarly the current work on Conjecture 1 involves (for an inductive argument) the determination of restricted nonsimple semisimple algebras with a standard C.s.a. of toral rank 2. The list is no longer finite, since, e.g., for every $r \geq 1$ it now includes certain algebras of derivations of $sl(2) \otimes F[x_1,\ldots,x_r]/(x_1^p,\ldots,x_r^p)$. Let us call the expected determination of these algebras Conjecture 2. It is hoped that the current work on Conjectures 1 and 2, when completed, can be used in work on restricted simple algebras of arbitrary rank in a somewhat similar way to that in which Theorems 1 and 2 have been used by Wilson in the case of algebras with a C.s.a. which is a torus. Thus a complete classification of the restricted simple algebras (over F) may be reasonably near.

We now indicate some of the ideas used in the proof of Theorem 1 (for the details, see [BW]) and how some of them are being extended in work on Conjecture 1.

Let L_0 be a maximal subalgebra of a simple Lie algebra L. A standard technique in investigations of simple Lie algebras is the use of a filtration of L associated with L_0, obtained by letting L_{-1} be minimal among subspaces of L properly containing L_0 and invariant under ad L_0; $L_{i+1} = \{x \in L_i | [L_{-1},x] \subseteq L_i\}$, $i \geq 0$; and $L_{-i-1} = [L_{-1},L_{-i}] + L_{-i}$, $i \geq 1$. Let $G = \Sigma G_i$ $(G_i = L_i/L_{i+1})$ be the associated graded algebra. There is a recognition theorem for Cartan type algebras (the result of work of Kostrikin and Šafarevič [KS], Kac [Kc1,Kc2] and Wilson [W4]) which states that if G satisfies certain properties, then L is of classical or Cartan type. Specifically, in the case in which L is restricted, if G_0 is a direct sum of restricted ideals which are either classical simple, **sl(pn)**, **gl(pn)**, or abelian, if the action of G_0 on G_1 is faithful, and if no nonzero ideal of the subalgebra of G generated by $\Sigma_{i<2}G_i$ is contained in $\Sigma_{i<0}G_i$, then L is either classical or Cartan (i.e., $X_n^{(2)}$ with $X = W,S,H$ or K). Unfortunately, even in algebras of Cartan type, not every maximal subalgebra L_0 gives a G satisfying these hypotheses. The proof of Theorem 1 above involves a new way of selecting a suitable L_0. Toward this end, let H be a 2-dimensional C.s.a. of the restricted simple algebra L. It follows from the classification [W5] of simple algebras of toral rank one that H is a torus. Let Δ be the additive subgroup of H^* generated by the set of roots. For $\alpha \in \Delta$ we set $K_\alpha = \{x \in L_\alpha | \alpha(x,L_{-\alpha}) = 0\}$, $T_\alpha = \{x \in L_\alpha | [x,L_{-\alpha}] = 0\}$, and $L^{(\alpha)} = H + \Sigma_i L_{i\alpha}$. Then $L^{(\alpha)}/\text{rad } L^{(\alpha)}$ is restricted semisimple of rank ≤ 1, hence (by [Kp]) $= 0$, sl(2) or W_1. We call $\alpha \in \Delta$ <u>proper</u> if $\alpha \neq 0$ and $L_{i\alpha} = K_{i\alpha}$ for some (hence most) i, $1 \leq i \leq p-1$; this is related to the choice of the maximal torus $x\frac{d}{dx}$ rather than $(x+1)d/dx$ in W_1. Cartan subalgebras can be switched via the exponentials of Winter [Wn] so as to send an improper root to a proper root. We call a rank 2 C.s.a. <u>optimal</u> if among all such it has the maximum number of $L^{(\alpha)}$ such that α is proper, and we call a maximal subalgebra L_0 <u>distinguished</u> if, for some optimal C.s.a. H, L_0 contains H and all (corresponding) T_α. We also prove (in [BW]) some useful bounds for the codimensions of the T_α, e.g., if $\alpha \in \Delta$ then $\dim L_\alpha/T_\alpha \leq 3$, and if α is proper then $T_{i\alpha} = L_{i\alpha}$ $(1 \leq i \leq p-1)$ except for $0,2$ or 4 values of i.

We then let L_0 be a distinguished maximal subalgebra, take a corresponding filtration, and use the above ideas in showing that the hypotheses for the recognition theorem for Cartan type algebras hold. The main task here is to determine the possible G_0. The argument splits into two cases: 1) the center $\mathfrak{z}(G_0) = 0$, and 2) $\mathfrak{z}(G_0) \neq 0$. If $\mathfrak{z}(G_0) = 0$ then G_0 is restricted semisimple of rank 2, and so, by Theorem 2 and an inductive hypothesis, G_0

is one of the algebras in the conclusions of Theorems 1 and 2, and detailed arguments then eliminate all of these algebras as possible G_0. If $z(G_0) \neq 0$ then $z(G_0)$ is 1-dimensional and $G_0/z(G_0)$ has rank 1. Very long arguments then show that either $G_1 = 0$ or the hypotheses of the recognition theorem for Cartan type algebras hold (in particular with $G_0 \cong sl(2) + Fz$). The proof in each of cases 1) and 2) makes use of Weisfeiler's theorem [Wf] on the structure of certain graded algebras. Finally the case $G_1 = 0$ is eliminated using an extension [B2] of the Mills-Seligman recognition theorem for algebras of classical type.

In considering a generalization of the above to the case of toral rank 2, we may assume that $H = T + I$ as above, and we then change the definition of T_α to: $T_\alpha = \{x \in L_\alpha | [x, L_{-\alpha}] \subseteq I\}$. Now $L^{(\alpha)}/\text{rad } L^{(\alpha)}$ has toral rank ≤ 1, and thus one of our first tasks is to determine the restricted semisimple Lie algebras with a standard C.s.a. of toral rank 1; in addition to $sl(2)$ and W_1, the list now includes some algebras of derivations of **two (one of them not** restricted) simple type H algebras of toral rank 1, for a total of **eight** algebras. We are then able to generalize to this situation the definition of proper root (and thus of distinguished maximal subalgebra) and the results on codimensions of the T_α.

It appears that the main steps in the proof of Theorem 1 can in principle be extended to the toral rank 2 case. Thus there is cause for optimism regarding a proof of Conjecture 1 (and 2); there remain however a mass of details to be worked out, and, perhaps, some surprises.

REFERENCES

[B1] Block, R. E., Determination of the differentiably simple rings with a minimal ideal, Ann. of Math. 90 (1969), 433-459.

[B2] ___________, On the Mills-Seligman axioms for Lie algebras of classical type, Trans. Amer. Math. Soc. 121 (1966), 378-392.

[BW] Block, R. E. and Wilson, R. L., The simple Lie p-algebras of rank two, Ann. of Math., **115 (1982), 93-168.**

[H] Humphreys, J., Restricted Lie algebras (and beyond). These proceedings, pp. **91-98.**

[J] Jacobson, N., Lie Algebras, Interscience, New York, 1962.

[Kc 1] Kac, V. G., The classification of the simple Lie algebras over a field of nonzero characteristic (Russian), Izv. Akad. Nauk SSSR 34 (1970), 385-408; English transl., Math. USSR-Izv. 4 (1970), 391-413.

[Kc 2] _________, Description of filtered Lie algebras with which graded Lie algebras of Cartan type are associated (Russian), Izv. Akad. Nauk SSSR 38 (1974), 800-838; Errata, 40 (1976), 1415; English transl. Math. USSR-Izv. 8 (1974), 801-835; 10 (1976), 1339.

[Kp] Kaplansky, I., Lie algebras of characteristic p, Trans. Amer. Math. Soc. 89 (1958), 149-183.

[KS] Kostrikin, A. I. and Šafarevič, I. R., Graded Lie algebras of finite characteristic (Russian), Izv. Akad. Nauk SSSR 33 (1969), 251-322; English transl. Math. USSR-Izv. 3 (1969), 237-304.

[S] Seligman, G. B., Modular Lie Algebras, Ergebnisse der Math., No. 40, Springer-Verlag, New York, 1967.

[Wf] Weisfeiler, B., On the structure of the minimal ideal of some graded Lie algebras in characteristic $p > 0$, J. Algebra 53 (1978), 344-361.

[W1] Wilson, R. L., The classification problem for simple Lie algebras of characteristic p, Algebra Carbondale 1980 Proceedings, in Lecture Notes in Math., No. 840, Springer-Verlag, Berlin, 1981.

[W2] ____________, Cartan subalgebras of simple Lie algebras, Trans. Amer. Math. Soc. 234 (1977), 435-446.

[W3] ____________, Restricted simple Lie algebras with toral Cartan subalgebras. These Proceedings, pp. 273-278.

[W4] ____________, A structural characterization of the simple Lie algebras of generalized Cartan type over fields of prime characteristic, J. Algebra 40 (1976), 418-465.

[W5] ____________, Simple Lie algebras of toral rank one, Trans. Amer. Math. Soc. 236 (1978), 287-295.

[Wn] Winter, D. J., On the toral structure of Lie p-algebras, Acta Math. 123 (1969), 70-81.

University of California
Riverside, California 92521

Contemporary Mathematics
Volume 13, 1982

RESTRICTED SIMPLE LIE ALGEBRAS WITH TORAL CARTAN SUBALGEBRAS

by

Robert Lee Wilson[1]

Dedicated to Nathan Jacobson, with admiration and gratitude, on the occasion of his retirement

In this paper we announce the classification of the finite-dimensional restricted simple Lie algebras over an algebraically closed field F of prime characteristic $p > 7$ which contain a toral Cartan subalgebra. (Recall that a Lie algebra over F is restricted (Jacobson [8]) if it has an extra mapping $x \mapsto x^p$ satisfying $(\text{ad } x)^p = \text{ad } x^p$; that an element x in a restricted Lie algebra is semisimple if $x \in \text{span}\{x^p, x^{p^2}, \ldots\}$; and that a torus is a subalgebra (necessarily abelian) every element of which is semisimple.) Such an L is either of classical type (i.e., an analogue over F of one of the finite-dimensional simple Lie algebras over $\mathbb{C}$) or a Jacobson-Witt algebra W_n (= Der $F[x_1,\ldots,x_n]/(x_1^p,\ldots,x_n^p)$) for some $n \geq 1$. We will outline the proof of this result here. Details will appear elsewhere.

This result verifies, for the special case of algebras with toral Cartan subalgebras, the conjecture of Kostrikin and Šafarevič [13] that all finite-dimensional restricted simple Lie algebras over F are either classical or of Cartan type. (The algebras of Cartan type are analogues (not necessarily restricted) over F of the infinite Lie algebras of Cartan over $\mathbb{C}$. The restricted Lie algebras of Cartan type form four infinite families ([10, Theorem 2]): W_n, $n \geq 1$; $S_n^{(1)}$, $n \geq 3$; $H_{2n}^{(2)}$, $n \geq 1$; and $K_{2n+1}^{(1)}$, $n \geq 1$. The algebras of types S, H, and K were originally discovered by M. Frank [1, 6, 7]. It is also conjectured [10, 12] that any (not necessarily

[1] The author gratefully acknowledges partial support from NSF grant MCS-8003000 and the Rutgers Faculty Academic Study Plan, the hospitality of the Institute for Advanced Study during part of the preparation of this work, and G. Seligman's **helpful comments on this work.**

0271-4132/82/0537/$02.50

restricted) finite-dimensional simple Lie algebra over F is classical or of Cartan type.)

An important technique in work on simple Lie algebras L is the construction of a filtration

$$L = L_{-k} \supseteq \dots \supseteq L_0 \supseteq \dots \supseteq L_r = (0)$$

determined by a maximal subalgebra L_0 of L [16]. Here L_{-1} is such that L_{-1}/L_0 is irreducible as an (ad L_0)-module,

$$L_i = [L_{i+1}, L_{-1}] + L_{i+1} \quad \text{for } i < -1,$$

and

$$L_{i+1} = \{x \in L_i \mid [L_{-1}, x] \subseteq L_i\} \quad \text{for } i \geq 0.$$

Let $G_i = L_i/L_{i+1}$ so that $G = \sum G_i$ is the associated graded algebra. The following "recognition theorem" (which combines results of Kostrikin-Šafarevič [14], Kac [9, 10] and Wilson [17]) characterizes algebras of Cartan type in terms of such a filtration:

THEOREM A: *Let* L *be a finite-dimensional simple Lie algebra over* F *and let* G *be as above. Assume that:* i) G_0 *is a direct sum of restricted ideals each of which is either classical simple,* $sl(n)$, $gl(n)$, *or* $pgl(n)$ *where* $p|n$, *or abelian;* ii) *the action of* G_0 *on* G_{-1} *is restricted; and* iii) *if* $x \in G_i$ *for* $i \leq 0$ *and* $[x, G_1] = (0)$, *then* $x = 0$. *Then* L *is either classical or of Cartan type.*

As mentioned above the restricted Lie algebras of Cartan type are just the W_n, $S_n^{(1)}$, $H_{2n}^{(2)}$, and $K_{2n+1}^{(1)}$. For each of these Demuskin [4, 5] has determined all conjugacy classes of Cartan subalgebras under the automorphism group. In particular, Demuskin has shown that all Cartan subalgebras of W_n are tori and that no Cartan subalgebra of $S_n^{(1)}$, $H_{2n}^{(2)}$ or $K_{2n+1}^{(1)}$ is a torus. Thus we obtain:

COROLLARY B: *Let* L *be as above and assume in addition that* L *is restricted and contains a toral Cartan subalgebra. Then* L *is either classical or* W_n *for some* $n \geq 1$.

Thus to prove our result it is sufficient to show that if L is a finite-dimensional restricted simple Lie algebra over F containing a toral Cartan subalgebra then L contains a maximal subalgebra L_0 for which the hypotheses of Theorem A are satisfied. To do this we consider in detail the rank one and two subalgebras of L. More precisely, if L has Cartan decomposition $L = H + \Sigma_{\gamma \in \Gamma} L_\gamma$ we define

$$L^{(\alpha)} = \Sigma_{i\in \mathbf{Z}} L_{i\alpha} \quad \text{for } \alpha \in \Gamma,$$

$$S^{(\alpha)} = \text{solv } L^{(\alpha)},$$

$$L^{(\alpha,\beta)} = \Sigma_{i,j\in \mathbf{Z}} L_{i\alpha+j\beta} \quad \text{for } \alpha,\beta \in \Gamma,$$

and

$$S^{(\alpha,\beta)} = \text{solv } L^{(\alpha,\beta)} .$$

Then $L^{(\alpha)}/S^{(\alpha)}$ is restricted semisimple of rank ≤ 1 and so is (0), or simple. Thus by Kaplansky's Theorem [11] it is (0), $sl(2)$, or W_1. Similarly $L^{(\alpha,\beta)}/S^{(\alpha,\beta)}$ is restricted semisimple of rank ≤ 2 and so is one of the algebras determined by Block and Wilson [3]. In particular, $L^{(\alpha,\beta)}/S^{(\alpha,\beta)}$ is either classical or isomorphic to a subalgebra of W_2.

Now define $K_\alpha = \{x \in L_\alpha | \alpha([x, L_{-\alpha}]) = (0)\}$ and say that α is <u>proper</u> if $L_{i\alpha} = K_{i\alpha}$ for some i, $1 \leq i \leq p-1$. It is easily seen (using Demuskin's result for W_1) that α is proper if and only if $L^{(\alpha)}/S^{(\alpha)}$ is is (0), $sl(2)$ or contains an ad H-invariant subalgebra of codimension 1.

For our purposes proper roots are easy to work with. Therefore our frist order of business is to show that we may assume all roots are proper.

<u>LEMMA 1</u>: <u>Let L be a finite-dimensional restricted Lie algebra over F containing a toral Cartan subalgebra. Then L contains a toral Cartan subalgebra for which every root is proper.</u>

To prove this let H be any toral Cartan subalgebra of L. For $x \in L_\alpha$ we define $E^x = \sum_{i=0}^{p-1} (\text{ad } x)^i/i!$ (Winter [18]). One sees that if $H' = E^x H$ then $E^x : H \to H'$ is a bijection. Thus we can define a map $' : H^* \to H'^*$ by

$$\beta'(E^x h) = \beta(h) + \alpha(h)\beta(x^p) \quad \text{for all } h \in H.$$

We show that if α is improper then $x \in L_\alpha$ can be chosen so that (with respect to H') α' is proper. We claim that the number of proper roots with respect to H' must be larger than the number with respect to H. Thus a toral Cartan subalgebra with the maximum number of proper roots must have all of its roots proper. An easy counting argument shows that it is sufficient to prove this claim for all $L^{(\alpha,\beta)}/S^{(\alpha,\beta)}$, and hence is sufficient to prove it for all the rank two restricted semisimple algebras. Since the claim is clearly true for classical algebras (for all roots are proper) it is sufficient to prove it for rank two semisimple subalgebras of W_2. Since a two-dimensional torus in W_2 is maximal Demuskin's results show that H can be assumed to be spanned by $\{x_1\partial/\partial x_1, x_2\partial/\partial x_2\}$, $\{(x_1 + 1)\partial/\partial x_1, x_2\partial/\partial x_2\}$, or $\{(x_1 + 1)\partial/\partial x_1, (x_2 + 1)\partial/\partial x_2\}$. The first case cannot occur (for all roots with respect to that torus are proper). The second case can be handled easily (for when E^x is applied all roots must become proper). The third case is handled by explicit calculation.

Now define an ad H-invariant subspace $Q = \Sigma Q_\alpha \subseteq L$ as follows: Let $Q^{(\alpha)}$ denote $\Sigma_{i\in\mathbb{Z}} Q_{i\alpha}$. Set $Q^{(\alpha)} = L^{(\alpha)}$ if $L^{(\alpha)}/S^{(\alpha)} = (0)$ or $sl(2)$ and let $Q^{(\alpha)}$ be the unique subalgebra of $L^{(\alpha)}$ of codimension one containing $H + S^{(\alpha)}$ otherwise (which exists as α is proper).

The main step in our proof is

LEMMA 2: Let L be a finite-dimensional restricted simple Lie algebra over F with a toral Cartan subalgebra H for which all roots are proper. Then Q is a subalgebra of L and $|\{\gamma \in \mathbb{Z}\alpha + \mathbb{Z}\beta \mid L_\gamma \neq Q_\gamma\}| \leq 2$.

The proof of this lemma also may be reduced to the rank two case. For example, to show that Q is a subalgebra it is sufficient to check $[Q_\alpha, Q_\beta] \subseteq Q_{\alpha+\beta}$ for all α and β. This may be checked in $L^{(\alpha,\beta)}$ and hence in $L^{(\alpha,\beta)}/S^{(\alpha,\beta)}$. However, the conclusion of the lemma does not hold for all rank two semisimple algebras. We must therefore show that not every rank two semisimple can occur as $L^{(\alpha,\beta)}/S^{(\alpha,\beta)}$ where L is simple. This depends on a number of detailed calculations.

There are now two cases, $L = Q$ and $L \neq Q$. The first is taken care of by the following characterization of classical algebras which is of independent interest.

PROPOSITION 3: Let L be a finite-dimensional restricted simple Lie algebra over an algebraically closed field of characteristic $p > 7$. Assume that for every α, $L^{(\alpha)}/S^{(\alpha)}$ is (0) or $sl(2)$. Then L is classical.

The proof divides into two parts. The first is to show that if every $L^{(\alpha,\beta)}/S^{(\alpha,\beta)}$ is classical semisimple then L is classical. This is fairly easy (essentially since the Mills-Seligman axioms [15] are stated in terms of pairs of roots). The more difficult part of the proof is devoted to showing that under the hypotheses of the proposition every $L^{(\alpha,\beta)}/S^{(\alpha,\beta)}$ must be classical semisimple. This depends on calculations similar to those in the proof of Lemma 2.

If $Q \neq L$ we let L_0 be a maximal subalgebra containing Q and construct a filtration and the associated graded algebra as above. Using Lemma 2 we immediately obtain that $L = L_{-1}$ and that if Γ_{-1} denotes the set of weights of G_0 acting on G_{-1} then for $\alpha, \beta \in \Gamma_{-1}$ we have $|\Gamma_{-1} \cap (\mathbb{Z}\alpha + \mathbb{Z}\beta)| \leq 2$ and each $\alpha \in \Gamma_{-1}$ has multiplicity one. We can then (using Block's characterization of the classical and Albert-Zassenhaus algebras [2]) prove:

LEMMA 4: $G_0^{(1)} / z(G_0^{(1)})$ is a direct sum of ideals each of which is classical simple.

We are now almost ready to apply Corollary B. (We need only eliminate one case (where $G_0^{(1)}$ has more than one summand and is a nonsplit central extension of $G_0^{(1)} / z(G_0^{(1)})$) which is not covered by Corollary B.) We then obtain:

THEOREM: Let L be a finite-dimensional restricted simple Lie algebra over an algebraically closed field of characteristic $p > 7$ containing a toral Cartan subalgebra. Then L is classical or isomorphic to some W_n, $n \geq 1$.

REFERENCES

[1] Albert, A. A. and M. S. Frank, Simple Lie algebras of characteristic p, Univ. e Politec. Torino. Rend. Sem. Mat. 14 (1954-55), 117-139.

[2] Block, R. E., On the Mills-Seligman axioms for Lie algebras of classical type, Trans. Amer. Math. Soc. 121 (1966), 378-392.

[3] Block, R. E. and R. L. Wilson, The simple Lie p-algebras of rank two, Ann. of Math., 115 (1982), 93-168.

[4] Demuskin, S. P., Cartan subalgebras of the simple Lie p-algebras W_n and S_n (Russian), Sibirsk. Mat. Z. 11 (1970), 310-325; English transl., Siberian Math. J. 11 (1970), 233-245.

[5] ______________, Cartan subalgebras of simple nonclassical Lie p-algebras (Russian), Izv. Akad. Nauk SSSR Ser. Mat. 36 (1972), 915-932; English transl., Math. USSR-Izv. 6 (1972), 905-924.

[6] Frank, M. S., A new class of simple Lie algebras, Proc. Nat. Acad. Sci. U.S.A. 40 (1954), 713-719.

[7] ___________, Two new classes of simple Lie algebras, Trans. Amer. Math. Soc. 112 (1964), 456-482.

[8] Jacobson, N., Restricted Lie algebras of characteristic p, Trans. Amer. Math. Soc. 50 (141), 15-25.

[9] Kac, V. G., The classification of the simple Lie algebras over a field with nonzero characteristic (Russian), Izv. Akad. Nauk SSSR Ser. Mat. 34 (1970), 385-408: English transl., Math. USSR-Izv. 4 (1970), 391-413.

[10] ________, Description of filtered Lie algebras with which graded Lie algebras of Cartan type are associated (Russian), Izv. Akad. Nauk SSSR Ser. Mat. 38 (1974), 800-838; Errata, 40 (1976), 1415; English transl., Math. USSR-Izv. 8 (1974), 801-835; Errata, 10 (1976), 1339.

[11] Kaplansky, I., Lie algebras of characteristic p, Trans. Amer. Math. Soc. 89 (1958), 149-183.

[12] Kostrikin, A. I., Variations modulaires sur un thème de Cartan, Actes Congres Intern. Math., Nice 1 (1970), 285-292.

[13] Kostrikin, A. I. and I. R. Šafarevič, Cartan pseudogroups and Lie p-algebras (Russian), Dokl. Akad. Nauk SSSR 168 (1966), 740-742; English transl., Soviet Math. Dokl. 7 (1966), 715-718.

[14] ________, Graded Lie algebras of finite characteristic (Russian), Izv. Akad. Nauk SSSR Ser. Math. 33 (1969), 251-322; English transl., Math. USSR-Izv. 3 (1969), 237-304.

[15] Seligman, G. B., Modular Lie Algebras, Ergebnisse der Math. und ihrer Grenzgebiete, Band 40, Springer-Verlag New York, Inc., New York, 1967.

[16] Weisfeiler, B. Ju., Infinite dimensional filtered Lie algebras and their connection with graded Lie algebras (Russian), Funkcional. Anal. i Prilozen. 2 (1968), 94-95.

[17] Wilson, R. L., A structural characterization of the simple Lie algebras of generalized Cartan type over fields of prime characteristic, J. Algebra 40 (1976), 418-465.

[18] Winter, D. J., On the toral structure of Lie p-algebras, Acta Math. 123 (1969), 70-81.

Department of Mathematics
Rutgers University
New Brunswick, NJ 08903

Contemporary Mathematics
Volume 13, 1982

COMPOSITION TRIPLES

by Kevin McCrimmon

I want to discuss a class of algebraic objects that arose in differential geometry, and which can be classified using the methods that Professor Jacobson developed for composition algebras. Throughout I will assume all quadratic forms and algebras live on a finite-dimensional vector space X over a field Φ of arbitrary characteristic. In the geometric case the field of interest is the real numbers, but the classification is valid over arbitrary fields. Let's begin the recalling the classical results concerning composition algebras.

1. QUADRATIC FORMS PERMITTING COMPOSITION

A nondegenerate quadratic form Q <u>permits composition</u> if $Q(x)Q(y) = Q(z(x,y))$, in which case $z(x,y) = x\cdot y$ defines the structure of a linear nonassociative algebra C on X so that

$$(1.1) \qquad Q(x\cdot y) = Q(x)Q(y) .$$

The possible quadratic forms permitting composition can be classified by classifying the corresponding <u>composition algebras</u>.

<u>STRUCTURE THEOREM</u>: (Hurwitz, Dickson, Albert, Schafer, Kaplansky, ...)

(1.2) <u>A general composition algebra</u> (C',Q') <u>is isotopic to a unital algebra</u> (C,Q): <u>there are maps</u> $(C',Q') \xrightarrow{f_i} (C,Q)$ <u>with</u> $f_0(x\cdot' y) = f_1(x)\cdot f_2(y)$, $Q(f_i(x)) = \alpha_i Q'(x)$, $\alpha_0 = \alpha_1\alpha_2 \neq 0$ <u>and there is</u> $u \in C$ <u>with</u> $u\cdot x = x\cdot u = x$ <u>(implying</u> $Q(u) = 1$).

---

* Research partially supported by NSF Grant. Full details will appear elsewhere.

0271-4132/82/0538/$03.00

(1.3) <u>A unital composition algebra is one of</u>

(I) Φ <u>of dimension</u> 1 (<u>or a purely inseparable field extension of exponent</u> 1 <u>in characteristic</u> 2)

(II) <u>a quadratic extension of dimension</u> 2 (<u>commutative, but with nontrivial involution</u>)

(III) <u>a quaternion algebra of dimension</u> 4 (<u>associative, but not commutative</u>)

(IV) <u>an octonion (Cayley) algebra of dimension</u> 8 (<u>alternative, but not associative</u>).

In his paper "Composition algebras and their automorphisms" (Rendi. Circ. Mat. Palermo 7 (1958), 55-80), Jacobson recovered this classification, and as in so many areas of algebra Jake showed us how to do it right: a composition algebra is built up by repeated application of the Cayley-Dickson doubling process:

(1.4) $C = C(B,\mu) = B \oplus B\ell$ for ANY nonsingular subalgebra B and $\ell \in B^{\perp}$ with $\mu = -Q(\ell) \neq 0$.

The importance of this process is not merely that it gives an elegant organic construction of the composition algebras, but it also gives a construction of isomorphisms. One particular application of this is

(1.5) If C, C' have the same norm and unit then C' is an elemental iso- or anti-isotope $C^{(t)}$ of C,

$$x\cdot' y = (xt^{-1})(ty) \quad \text{or} \quad (yt^{-1})(tx)$$

for some invertible t in C.

2. QUADRATIC FORMS PERMITTING TRIPLE COMPOSITION

In an algebraic investigation of isoparametric hypersurfaces in spheres, J. Dorfmeister and E. Neher encountered a space with a triple product (a trilinear map $X \times X \times X \to X$, denoted by $(x,y,z) \to \{x\ y\ z\}$) which <u>permitted triple composition</u> with respect to a nondegenerate quadratic form in analogy with (1.1):

(2.1) $Q(\{x\ y\ z\}) = Q(x)Q(y)Q(z)$.

It was important to be able to describe the possible <u>composition triples</u> up to (something slightly weaker than) isotopy; the hope was that there would be a "new" composition triple, leading to a new isoparametric hypersurface. As in (1.2), we can always pass to a unital isotope:

2.2. UNITALIZATION THEOREM: *A general composition triple* (T',Q') *is isotopic to a unital one* (T,Q): *there are maps* $(T',Q') \overset{f_i}{\to} (T,Q)$ *with* $f_0(\{x\ y\ z\}') = \{f_1(x) f_2(y) f_3(z)\}$, $Q(f_i(x)) = \alpha_i Q'(x)$, $\alpha_0 = \alpha_1\alpha_2\alpha_3 \neq 0$, *and there is* $u \in T$ *with* $\{u\ u\ x\} = \{u\ x\ u\} = \{x\ u\ u\} = x$ (*implying* $Q(u) = \pm 1$, *where we may normalize* Q *so that* $Q(u) = 1$). *Indeed, if* $\{u_1\ u_2\ u_3\} = u$ *for* $Q'(u) \neq 0$ *we can take* $Q(x) = Q'(u)^{-1} Q'(x)$ *and* $\{x\ y\ z\} = \{R(u_2,u_3)^{-1}x,\ M(u_1,u_3)^{-1}y,\ L(u_1,u_2)^{-1}z\}'$.

The next step is to show that a unital composition triple is inextricably linked to a composition algebra.

2.3. ASSOCIATED COMPOSITION ALGEBRA THEOREM: *Any unital composition triple* (T,Q,u) *lives in a unital composition algebra* $C(T,Q,u)$ *with the same unit* u *and norm* Q, *and product given by*

$$x \cdot_u y = \{x\ u\ y\} .$$

The products

$$x \cdot_{u,L} y = \{x\ y\ u\} , \quad x \cdot_{u,R} y = \{u\ x\ y\}$$

also determine composition algebras with the same unit and norm, thus by (1.5) *are elemental isotopes or anti-isotopes of* C. *We have* 4 *possible cases:*

(A) $\{x\ y\ u\} = (xt^{-1})(ty), \quad \{u\ y\ z\} = (ys^{-1})(sz)$

(B) $\{x\ y\ u\} = (xt^{-1})(ty), \quad \{u\ y\ z\} = (zs^{-1})(sy)$

(C) $\{x\ y\ u\} = (yt^{-1})(tx), \quad \{u\ y\ z\} = (ys^{-1})(sz)$

(D) $\{x\ y\ u\} = (yt^{-1})(tx), \quad \{u\ y\ z\} = (zs^{-1})(sy)$

for invertible $t,s \in C$.

We say (T,Q,u) has Type I-IV if the associated composition algebra $C(T,Q,u)$ does, and we henceforth write 1 for u. We begin by exhibiting the basic examples of "standard" composition triples living inside composition algebras.

3. STANDARD COMPOSITION TRIPLES

There is a natural way to get composition triples from composition algebras. If (C,Q) is a unital composition algebra then the left and right standard triple products

$$(3.1)\qquad \{x\ y\ z\}_L = (xy)z, \quad \{x\ y\ z\}_R = x(yz)$$

admit triple composition with Q, eg. $Q((xy)z) = Q(xy)Q(z) = Q(x)Q(y)Q(z)$ by double use of (1.1). The same applies to any permutation

$$(3.2)\qquad \{x_1\ x_2\ x_3\}_\pi = \{x_{\pi(1)}\ x_{\pi(2)}\ x_{\pi(3)}\}$$

of a composition triple. A standard triple is a permutation of a left or right standard triple. These are all unital composition triples, the unit being the unit of the underlying composition algebra.

In general there are 12 possible products $(x_{\pi(1)}x_{\pi(2)})x_{\pi(3)}$ and $x_{\pi(1)}(x_{\pi(2)}x_{\pi(3)})$, but because of the involution in a composition algebra there are only 6 possible nonisotopic products: $T \cong T^{op}$ since $\overline{(x_1x_2)x_3} = \bar{x}_3(\bar{x}_2\bar{x}_1)$. The important thing is that these products are in fact as nonisotopic as they can possibly be.

3.3. NONISOTOPY THEOREM: Let C be a unital composition algebra as in (1.3). If C is commutative associative (Types I-II) then there is only one standard product: xyz. If C is associative but not commutative (Type III) there are 3 nonisotopic standard products: $xyz \simeq zyx$, $yzx \sim xzy$, $zxy \simeq yxz$. If C is nonassociative (Type IV) there are 6 nonisotopic standard products:

$$\begin{array}{lll} x(yz) \sim (zy)x & y(zx) \simeq (xz)y & z(xy) \simeq (yx)z \\ (xy)z \cong z(yx) & (yz)z \simeq z(zy) & (zx)y \sim y(xz). \end{array}$$

I'll sketch the idea of the proof. Define x to be outside in a triple product $\{x\ y\ z\}$ if there exists a rational map $G : X \to \mathrm{End}(X)$ which cancels x,

$$(3.4)\qquad G(x)(\{x\ y\ z\}) = H(y,z)$$

for "injective" H (there is some u with $H(y,u) = 0 \Rightarrow y = 0$). Define x,y to be adjacent if there is a rational map $G : X \to X$ which cancels x when substituted for y,

$$(3.5) \qquad \{x\ G(x)\ z\} = H(z)$$

for some invertible H. These concepts are isotopy-invariant, and the pattern of outsideness and adjacenty determines which standard product you have: for example, with the right standard product $x(yz)$ we have

$$(3.6) \qquad \begin{array}{ll} x \text{ is outside in } x(yz) & (\text{via } x^{-1}\circ x(yz) = yz) \\ y \text{ is outside in } x(yz) & \text{iff } C \text{ is commutative} \\ z \text{ is outside in } x(yz) & \text{iff } C \text{ is associative} \\ x\text{-}y,\ y\text{-}z \text{ are adjacent in } x(yz) & (\text{via } x(x^{-1}z) = z,\ x(yy^{-1}) = x) \\ x\text{=}z \text{ are adjacent in } x(yz) & \text{iff } C \text{ is commutative} \end{array}$$

(since if $x(yG(x)) = H(y)$ as in (3.5) then $y = 1$ yields $xG(x) = t$, $G(x) = x^{-1}t$ for $t = H(1) = G(1)$, so $x = 1$ yields $yt = H(y)$, hence $x(y(x^{-1}t)) = yt$, $x = t$ forces $ty = yt$ for all y, and t lies in the center of C, thus $xyz^{-1} = y$ and C is commutative). Thus in the nonassociative case the 6 standard products have distinct outsideness and adjacency patterns, and therefore cannot be isotopic.

Having obtained a complete classification of standard triples into isotopy classes, the last (and surprisingly complicated) step is to show that any unital composition triple is standard.

4. GENERAL STRUCTURE THEORY

To show that every unital composition triple is standard, ie. a permutation of a left or right standard triple as in (3.1)-(3.2), it suffices to examine the 4 cases (A)-(D) of (2.3). We begin by reducing everything to case (A) where one of t, s disappears.

4.1. THEOREM: *Let T be a unital composition triple with associated composition algebra C.*

(i) *In Case (A) at least one of t,s is nuclear, in which case it can be deleted* ($t^{-1}t = s^{-1}s = 1$), *so we have 3 possibilities*

(A1) *if both t,s are nuclear, then T is strictly unital:* $\{x\ y\ 1\} = \{x\ 1\ y\} = \{1\ x\ y\} = xy$

(A2) *If s but not t is nuclear, then T is left-t-unital:* $\{x\ y\ 1\} = (xt^{-1})(ty)$, $\{x\ 1\ y\} = \{1\ x\ y\} = xy$ *for non-nuclear* $t \in C$.

(A3) If t but not s is nuclear then T is right-s-unital: $\{x\ y\ 1\} = \{x\ 1\ y\} = xy$, $\{1\ x\ y\} = (xs^{-1})(sy)$ for non-nuclear $s \in C$.
(naturally (A2), (A3) occur only when C is nonassociative of Type IV).

(ii) Cases (B), (C) are permutations of triples $\{x\ y\ z\}' = \{x\ z\ y\}$, $\{y\ x\ z\}$ of Case (A) whose associated algebras $C' = C^{(t)}$, $C^{(s)}$ are elemental isotopes of C.

(iii) Case (D) never occurs as a separate case: it is possible only when C is commutative, in which case it reduces to Case (A).

Here (ii) is a direct calculation, and (i) and (iii) are proven by showing that their failure would lead respectively to two Impossible Identities, which can be formulated succinctly in terms of the associator $[x,y,z] = (xy)z-x(yz)$.

4.2. FIRST IMPOSSIBILITY. Nonscalars $t,s \notin \Phi 1$ in an octonion algebra cannot satisfy $Q([xt,s,y], [x,t,sy]s\cdot y) = 0$ for all x,y.

4.3. REMARK. The linearization $Q([xt,s,y], [x,t,sy]) \equiv 0$ IS possible (for $s^{-1} = 2Q(t)1 - T(t)\bar{t}$, $t \notin \Phi 1$, $T(t) \neq 0$), so what is Possible and what is Impossible is not superficially obvious.

4.4. SECOND IMPOSSIBILITY. Invertible t,s in a quaternion or octonion algebra cannot satisfy $Q((yt)x, s^{-1}\{(s\bar{y})((yt)x-(tx)y)\}) = 0$ for all x,y.

It seems to be quite messy to show that 4.2, 4.4 are indeed Impossibilities. The calculations are fairly straightforward when t,s can be imbedded in a common quaternion subalgebra B, since then Jacobson's formula (1.4) for how C is built out of B makes the calculations easy. The general case where x,y,t,s, all have components in both B and $B\ell$ is too complicated for direct calculations, so as compromise one chooses B to at least contain t.

Having reduced the general unital case to 3 strictly unital cases 4.1 (A1-3), it suffices to show these are standard.

4.5. STRICTLY UNITAL THEOREM: (A1) If T is strictly unital then it is either left or right standard,

$$\{x\ y\ z\} = (xy)z \quad \text{or} \quad x(yz);$$

(A2) *If* T *is left-t-unital then*

$$\{x\ y\ z\} = ((xt^{-1})(ty))z;$$

(A3) *If* T *is right-s-unital then*

$$\{x\ y\ z\} = x((ys^{-1})(sz)).$$

In all cases, T *is isotopic to a standard triple.*

The idea of the proof for (A1) is as follows. We quickly get rid of Type I: here Q is injective, so $Q(\{x\ y\ z\}) = Q(x)Q(y)Q(z) = Q(xyz) \Rightarrow \{x\ y\ z\} = xyz$. Henceforth we may assume Q is nonsingular (which fails only in Type I of characteristic 2). We introduce the error terms

$$(4.6) \qquad E_L(x,y,z) = \{xyz\} - \{xyz\}_L,\ E_R(x,yz) = \{xyz\} - \{xyz\}_R$$

and attempt to show one of these vanishes. For these E's $Q(E(x,y,z),w)$ is an alternating form vanishing on 1, and in Types I-III the only alternating 4-linear form suppported on $1^{\perp}$ of dimension ≤ 3 is trivial, so $Q(E(x,y,z),w) = 0 \Rightarrow E(x,y,z) = 0 \Rightarrow \{xyz\} = \{xyz\}_L = \{xyz\}_R$.

Type IV is not so easy to dismiss (we expect one E to vanish and the other to equal the associator function). Here E is determined by a single scalar α; if $C = B \oplus B\ell$ for 1,u,v,uv an invertible basis of B with $T(u)T(v) = 0$, $(1,u,v)^{\perp} = \Phi w$, then necessarily E is generated by the value $E(u,v,\ell) = \alpha w\ell$. To show one of E_L, E_R vanishes we show their "product" (ie. the product of the scalars α_L, α_R) vanishes, hence one of the factors does:

$$\Phi\alpha_L\alpha_R = Q(E_L(u,v,\ell),E_R(u,v,B\ell)) = 0$$

by calculating with the defining identities (2.1), 4.1(A1).

Cases (A2), (A3) are handled similarly using the error terms

$$(4.7) \qquad E_t(x,y,z) = \{xyz\} - ((xt^{-1})(ty))z$$

$$E_s(x,y,z) = \{xyz\} - x((ys^{-1})(sz)).$$

Here the error term itself (not some product of it) vanishes; it suffices if $Q(E(x,y,z),\ (x^{-1}u)(uy)) = 0$ for all invertible x,y in C, and for $E = E_t,E_s$ this follows by choosing a B with basis 1,u,v,uv so that u = t or s. Here again it is crucial that we can build C out of any B via (1.4).

Summing up of chain of reductions, we have

4.8. CLASSIFICATION THEOREM: Every composition triple (T,Q) is isotopic to a standard composition triple on a unital composition algebra (C,Q):

(I) C field extension: $\{x\ y\ z\} = xyz$

(II) C quadratic extension: $\{x\ y\ z\} = xyz$

(III) C quaternion: $\{x\ y\ z\}$ is one of (i) xyz, (ii) xzy, (iii) yxz

(IV) C octonion: $\{x\ y\ z\}$ is one of (i) $(xy)z$, (ii) $(xz)y$, (iii) $(yx)z$, (iv) $x(yz)$, (v) $x(zy)$, (vi) $y(xz)$.

Two composition triples are isotopic iff they have similar norm forms and the same Type (Nn).

For example, over the complex numbers (or any algebraically closed field) there are precisely 11 classes of composition triples. Over the real numbers there are precisely 11 classes with positive definite form Q (the case of geometric interest).

Thus the techniques developed by Professor Jacobson for the study of composition algebras turn out to be powerful enough to classify composition triples as well. The fact that these triples turn out to be just the "known" ones arising naturally from composition algebras means geometrically that the resulting hypersurfaces are also "known" ones. This is yet another instance of Ecclesiastes' Principle that "there is no new exceptional thing under the sun".

Department of Mathematics
University of Virginia
Charlottesville, VA 22903

Contemporary Mathematics
Volume 13, 1982

TRIPLE PRODUCTS, MODULAR FORMS, AND HARMONIC VOLUMES

by Bruno Harris*

To Professor Nathan Jacobson

1. INTRODUCTION. Let A be an associative commutative algebra with a derivation e. For elements a_1, a_2, a_3 of A satisfying certain conditions we will construct a triple product $\langle a_1, a_2, a_3\rangle$ which is an element of A annihilated by e. We will be interested mainly in the case that the action of e on A extends to an action of the 3-dimensional simple Lie algebra, and more specifically A is the algebra over $\mathbb{C}$ generated by all C^∞ automorphic forms $f(z)$ of weights $2k$ $(k \in \mathbb{Z})$ with respect to a discrete subgroup Γ of $PSL(2,\mathbb{R})$ having fundamental domain of finite area: thus $f(z)(dz)^k$ is a Γ-invariant differential on the upper half plane. The action of e is:

$$e[f(z)(dz)^k] = y^2 \frac{\partial f}{\partial \bar{z}}(dz)^{k-1} \quad . \tag{1.1}$$

The triple product of 3 automorphic forms of weight $2k$ (satisfying some conditions) will then be a holomorphic automorphic form of weight $2k+2$, which is an alternating $\mathbb{R}$-trilinear function of its 3 arguments.

We will give an explicit formula for the triple product of holomorphic cusps forms for $\Gamma = PSL(2,\mathbb{Z})$.

Next, we will discuss a relation between triple products of holomorphic forms of weight 2, i.e. ordinary holomorphic differentials $f(z)dz$ for groups Γ with compact fundamental domain, and "harmonic volumes". Here, we will

*Supported by N.S.F. Grant No. MCS79-04905 to Brown University.

0271-4132/82/0539/$03.25

assume that the real parts of the three holomorphic 1-forms, i.e. three real harmonic 1-forms, have periods $\in \mathbb{Z}$ when integrated over any cycle on the Riemann surface X corresponding to Γ. Integration of these three harmonic 1-forms from a fixed point in the upper half-plane to a variable point then gives a map from the upper half plane mod Γ, i.e. the compact Riemann surface X, to the torus $\mathbb{R}^3/\mathbb{Z}^3$. The condition that had to be imposed on the original holomorphic 1-forms in order to define their triple product is equivalent to the fact that the image of X in the torus, i.e. a 2-cycle in a 3-dimensional manifold, actually bounds a 3-dimensional chain c_3. The volume of c_3 is now well-defined real number modulo $\mathbb{Z}$, which we refer to as a harmonic volume. This harmonic volume $\in \mathbb{R}/\mathbb{Z}$ depends on three integral cohomology classes of X (or of $H^1(\Gamma;\mathbb{Z})$), as well as on the complex-analytic structure of X, i.e. of the embedding $\Gamma \subset PSL(2,\mathbb{R})$. We now consider an infinitesimal deformation of the group Γ in $PSL(2,\mathbb{R})$, and compute how the harmonic volume varies. This directional derivative of the volume with respect to a vector in the tangent space to the moduli space is obtained by taking the triple product to the initial holomorphic one-forms, i.e. a holomorphic quadratic differential on X and so a complex linear function on the tangent space to moduli space, evaluating it on the given tangent vector, and finally taking the real part.

This last result can be put into the following framework. Consider the real vector space $H^1(\Gamma;\mathbb{R})$ (Γ acting trivially), form its 3rd exterior power over $\mathbb{R}$, $\Lambda^3H^1(\Gamma;\mathbb{R})$, and define a certain subspace $(\Lambda^3)'$ by the supplementary condition alluded to above. Then $(\Lambda^3)'$ is actually a complex vector space with positive definite hermitian inner product in the same way as $H^1(\Gamma;\mathbb{R})$ is. Furthermore $(\Lambda^3H^1(\Gamma;\mathbb{Z}))' = (\Lambda^3_{\mathbb{Z}})'$ is a lattice in $(\Lambda^3)'$ and $(\Lambda^3)'/(\Lambda^3_{\mathbb{Z}})'$ is an abelian variety. Harmonic volume is a homomorphism $(\Lambda^3_{\mathbb{Z}})' \to \mathbb{R}/\mathbb{Z}$, i.e. a point on a dual abelian variety. We have now attached to each analytic isomorphism class of compact Riemann surfaces (together with a given basis of $H^1(X;\mathbb{Z})$) i.e. to each point in Torelli space, an abelian variety and a point on this abelian variety, in other words a cross-section of a family of abelian varieties over Torelli space. The variational formula for harmonic volumes, namely the triple product, now gives us the differential of this cross-section.

2. TRIPLE PRODUCTS

Let A be an associative commutative algebra over $\mathbb{C}$ (usually infinite-dimensional), $a \to \bar{a}$ a $\mathbb{C}$-antilinear involution in A, e a $\mathbb{C}$-linear derivation of A. Let C,D be subspaces of A stable under e such that e is invertible on C and locally nilpotent on D (every $x \in D$ is annihilated by some e^k, $k \geq 1$).

For a,b in A, define

$$(a,b) = \frac{1}{2i}(a\bar{b} - b\bar{a}) = \mathrm{Im}(a\bar{b}) . \tag{2.1}$$

Form the third exterior power over $\mathbb{R}$: $\Lambda^3_{\mathbb{R}}(D)$ and map it to $A \otimes_{\mathbb{C}} D$ by

$$d_1 \wedge d_2 \wedge d_3 \longmapsto (d_1,d_2) \otimes d_3 + (d_2,d_3) \otimes d_1 + (d_3,d_1) \otimes d_2 . \tag{2.2}$$

Let $(\Lambda^3_{\mathbb{R}}D)'$ be the inverse image of $C \otimes D$ under this map. The triple product is then the following map $(\Lambda^3_{\mathbb{R}}D)' \to A \cap \ker e$

$$(\Lambda^3_{\mathbb{R}}D)' \to C \otimes D \xrightarrow{(e\otimes I+I\otimes e)^{-1}} C \otimes D \xrightarrow{\text{mult.}} A \tag{2.3}$$

I = identity map, $e \otimes I + I \otimes e = (e \otimes I)(I \otimes I + e^{-1} \otimes e)$ is invertible since e is invertible on C and locally nilpotent on D. The image of the map is annihilated by e since, if m = multiplication $A \otimes A \to A$, $e\, m = m(e \otimes I + I \otimes e)$ and finally m composed with the map (2.2) is zero (vanishing of a 3×3 determinant with entries d_1,d_2,d_3 as well as their real and imaginary parts). $\Lambda^3_{\mathbb{R}}(D)$ is a $\mathbb{C}$-module under $i(d_1 \wedge d_2 \wedge d_3) = id_1 \wedge id_2 \wedge id_3)$, and (2.2) is then $\mathbb{C}$-linear, so $(\Lambda^3_{\mathbb{R}}D)'$ is also a $\mathbb{C}$-module and (2.3) is $\mathbb{C}$-linear.

We will be interested in algebras A which have a positive definite hermitian inner product and on which the complexified Lie algebra of $SL(2,R)$ acts by derivations (those in the Lie algebra of $SL(2,\mathbb{R})$ being skew-adjoint). A basis for this Lie algebra L, as in Lang's "$SL(2,\mathbb{R})$" is W, E_+, E_-, with $[W,E_+] = 2iE_+$, $[W,E_-] = -2iE_-$, $[E_+,E_-] = -4iW$. We take $e = \frac{i}{4}E_-$. We will take the subspace C of A as the direct sum of all irreducible L-submodules of A having neither a highest nor a lowest weight vector, so $\frac{i}{4}E_-$ is invertible on it, and D as the direct sum of those irreducible submodules having a lowest weight vector, so $\frac{i}{4}E_-$ is locally nilpotent.

Next we specialize to the algebra A whose elements are all finite combinations of differentiable forms $f(z)(dz)^k$, $k \in \mathbb{Z}$, z in the upper half-plane, invariant under a discrete group $\Gamma \subset PSL(2,\mathbb{R})$ as in the introduction. The involution is $f(z)(dz)^k \to \overline{f(z)}\,\overline{(dz)}^k$ where $\overline{dz}$ is defined as $y^2(dz)^{-1}$ (and $z = x + iy$),

$$(2.4)\qquad \overline{f(z)(dz)^k g(z)(dz)^\ell} = f(z)\overline{g(z)}y^{2\ell}(dz)^{k-\ell}$$

The action of $e = \frac{i}{4}E_-$ is given by (1.1), and

$$(2.5)\qquad \frac{E_+}{4i}(f(z)(dz)^k) = \left(\frac{\partial f}{\partial z} + \frac{k}{iy}f(z)\right)(dz)^{k+1}.$$

The inner product is given by integration over a fundamental domain. The adjoint is given by $E_+^* = -E_-$. The Casimir operator is $\Delta = -\frac{1}{4}(E_+E_- + 2iW - W^2)$; on functions $(k = 0)$, $\Delta = -y^2(\frac{\partial^2}{\partial x^2} + \frac{\partial^2}{\partial y^2})$, the Laplace-Beltrami operator, and W is zero.

By assumption, Γ has fundamental domain of finite area. To define C and D we restrict ourselves first of all to the cusp forms (if there are no cusps, we just take the orthogonal complement of the constants).

D is then defined to be the union of the kernels of e^n for $n > 0$, and so is generated as L-module by the holomorphic cusp forms.

If Γ has compact fundamental domain then C is the L-module generated by all functions $f(z)$ orthogonal to 1. If Γ has cusps then we assume furthermore that the functions $f(z)$ are square-integrable on the fundamental domain and their 0th Fourier coefficient at each cusp vanishes. The Laplacian Δ is invertible on such functions, and since $\Delta = -\frac{1}{4}E_+E_- = -\frac{1}{4}E_-E_+$ on functions, to compute E_-^{-1} it suffices to compute Δ^{-1}. In the case of compact fundamental domain Δ^{-1} is determined by the following integral formula

$$(2.6)\qquad \int_X U_{q,q_0}(p)\Delta\varphi(p)\,dvol(p) = \varphi(q) - \varphi(q_0).$$

Here X is the compact Riemann surface (upper half-plane mod Γ), $U_{q,q_0}(p)$ is the harmonic function on X with logarithmic singularities at q,q_0 such that the analytic differential $dU + i * dU$ is the elementary differential of the third kind with simple poles at q,q_0 of residues $-\frac{1}{2\pi}$, $+\frac{1}{2\pi}$ respectively and with pure imaginary periods (Weyl, "The Concept of a Riemann Surface" §15, p. 120-121); q_0 is an arbitrarily fixed point on X, $dvol(p)$ denotes the volume form on X, and finally

$$(2.7)\qquad \Delta\varphi = f, \quad \varphi = \Delta^{-1}f.$$

(The proof of this classical formula (2.6) follows from Green's formulas.)

In order to use (2.6) one needs explicitly the elementary differentials of the third kind: e.g. on the Riemann sphere these are

$$\frac{1}{2\pi}\left(\frac{dz}{z-q_0} - \frac{dz}{z-q}\right) = \frac{d\log\left(\frac{z-q_0}{z-q}\right)}{2\pi}.$$

Much more care is needed if there are cusps, and we will only consider $\Gamma = PSL(2,\mathbb{Z})$, with only a sketch of the details. A $\mathbb{C}$-basis for holomorphic forms of weight $K = 2k \geq 12$ is provided by Poincare series

$$\text{(2.8)} \qquad g_K(z,m) = m^{K-1}\Sigma e^{2\pi i m\gamma(z)}\left(\frac{d(\gamma z)}{dz}\right)^k (dz)^k = \sum_{r=1}^{\infty} c_K(r,n)q^r$$

where $m \geq 1$, the first sum being over all coset representatives γ of Γ modulo Γ_∞, the stabilizer of the cusp, $(\Gamma = \cup\Gamma_\infty\gamma)$, and $q = e^{2\pi iz}$; the $c_K(r,m)$ are real. Thus $y^K(g_K(z,m)\overline{g_K(z,n)})$ has 0th Fourier coefficient $y^K\sum_r c_K(r,m)c_K(r,n)e^{-4\pi ry}$ which is real, and so its imaginary part is zero. $\mathrm{Im}(y^K g_K(z,m)\overline{g_K(z,n)})$ is furthermore bounded, and so square-integrable and finally a cusp form: let us denote it $g_{m,n}$ for short. A neighborhood of the cusp corresponds to a small q-disk around $q = 0$. If $|q| = r$, the hyperbolic metric $\frac{dzd\bar{z}}{y^2} = (4\pi)^2 \frac{dqd\bar{q}}{r^2(\log r)^2}$ blows up at $q = 0$; however we may consider the conformally equivalent metric obtained by multiplying by the factor $\mu(r) = r^2(\log r)^2$: this is just the Euclidean metric, and the new Laplacian is $\tilde{\Delta} = \frac{1}{\mu}\Delta$ = Euclidean Laplacian. The equation $\Delta\varphi = g_{m,n}$ is equivalent to $\tilde{\Delta}\varphi = \frac{g_{m,n}}{\mu} = \tilde{g}_{m,n}$. The expression for $g_{m,n}$ shows it to be of the order of $r^3(\log r)^K$ thus $\tilde{g}_{m,n}$ and its first partial derivatives are L^2 near $q = 0$, i.e. $\tilde{g}_{m,n}$ belongs to the Sobolev space H_1^{loc}. By the Sobolev lemma the solution φ of $\tilde{\Delta}\varphi = \tilde{g}_{m,n}$ (or $\Delta\varphi = g_{m,n}$) belongs to C^1, i.e. is continuously differentiable at $q = 0$. We can now apply the formula 2.6 in the new metric which is Euclidean near $q = 0$, and the new Laplacian $\tilde{\Delta}$, taking $q_0 = 0$. However the proof of (2.6) involves integrating only over the complement of small disks centered at q, q_0 and the integral is conformally invariant: $\Delta\varphi \, dvol = \tilde{\Delta}\varphi \, d\widetilde{vol}$, while U_{q,q_0} is also conformally invariant. The conclusion of this argument is that we may use (2.6) with $\Delta\varphi$ replaced by $g_{m,n}$ to calculate $\varphi = \Delta^{-1}g_{m,n}$. In fact the constant $\varphi(q_0)$ is irrelevant for us since we really want $E_-^{-1}(g_{m,n}) = E_+\Delta^{-1}g_{m,n}$. Using (2.5) with $k = 0$ and (2.6)

$$E_{-}^{-1}(g_{m,n}) = -i \int_{p\in X} \frac{\partial U_{q,q_0}(p)}{\partial q} dq\, g_{m,n}(p)dvol(p) .$$

The Riemann surface X has genus zero and is mapped $1-1$ onto the Riemann sphere by the Γ-invariant function $j(z)$ with Fourier expansion $j(z) = \frac{1}{q} + j_0 + \sum_{n=1}^{\infty} j_n q^n$. If q,Q corresponds to points z,Z in the fundamental domain then

$$(2.9) \qquad j(q) - j(Q) = \frac{Q-q}{Qq} e^{\sum_{n=1}^{\infty} \ell_n(Q)q^n}$$

where $\ell_1(Q) = -\sum_1^{\infty} j_n Q^n$, and the other $\ell_n(Q)$ can also be calculated as convergent power series in Q. The calculation of $E_{-}^{-1}(g_{m,n})$ now reduces to evaluation of the following integral

$$\int D \log(j(z) - j(Z))g_{m,n}(z) \frac{dxdy}{y^2}$$

over a fundamental domain, where D denotes the differential with respect to the variable Z, and the result of integration is a form of weight 2 in Z, (which is not analytic). We now choose (distinct) positive integers μ,ν,λ, put $m = \mu$, $n = \nu$ and multiply the form of weight 2 just obtained by the λth Poincare series then permute μ,ν,λ cyclically and add the three forms of weight $K + 2 = 2k + 2$ so obtained. This gives us the (holomorphic) triple product $\langle g_K(z,\mu),\ g_K(z,\nu),\ g_K(z,\lambda)\rangle$ for which we obtain the following formula

(on leaving out a factor $\frac{(4\pi)^{K-1}}{(K-1)!}$):

$$(2.10) \qquad \sum_{n\geq 1} [Q^{-n}+Q\ell_n'(Q)][(c_K(\nu-n,\mu) - c_K(\mu-n,\nu))g_K(Z,\lambda)$$
$$(c_K(\lambda-n,\nu) - c_K(\nu-n,\lambda))g_K(Z,\mu)$$
$$(c_K(\mu-n,\lambda) - c_K(\lambda-n,\mu))g_K(Z,\nu)](dZ)^{k+1}$$

($Q = e^{2\pi iZ}$ and $c_K(r,n) = 0$ for $r < 1$).

The coefficient of Q^n is a finite sum; for $n \leq 1$ it is zero as may be easily be checked from 2.10, since the terms $Q\ell_n'(Q)$ do not enter for $n \leq 1$.

The coefficient of Q^2 in (2.10) is a sum of 2×2 determinants (recall $c_K(r,n) = c_K(n,r)$): If $(\mu,\nu,\lambda) = (1,2,3)$ it is

$$\begin{vmatrix} c_K(1,1) & c_K(1,3) \\ c_K(3,1) & c_K(3,3) \end{vmatrix} - \begin{vmatrix} c_K(2,1) & c_K(2,2) \\ c_K(3,1) & c_K(3,2) \end{vmatrix}$$

(2.11)

$$+ j_1 \begin{vmatrix} c_K(2,1) & c_K(2,2) \\ c_K(1,1) & c_K(1,2) \end{vmatrix} - \begin{vmatrix} c_K(1,1) & c_K(1,2) \\ c_K(4,1) & c_K(4,2) \end{vmatrix}$$

$j_1 = 196884$.

For $K = 36$ (the first non-trivial case) and $\mu,\nu,\lambda = 1,2,3$ we get a non-zero multiple of the cusp form $E_{14}\Delta^2$ (of weight 38).

3. HARMONIC VOLUMES

Here we will only give a few details, leaving for another paper a complete exposition. Let X be a <u>compact</u> Riemann surface of genus $g \geq 3$. Our constructions will not use a metric but we could consider the Poincare metric if desired. Let H denote the real vector space of dimension $2g$ of real harmonic 1-forms α = Real part of the holomorphic form $\alpha + i * \alpha$. Let $H_{\mathbf{Z}}$ denote the lattice of harmonic 1-forms whose integrals over any 1-cycle on X belong to $\mathbf{Z}$. On H we have an alternating $\mathbf{R}$-valued bilinear form given by $\int_X \alpha\wedge\beta$, i.e. a map $\Lambda^2 H \to \mathbf{R}$, from which we obtain a map $\Lambda^3 H \to H$

$$(3.1) \qquad \alpha\wedge\beta\wedge\gamma \longmapsto (\int_X \alpha\wedge\beta)\gamma + (\int_X \beta\wedge\gamma)\alpha + (\int_X \gamma\wedge\alpha)\beta$$

We denote the kernel of this map by $(\Lambda^3 H)'$. The map (3.1) takes $\Lambda^3_{\mathbf{Z}} H_{\mathbf{Z}}$ to $H_{\mathbf{Z}}$ with kernel $(\Lambda^3 H_{\mathbf{Z}})' \subset (\Lambda^3 H)'$.

The alternating bilinear form on H is the real part of a positive definite hermitian form (where we let i act on H as $-*$); and $H/H_{\mathbf{Z}}$ is an abelian variety (Picard variety of X). The third exterior power of the alternating bilinear form on H is an alternating bilinear form on $\Lambda^3 H$ and $(\Lambda^3 H)'$ such that the hermitian form and complex structure are defined on $(\Lambda^3 H)'$ and make $(\Lambda^3 H)'/(\Lambda^3 H_{\mathbf{Z}})'$ an abelian variety.

We will now define a homomorphism $(\Lambda^3 H_{\mathbb{Z}})' \to \mathbb{R}/\mathbb{Z}$, in two equivalent ways, by iterated integrals and by harmonic volumes. In either case it suffices to start with $\alpha_1, \alpha_2, \alpha_3 \in H_{\mathbb{Z}}$ satisfying

(3.2) $$\int_X \alpha_i \wedge \alpha_j = 0 \qquad (i,j = 1,2,3)$$

it can be shown that such $\alpha_1 \wedge \alpha_2 \wedge \alpha_3$ generate $(\Lambda^3 H^1_{\mathbb{Z}})'$.

Let γ be a closed path with initial and terminal point $x_0 \in X$, which is Poincaré dual to α_3 : $\int_\gamma \beta = \int_X \beta \wedge \alpha_3$ for all $\beta \in H$.

By integrating α_1 along γ from x_0 to any point $x \in \gamma$, we obtain a function A_1 (defined on the unit interval if γ is a parametrized path); the product $A_1 \alpha_2$ is again a 1-form which may be integrated over γ, but the result $\int_\gamma A_1 \alpha_2$ is <u>not</u> a homotopy invariant of the path γ, since $A_1 \alpha_2$ is <u>not</u> a <u>closed</u> 1-form. Instead, from 3.2 it follows that

(3.3) $$\alpha_1 \wedge \alpha_2 = d\eta_{12}$$

where η_{12} is a 1-form on X, uniquely specified if we require it to be orthogonal to all closed 1-forms. We now consider the integral

(3.4) $$\int_\gamma (A_1 \alpha_2) - \eta_{12} \quad \underline{\text{taken modulo } \mathbb{Z}}$$

which is then seen to have the following properties:

(a) It depends only on the homology class of the closed path γ, namely the Poincare dual of α_3 and so is a function $I(\alpha_1, \alpha_2, \alpha_3)$ of the original three 1-forms.

(b) It is an alternating function of $\alpha_1, \alpha_2, \alpha_3$

(c) It defines a homomorphism

(3.5) $$I : (\Lambda^3 (H_{\mathbb{Z}})' \to \mathbb{R}/\mathbb{Z}$$

Thus I is a point on the <u>dual</u> abelian variety to $(\Lambda^3 H^1)'/(\Lambda^3 H^1_{\mathbb{Z}})'$. Since $\Lambda^3 H^1/\Lambda^3 H^1_{\mathbb{Z}}$ can be described as an intermediate Jacobian variety of $H^1/H^1_{\mathbb{Z}}$, we are led (as suggested to me by Langlands) to think in terms of volumes. Again starting with α_i, $i = 1,2,3 \in H_{\mathbb{Z}}$ satisfying (3.2) we may think of each α_i as the differential dh_i where $h_i : X \to \mathbb{R}/\mathbb{Z}$ is a harmonic function (defined only mod $\mathbb{Z}$). The 3 functions $h_i : X \to \mathbb{R}/\mathbb{Z}$ are the same as a single "harmonic map"

(3.6) $$h : X \to \mathbb{R}^3/\mathbb{Z}^3$$

(3.2) then states that any closed 2-form on $\mathbb{R}^3/\mathbb{Z}^3$ gives zero on integration over $h(X)$ (closed forms are cohomologous to invariant ones) which in turn implies that the 2-cycle $h(X)$ on $\mathbb{R}^3/\mathbb{Z}^3$ bounds a 3-chain c_3. If we normalize the volume of $\mathbb{R}^3/\mathbb{Z}^3$ to be 1, we can take the volume of c_3, only defined mod $\mathbb{Z}$ since c_3 is defined mod integral 3-cycles.

We can then check that the volume of c_3 mod Z is the same as (3.4), i.e. $I(\alpha_1,\alpha_2,\alpha_3)$. This is just an easy application of Green's theorem.

Finally we recall an elementary formula for volumes of closed surfaces in R^3 : if such a surface is given by an R^3-valued function $\vec{h}$ on X then the volume is $\frac{1}{6}\int_X \vec{h}\cdot(d\vec{h} \times d\vec{h})$ and under a variation $\delta\vec{h}$ of $\vec{h}$ the corresponding variation is

(3.5) $$\delta V = \frac{1}{6}\int_X \delta\vec{h}\cdot(d\vec{h} \times d\vec{h})$$

The formula (3.5) is still valid for surfaces in T^3 and allows us to calculate δV for a specific variation of $\vec{h}$ obtained by changing the conformal structure of X in a neighborhood of one point (Schiffer variation). Finally we obtain a formula described in the introduction valid for any infinitesimal variation: (up to an elementary constant) the variation of volume δV is a given direction $\xi \in$ dual space of quadratic differentials is obtained by taking $\alpha_i = \text{Re}\, f_i(z)dz$, $i = 1,2,3$ forming the triple product of weight 2 forms f_i, which is a quadratic differential, evaluating on ξ, and taking the real part of this complex number.

Contemporary Mathematics
Volume 13, 1982

BIVARIANTS OF TERNARY RINGS

by W. G. Lister

By a <u>ternary ring</u> we mean an additive group equipped with a triple product such that the system is isomorphic to an additive subgroup of a ring closed under triple ring products. These systems were characterized and their study begun in [1]. A related structure has been characterized by Loos in [2]. Any general significance of ternary rings will arise either from aspects not shared with rings or from contributing connections with ring theory. In this paper we introduce a type of distinctive structural transformation which determines a relationship among rings.

1. VARIANT MAPS AND VARIANTS

In a ternary ring (τ-ring hereinafter) T, a <u>variant map</u> is an additive endomorphism λ such that

(1) $$(xy\lambda z)\lambda = x\lambda yz\lambda, \quad \text{all } x,y,z \text{ in } T.$$

For any such λ define a new ternary product in T, by setting

(2) $$\langle xyz\rangle_\lambda = xy\lambda z.$$

The identity (1) implies that the product in (2) satisfies the associativity identities which make the new system $T_\lambda = (T,+,\langle\ \rangle_\lambda)$ a τ-ring, the <u>λ-variant</u> of T.

PROPOSITION 1. *For any τ-ring T the following are variant maps:*

i) *any additive endomorphism γ satisfying*

$$(xyz)\gamma = x\gamma yz = xy\gamma z = xyz\gamma,$$

ii) *any automorphism of order two,*

iii) *any map $t \to ata$ where a is an element of any enveloping ring of T and $aTa \subset T$.*

(Proofs requiring only straightforward verification will be omitted throughout.) Any τ-ring has an essentially unique imbedding in a <u>standard enveloping ring</u> $A = T \oplus T^2$ characterized by

$$w \in T^2 \text{ and } wT = Tw = 0 \Rightarrow w = 0. \qquad ([1], \text{ p. } 38).$$

0271-4132/82/0540/$03.00

The subring T^2 is the <u>complementary subring</u> for the imbedding. Unless otherwise stated "enveloping ring" and "complementary subring" will refer to the standard imbedding. In this paper attention will be limited to those variant maps which are bijective, the <u>bivariant maps</u>. Each of these determines, via (2), a bivariant of the given τ-ring. With reference to Proposition 1, the bivariant maps satisfying i) are <u>central</u>, those in class ii) are <u>automorphic</u> and those of the form iii) are the <u>inner</u> bivariant maps. Let

$\mathfrak{B}(T)$ denote the set of bivariant maps of T, and

$\mathfrak{C}(T)$ denote the set of central bivariant maps.

If $A = T \oplus T^2$, $c \in T^2$, and c is in the center of A then $t \to tc$ is in $\mathfrak{C}(T)$ but is not necessarily inner.

PROPOSITION 2. i) *If* $\beta \in \mathfrak{B}(T)$ *then* β^2 *and* β^{-1} *are in* $\beta(T)$.

ii) *If* β_1, β_2 *are in* $\mathfrak{B}(T)$ *and commute then* $\beta_1\beta_2 \in \mathfrak{B}(T)$.

iii) *If* β_1, β_2 *are in* $\mathfrak{B}(T)$ *then* $\beta_1\beta_2\beta_1 \in \beta(T)$.

iv) *If* $\beta \in \mathfrak{B}(T)$ *and* $\varphi \in \mathrm{Aut}(T)$ *then* $\varphi^{-1}\beta\varphi \in \mathfrak{B}(T)$.

If $\gamma \in \mathfrak{C}(T)$ *then* $\varphi^{-1}\gamma\varphi \in \mathfrak{C}(T)$.

For bivariant maps it will generally be convenient to replace the defining identity (1) by an equivalent:

$$(1)' \qquad (xyz)\beta = x\beta y\beta^{-1}z\beta$$

The converse of Proposition 2 i) and other properties relating to commuting bivariant maps require a mild non-degeneracy condition. An element x_0 in T is a <u>medial annihilator</u> provided $Tx_0T = 0$ and $x_0 \neq 0$. If T has no medial annihilators then

$$TTx_o = 0 \Rightarrow T^2x_oT^2 = 0 \Rightarrow T(Tx_oT)T = 0$$
$$\Rightarrow Tx_oT = 0 \Rightarrow x_o = 0.$$

Thus it follows that T also has neither <u>right annihilators</u> nor <u>left annihilators</u>.

PROPOSITION 3. *Suppose* T *has no medial annihilators.*

i) *For* α,β *in* $\mathfrak{B}(T)$, $\beta\alpha \in \mathfrak{B}(T) \Leftrightarrow \beta\alpha = \alpha\beta \Leftrightarrow \alpha \in \mathfrak{B}(T_\beta) \Leftrightarrow (T_\beta)_\alpha = T_{\beta\alpha}$.

ii) $\mathfrak{C}(T)$ *is an abelian group.*

<u>Proof</u>. As a sample computation we verify ii). Suppose γ,δ are in $\mathfrak{C}(T)$. In view of Proposition 2, we need only verify that $\gamma\delta = \delta\gamma$.

$$(xyz)\gamma\delta = xy\gamma\delta z = x\gamma y\delta z = (xy\delta z)\gamma = xy\delta\gamma z.$$

Now by the hypothesis, $y\gamma\delta - y\delta\gamma = 0$.

2. BASIC EXAMPLES

a) In any τ-ring T the map $-1: x \to -x$ is central and automorphic. The τ-ring T_{-1} is the "imaginary" bivariant of T. Specifically, if Φ is the real field and $T = \Phi_\tau$, the τ-ring derived from Φ by iterating the product in Φ, then $x \to ix$ is an isomorphism from T_{-1} to the imaginary numbers.

b) For any field θ of characteristic 0 and integers r,n with $0 < r < n$, the set T of n by n θ-matrices

$$\begin{bmatrix} 0 & A_1 \\ A_2 & 0 \end{bmatrix} \begin{matrix} (r) \\ (n-r) \end{matrix} = [A_1, A_2]$$
$$\quad (r) \quad (n-r)$$

is a simple τ-algebra over θ since its enveloping algebra is θ_n ([1], p. 51). Here the map $[A_1, A_2] \to [A_1, -A_2]$ is an isomorphism from T_{-1} to T.

c) With T as in b) and $n = 2r$, the map

$$\sigma: [A_1, A_2] \to [A_2, A_1]$$

is automorphic bivariant. Since

$$<[A_1, A_2][B_1, B_2][C_1, C_2]>_\sigma = [A_1B_1C_1, A_2B_2C_2],$$

T_σ has mutually annihilating ideals $\{[A_1, 0]\}$, $\{[0, A_2]\}$ and is therefore not simple and so not isomorphic to T.

d) Any central variant of a commutative τ-ring is commutative, but this is not true of variants in general. Convert the complex field P to the τ-ring $T = P_\tau$ and let $\sigma: x \to \bar{x}$. In T_σ $<xyz>_\sigma = x\bar{y}z$ while $<yxz>_\sigma = \bar{x}yz$.

e) As an instance of the "natural" occurrence of a bivariant consider a finite dimensional complex vector space V and the real τ-algebras $\mathcal{L}_\tau$ and $\mathcal{L}'_\tau$ of all linear and conjugate linear transformations in V respectively. Fix a basis $v_1, v_2, \ldots, v_n$ for V and use the basis $v_1, \ldots, v_n, iv_1, \ldots, iv_n$ to identify elements of $\mathcal{L} (= \mathcal{L}_\tau)$ and $\mathcal{L}' \; (= \mathcal{L}'_\tau)$ with matrices.

$$L \in \mathcal{L} \Rightarrow L = \begin{bmatrix} A & -B \\ B & A \end{bmatrix},$$

where A,B are n by n, and

$$M' \in \mathcal{L}' \Rightarrow M' = \begin{bmatrix} C & D \\ D & -C \end{bmatrix}.$$

Now define

$$L\rho = \begin{bmatrix} A & B \\ -B & A \end{bmatrix} \quad \text{and} \quad L\varphi = \begin{bmatrix} A & B \\ B & -A \end{bmatrix}.$$

The map ρ is automorphic bivariant in $\mathcal{L}$, φ maps $\mathcal{L}$ into $\mathcal{L}'$ and calculation shows φ is an isomorphism from $\mathcal{L}_\rho$ to $\mathcal{L}'$. Note also that

$$J = \begin{bmatrix} I & 0 \\ 0 & -I \end{bmatrix} \Rightarrow L\rho = JLJ \text{ and } L\varphi = LJ.$$

To put this in context let A be the algebra of all real linear transformations in V regarded as real space. Then

$$A = \mathcal{L}' \oplus (\mathcal{L}')^2 = \mathcal{L}' \oplus \mathcal{L}$$

is the standard enveloping algebra of $\mathcal{L}'$. Thus $J \in (\mathcal{L}')^2$ and ρ is inner.

3. ISOMORPHISMS

PROPOSITION 4. <u>For any</u> α,β <u>in</u> $\mathcal{B}(T)$ <u>and</u> φ <u>in</u> $\mathrm{Aut}(T)$,

i) α <u>is an isomorphism from</u> $T_{\alpha\beta\alpha}$ <u>to</u> T_β,

ii) φ^{-1} <u>is an isomorphism from</u> $T_{\varphi^{-1}\beta\varphi}$ <u>to</u> T_β.

Because the set $\mathfrak{C}(T)$ of central bivariant maps is an abelian group, if γ,δ are in $\mathfrak{C}(T)$ then $T_{\gamma\delta^2} \cong T_\gamma$. In particular $T_{\delta^2} \cong T$ and $T_{-\delta^2} \cong T_{-1}$. Although a τ-ring T and any bivariant T_β are closely related, they do not necessarily have the same standard enveloping algebra. (See examples c), d), e) of Section 2). They do, however, have a common complementary subalgebra.

THEOREM 1. <u>For</u> $\beta \in \mathcal{B}(T)$

$$\beta_r : \Sigma x_i y_i \to \Sigma \langle x_i y_i \beta^{-1}\rangle_\beta, \qquad x_i, y_i \text{ in } T,$$

<u>and</u>

$$\beta_\ell : \Sigma x_i y_i \to \Sigma \langle x_i \beta^{-1} y_i\rangle_\beta$$

<u>each define an isomorphism from</u> T^2 <u>to</u> T_β^2.

<u>Proof</u>. Since the formula claimed for β_r is additive, to prove that it defines a map in T requires only that if $\Sigma x_i y_i = 0$ then $\Sigma\langle x_i y_i \beta^{-1}\rangle_\beta = 0$. Now

$$\begin{aligned} \Sigma x_i y_i = 0 &\Rightarrow \Sigma x_i y_i t = 0, \quad \text{all } t \text{ in } T, \\ &\Rightarrow \Sigma\langle x_i y_i \beta^{-1} t\rangle_\beta = 0. \end{aligned}$$

Similarly,

$$\begin{aligned} \Sigma x_i y_i = 0 &\Rightarrow \Sigma t x_i y_i = 0, \quad \text{all } t \text{ in } T, \\ &\Rightarrow \Sigma (t x_i y_i)\beta^{-1} = 0 = \Sigma t\beta x_i \beta y_i \beta^{-1} \\ &\Rightarrow \Sigma t x_i y_i \beta^{-1} \rangle_\beta = 0, \quad \text{all } t \text{ in } T. \end{aligned}$$

Thus if $w = \Sigma\langle x_i y_i \beta^{-1}\rangle_\beta \langle wT\rangle_\beta = \langle Tw\rangle_\beta = 0$ in the standard enveloping algebra of T_β, hence $w = 0$. Since

$$(\beta_r)^{-1} : \Sigma\langle x_i y_i\rangle_\beta \to \Sigma x_i y_i \beta,$$

the argument above shows β_r is a bijection. Because β_r is additive, the following establishes that it is an isomorphism: For x,y,z,t in T

$$[(xy)(zt)]\beta_r = [(xyz)t]\beta_r = <<xy\beta^{-1}z>_\beta t\beta^{-1}>_\beta$$
$$= <xy\beta^{-1}zt\beta^{-1}>_\beta$$
$$(xy)\beta_r(zt)\beta_r = <xy\beta^{-1}>_\beta<zt\beta^{-1}>_\beta$$
$$= <xy\beta^{-1}zt\beta^{-1}>_\beta.$$

That β_ℓ is also an isomorphism follows in the same way or from an independent verification of the next result.

COROLLARY. *For* $\beta \in \mathcal{B}(T)$,

i) $\beta_+ : \Sigma x_i y_i \to \Sigma x_i\beta y_i\beta^{-1}$, x_i, y_i in T,

defines an automorphism of T^2, *and*

ii) β_+ *is the identity if and only if* $\beta \in \mathfrak{C}(T)$.

Proof. The assertion i) is immediate from $\beta_+ = \beta_r\beta_\ell^{-1}$. As for ii),

$$\beta_+ = I \Leftrightarrow x\beta y\beta^{-1}t\beta = xyt\beta \text{ and } t\beta x\beta y\beta^{-1} = t\beta xy$$
$$\Leftrightarrow (xyt)\beta = xyt\beta \text{ and } (txy)\beta = t\beta xy,$$

and this pair of identities implies that β is in $\mathfrak{C}(T)$, for

$$xy\beta t = x\beta\beta^{-1}y\beta t\beta\beta^{-1} = (x\beta yt\beta)\beta^{-1} = (xyt)\beta^2\beta^{-1} = (xyt)\beta.$$

Observe that if $\beta : t \to btb$ for some b in $A = T \oplus T^2$, then $\beta_+ : w \to bwb^{-1}$, $w \in T^2$.

4. MODULES AND THE RADICAL.

Definitions and properties used in this section will be found in Sections 2, 3, and 4 of [1].

PROPOSITION 5. *For* $\beta \in \mathcal{B}(T)$ *and* U *a right (left) ideal of* T,

i) $U\beta$ *is a right (left) ideal of* T,

ii) U *is a right (left) ideal of* T_β.

iii) *If* V *is a* β-*invariant ideal of* T *then* V *is an ideal of* T_β.

PROPOSITION 6. *For* $\beta \in \mathcal{B}(T)$ *and* M *a* T-*module set* $<mxy>_\beta = m(x\beta)y$ $(m \in M,\ x,y \in T)$. *This defines a* T_β-*module on* $(M,+)$, *which will be denoted by* M_β.

Proof. The requirement is that

$$\langle\langle muv\rangle_\beta xy\rangle_\beta = \langle mu\langle vxy\rangle_\beta\rangle_\beta = \langle m\langle uvx\rangle_\beta y\rangle_\beta .$$

This translates to the identity

$$(mu\beta v)x\beta y = mu\beta(vx\beta y) = m(uv\beta x)\beta y\beta .$$

For any T-module M, $K(M) = \{x \mid MxT = 0\}$ is a left-right ideal, and the radical $R(T)$ of T is defined by

$$R(T) = \bigcap K(M), \quad M \text{ irreducible}.$$

PROPOSITION 7. <u>For</u> $\beta \in \mathcal{B}(T)$ <u>and</u> M <u>a</u> T-<u>module</u>,

i) <u>the map</u> $m \to m(x\beta)(y\beta^{-1})$ <u>defines a</u> T-<u>module on</u> $(M,+)$ <u>denoted by</u> M^β,

ii) $K(M_\beta)\beta = K(M)$,

iii) $K(M^\beta) = K(M_\beta)$.

These are immediate, with the verification of i) similar to that of Proposition 6.

THEOREM 2. <u>For</u> $\beta \in \mathcal{B}(T)$, $R(T_\beta) = R(T)$.

<u>Proof</u>. Note first that the submodules of M, M^β, M_β coincide. Next observe that if M^* is a T_β-module then $M^*_{\beta^{-1}}$ is a $(T_\beta)_{\beta^{-1}}$ = T-module. By Proposition 7, ii) and iii), β permutes the kernels of irreducible T-modules. Thus if $K(M) = K(N)\beta^{-1}$ then $K(M) = K(N_\beta)$. On the other hand if M^* is an irreducible T_β-module then the above applies to β^{-1} and produces a $(T_\beta)_{\beta^{-1}}$-module $N^*_{\beta^{-1}}$ with $K(M^*) = K(N^*_{\beta^{-1}})$.

5. BIVARIANTS OF SEMISIMPLE ARTINIAN τ-RINGS.

A natural test problem is to determine all bivariant maps and then all bivariants of each τ-ring in this well known class. Here we begin such an effort.

THEOREM 3. <u>If</u> T <u>is a</u> τ-<u>division ring and</u> $\beta \in \mathcal{B}(T)$ <u>then</u> T_β <u>is a</u> τ-<u>division ring</u>.

<u>Proof</u>. The elements x,y of T are inverse if and only if $xyt = txy = t$ for every t in T. The decisive fact is that if x and y are inverse in T then x and $y\beta^{-1}$ are inverse in T_β. This follows from

$$\langle xy\beta^{-1}t\rangle_\beta = xyt \quad \text{and} \quad \langle t\beta^{-1}xy\beta^{-1}\rangle_\beta = txy.$$

A simple Artinian T may produce a non-simple T_β, as example c) of Section 2 shows. This can occur only if T has a proper left-right ideal, in which case

$T = U_1 \oplus U_2$ where U_1, U_2 are the unique left-right ideals in T, and $T^2 = U_1U_2 \oplus U_2U_1$ with U_1U_2 and U_2U_1 simple ideals. If T has no proper left-right ideal, T^2 is simple. (See Theorem 16 of [1]).

THEOREM 4. _A necessary and sufficient condition that a simple_ T _have a non-simple bivariant_ T_β _is that_ T _have left-right ideals_ U_1 _and_ U_2 _exchanged by_ β, _in which case_ U_1 _and_ U_2 _are simple ideals in_ T_β.

Proof. If T_β is not simple neither is T_β^2 and therefore T has proper left-right ideals U_1, U_2. Since $U_1\beta$ is also a left-right ideal in T and U_1 is not an ideal in T, $U_1\beta = U_2$. Similarly $U_2\beta = U_1$. Conversely, if $T = U_1 \oplus U_2$ and β exchanges the left-right ideals U_1, U_2, then each is a left-right ideal in T_β and $\langle TU_1T\rangle_\beta = TU_2T = U_1$ because TU_2T is a left-right ideal in T and U_2 is not an ideal in T.

The preceding result may suggest that determination of the bivariant maps of a semisimple T reduces to the case in which there are at most two simple components. As an instructive counterexample consider a sum T of three isomorphic simple ideals. Then

$$T = U \oplus U\psi \oplus U\psi^2,$$

where $\psi \in \text{Aut } T$, $\psi^3 = I$, and U is a simple ideal. In addition suppose U has proper left-right ideals; say

$$U = U_1 \oplus U_2.$$

Now define a bijection β of T as follows:

Restricted to $U_1 \oplus U_1\psi \oplus U_1\psi^2$, β is ψ.

Restricted to $U_2 \oplus U_2\psi \oplus U_2\psi^2$, β is ψ^2.

Computation shows that $\beta \in \mathcal{B}(T)$.

For T semisimple Artinian, $A = T \oplus T^2$ and $\alpha: t \to ata$, $a \in A$, an inner bivariant map, the associated bivariants can be identified. The problem reduces immediately to the case T simple and there the result, without proof, is the following:

THEOREM 4. _Suppose that_ T _is Artinian and simple_, $A = T \oplus T^2$, $a \in A$ _is invertible, and_ $\alpha: t \to ata$ _is in_ $\beta(T)$. _Then_

i) _either_ $a \in T$ _or_ $a \in T^2$, _and_

ii) _if_ $a \in T$, $t \to ta$ _is an isomorphism from_ T_α _to_ $(T^2)_\tau$, _while if_ $a \in T^2$, $t \to ta$ _is an isomorphism from_ T_α _to_ T.

6. TERNARILY PAIRED RINGS

The bivariant relationship between τ-rings determines a relationship between rings via standard enveloping algebras. Call β a ternary pairing of $A = T \oplus T^2$ and $B = T_\beta \oplus T_\beta^2$. One route of investigation is to explore this relationship, which can be viewed as a method of constructing B from A. We cite one result.

THEOREM 5. *Every finite dimensional simple real algebra is ternarily paired to an algebra of all* n *by* n *real matrices,* Φ_n, *or to* $\Phi_n \oplus \Phi_n$.

REFERENCES

1. W. G. Lister, Ternary Rings, Trans. Amer. Math. Soc. 154 (1971), 37-55.

2. Ottmar Loos, Assoziative Tripelsysteme, Manuscripta Math. 7 (1972), 103-112.

State University of New York
at Stony Brook
Stony Brook, New York 11794

Contemporary Mathematics
Volume 13, 1982

SEPARABLE ALTERNATIVE ALGEBRAS OVER COMMUTATIVE RINGS

by

Robert Bix

Let A be a unital alternative algebra over a commutative ring R, and let $U_R(A)$ be the unital universal multiplication envelope of A. A is called separable over R if $U_R(A)$ is a separable associative R-algebra. A unital nonassociative algebra C over a commutative ring S is called octonion if C is a finitely spanned projective S-module of rank 8 having a quadratic form Q such that the associated bilinear form is nondegenerate and $Q(xy) = Q(x)Q(y)$ for all $x, y \in C$. It is proved that A is separable over R if and only if $A \cong B \oplus C$, where B is a separable associative R-algebra and C is octonion over a commutative separable associative R-algebra S. Details will appear elsewhere.

The University of Michigan-Flint
Flint, MI. 48503

0271-4132/82/0541/$01.25

Contemporary Mathematics
Volume 13, 1982

EXCEPTIONAL JORDAN DIVISION ALGEBRAS

Holger P. Petersson

and

Michel Racine*

A notre maître et ami, Nathan Jacobson.

1. INTRODUCTION

The first examples of exceptional Jordan division algebras were constructed by Albert [1] as fixed point sets of a semilinear automorphism of reduced exceptional simple Jordan algebras. In [1] Albert obtains all exceptional division algebras containing a given cyclic cubic subfield and in [3] those containing a given noncyclic cubic subfield. Thus all finite-dimensional exceptional Jordan division algebras over a field of characteristic not 2 were in some sense determined.

Tits gave two constructions of finite dimensional central exceptional linear Jordan algebras containing A^+, A a central simple associative algebra of degree 3, respectively $H(B,*)$, B a central simple associative algebra of degree 3 with involution $*$ of the second kind, and proved that all finite-dimensional exceptional division algebras are obtainable from one or the other construction (char. $\neq 2$) [4].

McCrimmon [6] gave a beautiful formulation of the Tits constructions in the context of generically algebraic algebras of degree 3 over fields of arbitrary characteristic and proved that all forms of $H(O_3)$, O an octonion algebra, are obtained by one or the other of the Tits constructions, irrespective of the characteristic [8].

The question of isomorphism is left open. In fact it was not known whether the Tits constructions are disjoint for division algebras or whether every exceptional Jordan division algebra contains a cyclic extension of the center.

The recent results of Zelmanov [12], which tell us that there are no infinite dimensional exceptional Jordan division algebras add to the interest of these questions. In fact it is now even more apparent that the simple exceptional Jordan algebras of dimension 27 play a role in the structure theory of Jordan algebras which is analogous to that of octonion algebras in the structure theory of alternative algebras.

*The research of the second author was supported in part by an NSERC grant and by a von Humboldt Fellowship at the University of Münster.

The results announced here grew out of an attempt to answer the questions mentioned above. Proofs will appear elsewhere.

2. THE TITS PROCESS

Let Φ be a unital commutative associative ring and J a Φ-module possessing a cubic form N with values in Φ, a quadratic mapping $x \to x^{\#}$ in J and a distinguished element $1 \in J$ satisfying

(1) $x^{\#\#} = N(x)x$

(2) $N(1) = 1$

(3) $T(x^{\#},y) = \Delta_x^y N$, the directional derivative of N in the direction y, evaluated at x; $T(x,y) = -\Delta_1^x \Delta^y \log N$,

(4) $1^{\#} = 1$,

(5) $1 \times y = T(y)1 - y$, where $T(y) = T(y,1)$ and $x \times y = (x + y)^{\#} - x^{\#} - y^{\#}$.

If these hold for every scalar extension of Φ then McCrimmon [6] has shown that

(6) $yU_x = T(x,y)x - x^{\#} \times y$

defines a quadratic Jordan algebra structure $(J,U,1)$, which we denote $J(N,\#,1)$. Every element of J satisfies

(7) $x^3 - T(x)x^2 + S(x)x - N(x)1 = 0$, where $S(x) = T(x^{\#})$.

Next we recall the Tits constructions. Let A be a separable associative algebra of degree 3 over a field F, μ a non-zero element of F. Let $J(A,\mu) = A_0 \oplus A_1 \oplus A_2$, $A_i = A$; $x = (a_0,a_1,a_2)$, $y = (b_0,b_1,b_2) \in J(A,\mu)$. The cubic form

(8) $N(x) = N(a_0) + \mu N(a_1) + \mu^{-1}N(a_2) - T(a_0a_1a_2)$,

the quadratic map

(9) $x^{\#} = (a_0^{\#} - a_1a_2, \mu^{-1}a_2^{\#} - a_0a_1, \mu a_1^{\#} - a_2a_0)$,

and the trace form

(10) $T(x,y) = T(a_0,b_0) + T(a_1,b_2) + T(a_2,b_1)$,

where $\#$, N, T are the adjoint, norm and trace maps of A where appropriate, satisfy (1) - (5) and so give $J = J(A,\mu)$ a Jordan algebra structure. J is a division algebra if and only if A is an associative division algebra and $\mu \notin N(A)$. This is known as the <u>first Tits construction</u>.

Let B be a separable associative algebra of degree 3 over the field K with involution $*$ of the second kind such that $K^* = K$. Let $A = H(B,*)$, the symmetric elements of B, $F = H(K,*)$, u an element of A such that

$N(u) = \mu\mu^*$, μ a non-zero element of K. The following norm form, quadratic map and trace form define a Jordan algebra structure on $J = J(B,u,\mu,*) = A \oplus B$. Let $x = (a,b)$, $y = (c,d) \in J$.

(11) $N(x) = N(a) + \mu N(b) + \mu^* N(b^*) - T(abub^*)$,

(12) $x^{\#} = (a^{\#} - bub^*, \mu^* b^{*\#} u^{-1} - ab)$,

(13) $T(x,y) = T(a,c) + T(ub^*,d) + T(bu,d^*)$,

where $\#$, N, T are those of B where appropriate. This is the <u>second Tits construction</u>. J is a division algebra if and only if B is an associative division algebra and $\mu \notin N(B)$. It was remarked that $J(B,u,\mu,*)$ is the subalgebra of fixed points of $*$ which sends $(b_0,b_1,ub_2) \in J(B,\mu)$ to (b_0^*,b_2^*,ub_1^*). In fact if we allow K to be a ring then the first constructions are second constructions.

LEMMA 1. *Let A be a separable associative algebra of degree 3, $B = A \oplus A^{op}$, where A^{op} denotes the opposite algebra of A. Then, for any invertible $\mu \in F$, $J(A,\mu)$ is isomorphic to $J(B,1,(\mu,\mu^{-1}),*)$, where $*$ is the exchange involution of B.* □

In fact, we introduce a somewhat more general construction. Let B be an associative Φ-algebra with involution $*$. Recall that a Φ-submodule A of $H(B,*)$ is called <u>ample</u> if bb^*, $b + b^*$ and $bAb^* \subset A$ for all $b \in B$ [7]. If $\frac{1}{2} \in \Phi$ or if $*$ is of the second kind then $A = H(B,*)$.

THEOREM 2. *If B is an associative Φ-algebra with involution $*$ such that $B^+ = J(N,\#,1)$, A an ample Φ-subalgebra of $H(B,*)$ and u an invertible element of A such that $N(u) = \mu\mu^*$, μ a central element of B, then equations (11) - (13) define a norm, adjoint and trace form on $J = J(B,*,u,\mu,A) = A \oplus B$ which give J a Jordan algebra structure. J is a Jordan division algebra if and only if A is a Jordan division algebra and $\mu \notin N(B)$.* □

Theorem 2 is proved directly. We will call this construction the <u>Tits process</u> and drop the A when $A = H(B,*)$. The Tits process plays a role in Jordan theory similar to the Cayley-Dickson process in alternative theory.

THEOREM 3. *If $J = J(N,\#,1)$ is a simple F-algebra then J is obtained by repeated applications of the Tits process unless J is a separable cubic extension over a field of characteristic 3 or $H(Q_3,J_\gamma)$, Q a quaternion algebra over a field of characteristic 2.* □

The key to the proof is the following

LEMMA 4. ([8]). *Let A be a proper subalgebra of $J = J(N,\#,1)$ on* *which* $T(\ ,\ )$ *is non-degenerate. Then* $\mathcal{J} = A \oplus A^{\perp}$, $A^{\perp}$ *the orthogonal complement of* A *with respect to* $T(\ ,\ )$ *and* $g: A \to (\mathrm{End}_F(A^{\perp}))^{+}$ *given by*

(14) $\quad g(a)z = -a \times z,\ a \in A,\ z \in A^{\perp},$

is a homomorphism of Jordan algebras. □

This map factors through $su(A)$ the special universal envelope of A.

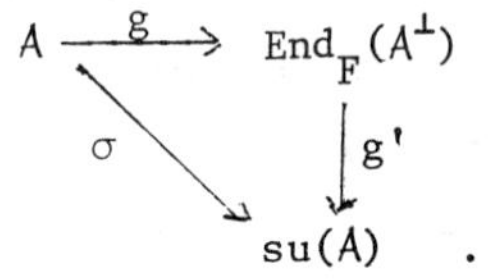

Given $z \in A^{\perp}$, define $q(z) = q_A(z)$, respectively $r(z) = r_A(z)$, to be the projections of $z^{\#}$ on A, respectively $A^{\perp}$, so that

(15) $\quad z^{\#} = q(z) + r(z),\ q(z) \in A,\ r(z) \in A^{\perp}.$

The proof of Theorem 3 is essentially rational and independent of structure theory, in so far as

THEOREM 5. *Let* $\mathcal{J} = \mathcal{J}(N,\#,1)$ *be a unital quadratic Jordan algebra over a field* F. *If* $\mathcal{J}$ *has no absolute zero divisors then* $\mathcal{J}$ *is either* i) *a Jordan division algebra,* ii) *a direct sum* $F \oplus \mathcal{J}(Q,1')$, *where* $\mathcal{J}(Q,1')$ *is the Jordan algebra of the quadratic form* Q *with base point* $1'$, *or* iii) $\mathcal{H}(C_3, J_\gamma)$, C *a composition algebra over* F,

was proved in [9] using only the axioms (1) - (5).

3. THE ALGEBRAS OF DIMENSION 9

Since an exceptional Jordan division algebra is a first Tits construction if and only if it contains a nine-dimensional subalgebra of the form A^{+} and a second Tits construction if and only if it contains a nine-dimensional subalgebra of the form $\mathcal{H}(\mathcal{B},*)$, $*$ an involution of the second kind, we study the simple algebras of dimension 9 over a given field F.

LEMMA 6. *Every separable division algebra* $\mathcal{J} = \mathcal{J}(N,\#,1)$ *over* F *contains a separable cubic extension* E *of* F. □

E/F is either cyclic or has a splitting field L, with $\mathrm{Gal}(L/F) = \mathcal{D}_6$ the dihedral group generated by σ,τ with $\sigma^3 = 1 = \tau^2$ and $\sigma\tau = \tau\sigma^2$; $E = L^{\{1,\tau\}}$ and let $K = L^{\{1,\sigma,\sigma^2\}}$. Of course L/K is cyclic. Albert [2] proves that if $\mathcal{B}$ is a central associative division algebra of degree 3 over a field K of characteristic not 2 and $\mathcal{B}$ has an involution $*$ of the second kind then $H(B,*)$ contains a separable cubic extension E of $F = H(K,*)$. He gives an explicit description of the involution according as E/F is cyclic or not and refers to the <u>abelian</u> and <u>non-abelian case</u>.

THEOREM 7. *If E is a separable cubic extension of F contained in a division algebra $J = J(N,\#,1)$ and y a non-zero element of $E^{\perp}$, then E and y generate a 9-dimensional subalgebra $E \oplus E \times y \oplus E \times r_E(y)$ and this algebra is a first Tits construction $J(E,\mu)$ if and only if $1 + N(y)^{-2}N(q_E(y)) \in F^2$ (char $F \neq 2$) or $1 + N(y)^{-2}N(q_E(y)) \in \{\alpha + \alpha^2 | \alpha \in F\}$ (char $F = 2$).* □

A closer look at the two Tits construction yields

Separable Extension	First Tits Construction	Separable Extension		Second Tits Construction
F / E cyclic	A^+	E — L / F — K cyclic	F / E cyclic	$H(B,*)$ (abelian)
			F / E non-cyclic	A^+
F / E non-cyclic	$H(B,*)$ (non-abelian)	E — L / F — K non-cyclic		$H(B,*)$ (???)

Albert [3] has an example of an exceptional division algebra which is not a first construction, and we have an example of an exceptional division algebra which is not a second construction. As is apparent from the above table, there exist exceptional Jordan division algebras which are obtainable from both Tits constructions.

4. THE QUADRATIC FORM OF A MAXIMAL TORUS

From now on, our base field F will be of characteristic not two. Adapting the terminology of Loos [5] to the present setup, we understand by a <u>torus</u> a separable commutative associative algebra of finite dimension over F. Each torus E admits an important numerical invariant, its discriminant d_E, uniquely determined up to a non-zero square factor in F. Recall that the isomorphism class of a two-dimensional torus is uniquely determined by its discriminant and that a torus of dimension 3 has discriminant 1 iff it is either a cyclic field extension or the direct sum of three copies of the base field.

We fix a unital Jordan algebra $J = J(N,\#,1)$ of degree 3 over F.

Every torus E, being an associative F-algebra, carries canonically an induced Jordan algebra structure E^+, which, since we are in characteristic not two, may be identified with E. Hence the notion of a maximal torus in J makes good and obvious sense. Evidently, all maximal tori in J have dimension 3.

The following proposition is due to Springer [11].

PROPOSITION 8. *Let E be a maximal torus in J and write $E^\perp$ for its orthogonal complement relative to the generic trace.*

(i) *The rule*

$$(a,v) \to a\cdot v := -a \times v \qquad (a \in E, v \in E^\perp)$$

defines the structure of a unital left E-module on $E^\perp$ which is compatible with its vector space structure over F.

(ii) *There is a unique map $q_E : E^\perp \to E$ satisfying*

$$T(q_E(v),a) = T(v^{\#},a)$$

for $v \in E^\perp$, $a \in E$, and, when $E^\perp$ is regarded as an E-module according to (i), *q_E is a quadratic form over E.*

(iii) *If J is separable, q_E is non-degenerate.*

Up to a certain point, this is just a special case of Lemma 4.

Assume now that J is central simple. Then, if E is a maximal torus in J and at the same time a field (so, altogether, a separable cubic subfield of J), the scalar extension $J_E = J \otimes_F E$ obviously is a reduced simple Jordan algebra of degree 3 over E. As such it admits a unique coordinate algebra, whose generic norm is a quadratic form over E. It is natural to ask how this quadratic form relates to q_E as defined in Proposition 8. The answer is provided by

THEOREM 9. *Let J be central simple and E a separable cubic subfield of J. Write C for the coordinate algebra of the scalar extension J_E, q for the generic norm of C and q_o for the restriction of q to the elements of C having trace zero. Then*

$$q_E \text{ and } \langle d_E \rangle \perp q_o$$

are similar quadratic forms over E.

Here $\langle\alpha\rangle$, for a non-zero scalar α, is the one-dimensional quadratic form $\xi \to \alpha\xi^2$, and $\perp$ denotes the orthogonal sum. Two quadratic forms are said to be *similar* if they are isometric up to a non-zero scalar factor. The proof of Theorem 9 is fairly straightforward in case E/F is cyclic but becomes quite messy otherwise. The cyclic case has also been discussed in

Springer [11], where $E^{\perp}$ is shown to carry the structure of a twisted composition algebra.

5. EXCEPTIONAL ALGEBRAS

We now wish to apply our results to an exceptional Jordan division algebra $\mathcal{J}$ over F. To this end, the following proposition turns out to be quite useful.

PROPOSITION 10. *Suppose there exists a field extension* K/F *of odd degree which splits* $\mathcal{J}$ *(that is, the scalar extension* $\mathcal{J}_K$ *is split). Then* $\mathcal{J}$ *is split by every cubic subfield of* $\mathcal{J}$. □

The proof of this is based upon a well known theorem of Springer [10] according to which two non-degenerate quadratic forms over F which become isometric after performing an odd degree field extension must have been isometric over F to begin with.

We wish to obtain criteria for $\mathcal{J}$ to be a first Tits construction, that is, to contain a subalgebra of the form D^+, D a central associative division algebra of degree 3. As a preliminary step, we first present a criterion for $\mathcal{J}$ to contain a (nine-dimensional) first Tits construction.

THEOREM 11. *Given a cubic separable subfield* E *of* $\mathcal{J}$, *the following statements are equivalent.*

(i) *There exists a non-zero element* $\mu \in F$ *such that the first Tits construction* $\mathcal{J}(E,\mu)$ *becomes a subalgebra of* $\mathcal{J}$.

(ii) *The quadratic form* q_E *over* E *is isotropic.*

(iii) *The Galois closure of* E/F *is a splitting field of* $\mathcal{J}$. □

It is well known that a reduced first Tits construction exceptional simple Jordan algebra is necessarily split (cf. Jacobson [4, IX Theorem 20]). Hence Theorem 11 yields

COROLLARY 1. *Suppose* $\mathcal{J}$ *is a first Tits construction. Then, given a cubic separable subfield* E *of* $\mathcal{J}$, *there exists a non-zero element* F *such that the first Tits construction* $\mathcal{J}(E,\mu)$ *becomes a subalgebra of* $\mathcal{J}$. □

Since the first Tits constructions starting from a cyclic cubic field have the form A^+, A a central simple associative algebra of degree 3 (see table in §3), Corollary 1 implies

COROLLARY 2. *Suppose* $\mathcal{J}$ *is a first Tits construction. Then every cyclic cubic subfield of* $\mathcal{J}$ *imbeds into a subalgebra of* $\mathcal{J}$ *having the form* D^+,

D a central associative division algebra of degree 3.

The situation becomes particularly nice in the presence of the third roots of unity. We have

THEOREM 12. *If F has characteristic not 3 and contains the third roots of unity, the following statements are equivalent.*

(i) *$\mathcal{J}$ is a first Tits construction.*

(ii) *Every reducing field of $\mathcal{J}$ is a splitting field of $\mathcal{J}$.*

(iii) *There exists a cubic subfield of $\mathcal{J}$ which splits $\mathcal{J}$.*

(iv) *The quadratic form S on $\mathcal{J}$ defined by $S(u) = T(u^{\#})$ for $u \in \mathcal{J}$ has maximal Witt index.* □

It is generally believed that first Tits constructions are much easier to handle than second Tits constructions. This philosophy is supported by the following result.

THEOREM 13. *Let $\mathcal{J}$ be a first Tits construction exceptional Jordan division algebra. Then every isotope of $\mathcal{J}$ is isomorphic with $\mathcal{J}$.* □

6. OPEN QUESTIONS

With regard to Corollary 2 of Theorem 11, it is natural to conjecture the following.

CONJECTURE I. *Every separable cubic subfield (not necessarily cyclic) of a first Tits construction exceptional Jordan division algebra imbeds into a subalgebra of the form D^+, D a central associative division algebra of degree 3.*

Another natural conjecture is

CONJECTURE II. *Theorem 12 holds irrespective of characteristics and the presence of third roots of unity.*

In connection with Corollary 1 of Theorem 11, the following question presents itself:

Is every nine-dimensional subalgebra of a first Tits construction exceptional Jordan division algebra itself a first Tits construction?

The answer to this question is negative. In fact, combining the Amitsur technique of generic matrices with formally real Jordan algebras, we can construct a first Tits construction exceptional Jordan division algebra $\mathcal{J}$ containing a nine-dimensional subalgebra J' such that the quadratic form S, $u \to T(u^{\#})$, restricted to the elements of trace zero in J' is anisotropic. Under these circumstances, J' obviously cannot be a first Tits construction.

One of the most significant questions in this context has been raised by Albert [3]: Does every exceptional Jordan division algebra contain a cyclic cubic field extension of the center? We conjecture:

CONJECTURE III. *There exist exceptional Jordan division algebras which do not contain a cyclic cubic field extension of the center.*

Since, by a classical result of Wedderburn, every central associative division algebra of degree 3 is cyclic, such an exceptional division algebra could not be a first construction. The canonical candidates are of course the (appropriately defined) generic exceptional division algebras. Conjecture III is supported by the fact that there exist nine-dimensional first Tits construction Jordan division algebras which do not contain cyclic cubic subfields; our examples live over the field of iterated Laurent series in sufficiently many variables having real coefficients.

REFERENCES

[1] Albert, A. A., A construction of exceptional Jordan division algebras, Ann. of Math. 67 (1958), 1-25.

[2] ____________, On involutorial associative division algebras, Scripta Math. 26 (1963), 309-316.

[3] ____________, On exceptional Jordan division algebras, Pacific J. Math. 15 (1965), 377-404.

[4] Jacobson N., Structure and Representations of Jordan Aglebras, AMS Colloquium Pub. Vol. XXXIX, Providence, 1968.

[5] Loos, O., Jordan pairs. Lecture Notes in Mathematics No. 460. Berlin-Heidelberg-New York: Springer 1975.

[6] Mc Crimmon, K., The Freudenthal-Springer-Tits constructions of exceptional Jordan algebras, Trans. AMS 139 (1969), 495-510.

[7] ____________, On Herstein's theorems relating Jordan and associative algebra, J. of Algebra 13 (1969), 382-392.

[8] ____________, The Freudenthal-Springer-Tits constructions revisited, Trans. AMS, 148 (1970), 293-314.

[9] Racine, M. L., A note on quadratic Jordan algebras of degree 3, Trans. AMS, 164 (1972), 93-103.

[10] Springer, T. A., Sur les formes quadratiques d'indice zéro, C. R. Acad. Sci. Paris 234 (1952), 1517-1519.

[11] ____________, Oktaven, Jordan-Algebren und Ausnahmegruppen, Lecture Notes, Gottingen 1963.

[12] Zel'manov, E. I., Jordan Division Algebras, Algebra y Logika, 18 (1979), 286–310.

H. Petersson
Fachbereich Mathematik
Fernuniversitaet-Gesamthochschule
Postfach 940, D-58 Hagen
Bundesrepublik Deutschland

M. Racine
Fachbereich Mathematik
Universitaet Muenster
Einsteinstrasse 64
D-44 Muenster
Bundesrepublik Deutschland

and

Department of Mathematics
The University of Ottawa
Ottawa, Ontario K1N9B4
CANADA

Contemporary Mathematics
Volume 13, 1982

AN APOLOGY FOR JORDAN ALGEBRAS IN QUANTUM THEORY

By John R. Faulkner

Already with the introduction of the algebras which came to bear his name, Jordan [1] stressed the physical significance of taking powers of observables. More generally, if A is an observable and u is a well behaved real valued function, one can interpret u(A) as the observable obtained by first measuring A and then applying u to each value of A obtained (as opposed to applying u to the expected value of A). Jordan also assumed the existence of the <u>sum</u> A+B of two observables A and B satisfying

(1) $$E_x(A+B) = E_x(A) + E_x(B)$$

for all states x, where $E_x(A)$ is the expected value of A on state x. Also, implicitly assumed by Jordan was the condition that the powers A^n come from an algebra structure; in particular

(2) $$A\cdot B = \frac{1}{2}((A+B)^2-A^2-B^2) \text{ is bilinear.}$$

One is led in this way to a formally real ($\Sigma A_i^2 = 0$ implies $A_i = 0$), power associative algebra which was shown by Jordan, von Neumann, and Wigner [2], under a finite dimensionality assumption, to be a Jordan algebra, i.e.

(3) $$A^2\cdot(B\cdot A) = (A^2\cdot B)\cdot A \ .$$

Among the difficulties with this approach is the assignemt of physical meaning to A+B, for observables A and B which are not "compatible". In particular, while (1) holds, it is possible for A+B to take values which are not among the sums of values taken by A and by B. Similarly, one might try to interpret $E_x(A\cdot B)$ for $E_x(A) = E_x(B) = 0$ as a "covariance", but what does this mean for noncompatible observables? Finally, the assumption of finite dimensionality is an undesirable limitation.

0271-4132/82/0543/$02.00

In this note, we give an axiomatic formulation of the measuring process which avoids the above objections and which obtains Jordan algebras as the underlying algebraic structure. Full details, including proofs, will appear elsewhere.

Preliminary to the axiomatization, we consider the "classical" measuring process which may be viewed as a set Ω of <u>states</u> and the set $\mathfrak{A}$ of (bounded) functions on Ω called <u>observables</u>. The value of the observable f in state p is simply $f(p)$. Passing to a "statistical" measuring process, we consider a measure space (Ω,μ) with $\mu(\Omega) < \infty$ and take <u>states</u> to be probability measures on Ω which are absolutely continuous with respect to μ. The <u>observables</u> are the set $\mathfrak{A}_\Omega = L^\infty(\Omega,\mu)$ and $E_\Omega(f,\nu) = \int_\Omega f d\nu$ is the <u>expectation</u> of $f \in \mathfrak{A}_\Omega$ in state ν. Expanding slightly on this formalism, let E_Ω denote the set of nonnegative, nonzero, finite measures ν on Ω which are absolutely continuous with respect to μ. Thus, if $\nu \in E_\Omega$, then $\nu = a\nu_0$, $a > 0$, for some probability measure (state) ν_0. We may interpret ν as an <u>ensemble</u> of size $\nu(\Omega) = a$ of objects in state ν_0. Identifying the state ν_0 with the ray $\mathbb{R}^+\nu_0 = \mathbb{R}^+\nu \equiv \bar{\nu}$, and defining the <u>value</u> of $f \in \mathfrak{A}_\Omega$ on the ensemble ν to be $\langle f,\nu\rangle_\Omega \equiv \int_\Omega f d\nu$, we obtain

$$E_\Omega(f,\bar{\nu}) = \frac{\langle f,\nu\rangle_\Omega}{\langle 1,\nu\rangle_\Omega}$$

independent of the representative ν of $\bar{\nu}$.

Note that in both the classical and statistical processes, if u is a suitable real value function and f is an observable, then $u \circ f$ is also an observable so one can take functions of observables.

We also note that associated with Ω in the classical set-up is another set Ω' and a bijection $\theta : \Omega \to \Omega'$ where the state $\theta(p)$ in Ω' is interpreted as a certain number, say $h(p)$, of copies of the state p in Ω. The observable $f \in \mathfrak{A}$ then corresponds to the observable $f' \in \mathfrak{A}'$ given by $f'(\theta(p)) = h(p)f(p)$. Extending this association to the statistical set-up, we may interpret the ensemble ν' in Ω' given by $d\nu' = g d\mu'$ as the corresponding mixture $(g \circ \theta)d\mu$ of $h(p)$ copies of p, i.e. ν' is interpreted in Ω as ν where $d\nu = h(g \circ \theta)d\mu$. Identifying Ω with Ω' via θ, we have maps $\omega_h : E_{\Omega'} \to E_\Omega$, $\omega_h(\nu') = \nu$ where $d\nu' = g d\mu$, $d\nu = hg d\mu$ and $\omega_h^* : \mathfrak{A}_\Omega \to \mathfrak{A}_{\Omega'}$ with $\omega_h^*(f) = hf$. Note $\omega_h(E_\Omega) = E_\Omega$ provided $h, h^{-1} \in \mathfrak{A}_\Omega$, $h > 0$. We intepret (ω_h, ω_h^*) as giving the change in the statistical set-up resulting from a change in the choice of which ensembles are fundamental events.

We shall formulate the desired axoimatization in terms of the following definitions.

Let E and $\mathfrak{A}$ be sets and $<,> : \mathfrak{A} \times E \to \mathbb{R}$ be a nondegenerate pairing; i.e.,

$$<A,x> = <B,x> \quad \text{for all} \quad x \in E \quad \text{implies} \quad A = B,$$

and

$$<A,x> = <A,y> \quad \text{for all} \quad A \in \mathfrak{A} \text{ implies} \quad x = y.$$

If $(E', \mathfrak{A}', <,>')$ is another such triple and if $\theta : E \to E'$, $\varphi : \mathfrak{A}' \to \mathfrak{A}$ satisfy

$$<A', \theta(x)>' = <\varphi(A'), x> \quad \text{for} \quad A' \in \mathfrak{A}', \; x \in E,$$

then we say (θ, φ) is a homomorphism. Note φ is uniquely determined by θ and write $\varphi = \theta^*$. As an example, we note that for $h, h^{-1} \in \mathfrak{A}_\Omega$, $h > 0$, (ω_h, ω_h^*) is an automorphism of $(E_\Omega, \mathfrak{A}, <,>_\Omega)$. A homomorphism $\theta: E \to E_\Omega$ is a representation of $(E, \mathfrak{A}, <,>)$ on (Ω, μ). Representations $\theta : E \to E_\Omega$ and $\theta' : E \to E_\Omega$ are equivalent if $\theta'(x) = \theta(x) \circ \alpha$ for a measure space isomorphism $\alpha : \Omega' \to \Omega$.

A measurement system $M = (E, \mathfrak{A}, <,>, W, I, x_I, T)$ consists of a set E of ensembles, a set $\mathfrak{A}$ of observables, a nondegenerate pairing $<,> : \mathfrak{A} \times E \to \mathbb{R}$ a group W of automorphisms of $(E, \mathfrak{A}, <,>)$ representing a change in the fundamental events, a fixed observable $I \in \mathfrak{A}$ called the counting observable, a fixed ensemble $x_I \in E$ called the fundamental ensemble, and a family T of representations of $(E, \mathfrak{A}, <,>)$ satisfying

(MS 1) $\mathfrak{A} = \bigcup_{\theta \in T} \theta^*(\mathfrak{A}_{\Omega_\theta})$,

(MS 2) $\theta(x_I) = \mu_\theta, \theta^*(1) = I$ for all $\theta \in T$,

(MS 3) If $\theta_1, \theta_2 \in T$, if $\theta_i(f_i) = A$ for $f_i \in \mathfrak{A}_{\Omega_{\theta_i}}$,

and if $u : \mathbb{R} \to \mathbb{R}$ has $u \circ f_1 \in \mathfrak{A}_{\Omega_{\theta_1}}$, then $u \circ f_2 \in \mathfrak{A}_{\Omega_{\theta_2}}$ and $\theta_1^*(u_1 \circ f_1) = \theta_2^*(u_2 \circ f_2)$. We denote the common value by $u_I(A)$.

(MS 4) Given $\theta \in \mathcal{T}$ and $h \in \mathfrak{A}_{\Omega_\theta}$, then $\theta(h) = W^*(I)$ for some $W \in \mathcal{W}$ if and only if $h^{-1} \in \mathfrak{A}_{\Omega_\theta}$, $h > 0$. In this case, there is a $\tilde{\theta} \in \mathcal{T}$ with $\tilde{\theta} \circ W$ equivalent to $\omega_h \circ \theta$.

(MS 5) If $W \in \mathcal{W}$ and $j:\mathbb{R}\setminus\{0\} \to \mathbb{R}$ is $j(t) = t^{-1}$, then there is $\hat{W} \in \mathcal{W}$ with $j_I W^* = \hat{W}^* j_I$ where j_I is defined.

<u>Main Theorem</u>: If $\mathcal{M}$ is a measurement system then $\mathfrak{A}$ has the structure of a formally real Jordan algebra and a normed linear space with continuous operations.

In the proof of the theorem, one introduces the Jordan quadratic operator U_A. Indeed, one shows that for $W \in \mathcal{W}$, $U_A \equiv W^*\hat{W}^{*-1}$ and $x_A \equiv \hat{W}^{-1}x_I$ depend only on $A = W^*(I)$. Setting $\langle B,x\rangle_A = \langle U_A^{-1}B,x\rangle$ and $\mathcal{T}_{\mathcal{W}} = \{\theta \circ W \mid \theta \in \mathcal{T}\}$ one shows the following

<u>Lemma</u>. $\mathcal{M}' = (E, \mathfrak{A}, \langle,\rangle_A, \mathcal{W}, A, x_A, \mathcal{T}_W)$ is a measurement system uniquely determined by $A = W^*(I)$ up equivalence of the representations.

REFERENCES

[1] P. Jordan, "Uber die Multiplikation quantenmechanischer Grossen", Z. Phys., 80 (1933), 285-291.

[2] P. Jordan, J. von Neumann and E. Wigner, "On an algebraic generalization of the quantum mechanical formalism", Ann. of Math. 35 (1934), 29-64.

University of Virginia
Charlottesville, VA 22903

Contemporary Mathematics
Volume 13, 1982

IDENTITIES OF NONASSOCIATIVE ALGEBRAS STUDIED BY COMPUTER

Leslie Hogben
Department of Mathematics
Iowa State University
Ames, Iowa 50011

Group representations offer an efficient means of representing identities of nonassociative algebras by matrices [1]. These matrices can be row reduced to determine whether one identity is a consequence of other identities. The entire process can be done by a computer.

Let V be a variety of nonassociative algebras over a field, defined by a multilinear identity g of degree n. To determine whether a given identity f is a consequence of g (i.e., whether it is true in V) we construct the group ring R on the symmetric group S_n over F. We then consider a free (left) R module of rank equal to the number of distinct associations in the identities. For example, the variety of right alternative algebras is defined by

$$(x,y,y) = (xy)y - x(yy) = 0. \tag{1}$$

Linearizing this we obtain

$$(xy)z + (xz)y - x(yz) - x(zy) = 0. \tag{2}$$

This identity has 2 association types, $(ab)c$ and $a(bc)$ (the only two possible types in degree 3). Thus to determine whether another identity of degeee 3 is true for right alternative algebras we would use a free R module of rank 2, (R,R), with the first coordinate corresponding to $(ab)c$ and the second to $a(bc)$, where R is the group ring on S_3. The identity (2) would be represented by the element $(I+(23), - I - (23))$.

Given an identity, permutations of the variables, and sums of these, yield consequences of the identity. Thus the consequences of an identity form a (left) submodule; i.e., we can multiply by any linear combination of permutations. Note: permutations are acting on the right. (2) was $I+(23)$ applied to $(xy)z - x(yz)$. If we multiply on the left by the transposition (12) we obtain $(12) + (12)(23) = (12) + (132)$. Applying this to $(xy)z - x(yz)$ gives $(yx)z + (yz)x - y(xz) - y(zx)$ which is clearly a consequence of (2). Right multiplication by elements of R does not work: $(I+(23))(12) = (12) + (123)$ which yields $(yx)z + (zx)y - y(xz) - z(xy)$ which is not a consequence of (2).

0271-4132/82/0544/$02.00

More generally, we need not restrict the variety to one defining identity. We can consider the submodule generated by several identities. Any identity contained in this submodule is a consequence of the defining identities, and any identity not in the submodule is not a consequence.

The advantage of this description of identities is that we can now study them by means of representations of S_n as matrices. (Left) multiplication by an element of R corresponds to (left) multiplication by the appropriate matrix and vice versa. However, the matrix A can be expressed as CB for some matrix C if and only if every row of A is in the span of the rows of B. Thus, an identity f is a consequence of identities $g_1,\ldots,g_m$ if and only if the element of the group ring module associated with f is in the submodule generated by $g_1,\ldots,g_m$, if and only if in every irreducible representation of S_n the rows of the matrix of f are in the span of the rows of the matrix of $g_1,\ldots,g_m$.

Thus to determine if identity f is a consequence of identities $g_1,\ldots,g_m$ we could, in a fixed representation, construct the matrix of $g_1,\ldots,g_m$, compute its rank (by reducing to reduced row echelon form), construct the matrix of $g_1,\ldots,g_m$, f and compute its rank. If the ranks are equal, the rows of f are in the span of $g_1,\ldots,g_m$. If this is true in every representation f is a consequence of $g_1,\ldots,g_m$.

More efficiently, we can construct the matrix of $g_1,\ldots,g_m$, reduce it to reduced row echelon form and store the columns which do not contain a leading one, and their locations. Then to test any identity f we construct its representation and compare it to the (easily reconstructed) reduced row echelon form of $g_1,\ldots,g_m$ to see if it is in the span.

The preceding discussion is predicated on the assumption that all the identities $g_1,\ldots,g_m$ and f have the same degree. However this is not always the case. One frequently wishes to determine whether an identity f of higher degree is a consequence of the identities of the variety. In order to do this, we need to generate higher degree identities from the defining identities. In general, an identity of degree 3 will yield 5 identities of degree 4, by left multiplication, right multiplication, and substitution of a product for each of the three variables. However, if the identity has some symmetry it will yield fewer (not redundant) identities. For example, the right alternative law, (2), being symmetric in y and z, yields only 4 identities of degree 4, because substitutions for y and z yield the same identity. Also if one expands an identity of degree 3 to degree 4, and then expands each of these to degree 5, one obtains $5 \cdot 6 = 30$ identities. However 9 of these are redundant, as direct expansion to degree 5 shows.

Because of limitations of computer time and storage space, it is desirable to avoid redundancies. Partly because of the difficulties this introduces into the expansion process, at the present time we are doing this expansion by hand.

Unfortunately this offers many opportunities for human error and it is necessary to check and recheck the data entered into the computer.

The statement "reduce the matrix to reduced row echelon form" glosses over a great many difficulties. There are many techniques of numerical linear algebra and canned computer routines available for doing this efficiently, minimizing round off errors. Such errors can create a serious problem with large (100×100) matrices. To avoid this we have chosen to write our own programs which work entirely with rational matrices (computers work only with reals and integers; thus a rational number is actually stored as 2 integers). Although this presents occasional problems with integer overflow (numbers too big for the computer to handle) such problems can usually be avoided by a little forsight, such as clearing common factors from rows. By working with rational numbers all results are exact. We also keep track of what numbers we divide by (what characteristics to exclude) and check row operations by matrix multiplication. We have programs to linearize identities, to represent identities by a matrix and row reduce the matrix, and to compare the reduced row echelon form to the representation of another identity.

Assuming the data is accurate, we are thus able to prove on the computer that a given identity is true in a given variety. However, we are currently using this only as a test; we prove the identity by hand after determining it is true. Although it is theoretically possible to recover the proof of the identity from the matrix reduction (in each representation) we don't have a program to do this and it is a rather arduous task to undertake by hand.

Despite the speed of modern computers, considerations of computer time and storage space still limit the degrees of identities we can study. Unfortunately the size of the matrices involved increases drastically with the degree. The matrix to be row reduced has (number of identities) × (dimension of the representation) rows and (number of associations) × (dimension of the representation) columns. As remarked earlier one defining identity of degree 3 may yield 5 identities of degree 4, 21 of degree 5, 84 of degree 6, etc. There are 2 associations of degree 3, 5 of degree 4, 14 of degree 5, and 42 of degree 6. The maximum dimension of a representation is 2 for degree 3, 3 for degree 4, 6 for degree 5, and 16 for degree 6. Additionally, there are 3 representations of degree 3, 5 of degree 4, 7 of degree 5, and 11 of degree 6. Thus one defining identity of degree 3 (with no symmetry) would yield a 1344×672 matrix to be row reduced (in the largest representation) in degree 6. Since the work involved in the row reduction of an $n \times n$ matrix increases with the cube of n, it is easy to see that the matrices soon become too large. And no account has been taken of the work needed to construct the representation, although this is usually less significant than the row reduction. The work can be greatly reduced by reducing the number of identities and/or by reducing the number of associations used. For example, if the variety is commutative, only 6 associations are needed in degree 6. At the present time we are able to handle identities of degree less than or equal to 5, and those of degree 6 which do not involve too many identities and associations. Additional savings may be achieved by checking only those representations where the identity being tested is nonzero. When the identity being tested is the linearization of an identity with several elements equal, this is significant. For degree 5 and up we use the more efficient method of comparing the identity being tested with the previously constructed reduced row echelon form, rather than redoing the entire row reduction.

Several colleagues at Iowa State University and I have used these programs to study numerous nonassociative algebras including Jordan algebras, alternative algebras, right alternative algebras, flexible derivation alternator algebras, and Lie admissable algebras.

Recently Irvin Hentzel, Harry Smith, and I have been studying algebras in the variety defined by

$$(yz,x,x) = y(z,x,x) + (y,x,x)z = 0 \tag{3}$$

and the Lie admissable identity

$$(x,y,z) + (y,z,x) + (z,x,y) = 0. \tag{4}$$

We hoped to establish the identity

$$((y,x,x),x,x) = 0 \tag{5}$$

so we expanded (3) and (4) to degree 5, obtaining 5 and 11 identities respectively. Unfortunately (5) proved to be false. However we were able to determine that

$$(x,(x,y,x),x) = 0 \tag{6}$$

is a consequence of (3) and (4), which was subsequently proved by Smith. We also found several other useful consequences of (3) and (4). The value of negative results should not be overlooked, as a great deal of time can be saved by not attempting to prove false identities.

REFERENCES

1. I. R. Hentzel, Processing Identities by Group Representation, in Computers in Nonassociative Rings and Algebras, (R. E. Beck and B. Kolman, Eds.), pp. 13-40, Academic Press, New York, 1977.

2. I. R. Hentzel, Alternators of a Right Alternative Algebra, Trans. Amer. Math. Soc., 242(1978), 141-156.

3. I. R. Hentzel, L. Hogben, and H. F. Smith, Flexible Derivation Alternator Rings, Comm. Algebra 8(1980), 1997-2014.

4. I. R. Hentzel, H. F. Smith, and L. Hogben, Lie Admissable Right Derivation Alternator Rings, (to appear).

5. G. M. Piacentini and I. R. Hentzel, Right Alternative Algebras and Alternators, (to appear).

Contemporary Mathematics
Volume 13, 1982

ON REGULARLY CLOSED FIELDS

Tsuneo Tamagawa

Let K be a field of characteristic p. If for every regular extension $L = K(x_1 \dots x_n)$ of K there exists a morphism φ of the affine ring $K[x_1, \dots, x_n]$ onto K, the field K is called regularly closed. Geometrically K is regularly closed if every affine variety V defined over K has a K-rational point. If K is regularly closed then the set V_K of all K-rational points on V is Zariski-dense in V. The term "regularly closed field" or RC field is introduced by G. Cherlin-L. van den Dries-A. Macintyre in their paper "The Elementary Theory of Regularly Closed Fields" and it becomes necessary to prove the following two theorems in the case where $p > 0$.

THEOREM 1. *Let* k *be a regularly closed field. If* K/k *is a purely inseparable extension, then* K *is also regularly closed.*

THEOREM 2. *Suppose that* K *is regularly closed. Let* V *be a variety of dimension* > 0 *defined over* K *and* $(x) = (x_1, \dots, x_n)$ *a generic point of* V *over* K. *If* $z_1, \dots, z_m \in K[x]$ *are p-independent in* $K(x)$ *and* $[K:K^p] \geq p^m$, *then there exist infinitely many K-rational points* $(\xi) = (\xi_1, \dots, \xi_n)$ *on* V *such that* $\varphi_{(\xi)}(z_1) \dots \varphi_{(\xi)}(z_m)$ *are p-independent in* K *where* $\varphi_{(\xi)}$ *denotes the morphism of* $K[x]$ *onto* K *associated to the specialization* $(x) \to (\xi)$ *over* K.

In this note, we will give proofs of these theorems. Throughout this note we adopt terminologies of A. Weil's monumental book "Foundations of Algebraic Geometry". Thus fields are contained in a fixed "universal domain" $\mathbb{K}$ of characteristic $p > 0$ and the degree of transcendency of $\mathbb{K}$ over any "field" is infinite. Quantities are elements of $\mathbb{K}$ and the dimension of an extension $K(x_1, \dots x_n)$ over K is the degree of transcendency of the extension. (cf. A. Weil, FAG. Chap. I).

The theory of derivations provides a powerful tool to study fields of characteristic $p > 0$, especially problems concerning with separability and

---

Written with the partial support of NSF grant MCS79-04470

p-independency, and plays the central role in this note. We are greatly indebted to Professor Nathan Jacobson who laid foundations of the theory.

Let K be a field. A derivation D of K will be called standard if $D \neq 0$ and $D^p = 0$. The constant field K^D is the subfield of all $a \in K$ with $Da = 0$. A system of standard derivations $D = \{D_1, \ldots, D_m\}$ will be called P.D. system if $D_1, \ldots, D_m$ are linearly independent over K and $D_iD_j = D_jD_i$ for all i, j. By Jacobson's theorem (cf. Jacobson p. 186) the constant field $k = K^{D_1} \cap \ldots \cap K^{D_m}$ is a subfield of K such that $K \supset k \supset K^p$ and $[K:k] = p^m$. Let I denote the set of all m-tuples of integers $(\nu) = (\nu_1, \ldots, \nu_m)$ such that $0 \leq \nu_j \leq p-1$. For $(\nu) = (\nu_1, \ldots, \nu_m)$ put $D^\nu = D_1^{\nu_1} \cdots D_m^{\nu_m}$, $\alpha^{(\nu)} = D^\nu \alpha$ $(\alpha \in K)$ and $|\nu| = \Sigma\, \nu_j$. Put $D^{p-1} = (D_1 \cdots D_m)^{p-1}$. Then $\alpha \to D^{p-1}\alpha$ is a non-trivial k-linear function on K and we have $D^{p-1}(\alpha\beta) = \Sigma_\nu (-1)^{|\nu|} \alpha^{(\nu)} \beta^{(p-1-\nu)}$ where $(p-1-\nu)$ denotes $(p-1-\nu_1, \ldots, p-1-\nu_m)$ for $(\nu) = (\nu_1 \cdots \nu_m)$. Put $q = p^m$. If $\omega_1, \ldots, \omega_q$ are a base of K/k then the complementary base $\omega_1^*, \ldots, \omega_q^*$ are defined by $D^{p-1}(\omega_i\, \omega_j^*) = \delta_{ij}$ and for $\alpha \in K$ we have

$$\alpha^{(\nu)} = \Sigma\, a_\mu\, \omega_\mu^{(\nu)}, \quad a_\mu = D^{p-1}(\alpha\, \omega_\mu^*) = \Sigma\, (-1)^{|\nu|}\, \alpha^{(\nu)}\, \omega_\mu^{*\,(p-1-\nu)}$$

We fix an ordering of I lexicographically. If $\alpha_1, \ldots, \alpha_t \in K$ then they are p-independent over k (cf. Jacobson p. 174) if and only if the Jacobian matrix $(D_i\, \alpha_j)$ is of rank t. There exist $\gamma_1, \ldots, \gamma_m \in K$ such that $D_i\, \gamma_j = \delta_{ij}$ and derivations $D_1, \ldots, D_m$ are the partial derivations

$$\frac{\partial}{\partial\gamma_1}, \ldots, \frac{\partial}{\partial\gamma_m}.$$

Let F be a field $\supset K$ and $D' = \{D_1', \ldots, D_m'\}$ a P.D. system of F such that $D_j'|_K = D_j$, $1 \leq j \leq m$. Then we have $F^{D'} \cap K = k$ and $F^{D'}K = F$. $F^{D'}$ and K are linearly disjoint over k.

Let V be a variety of dimension $d > 0$ defined over K and $(x) = (x_1, \ldots, x_m)$ a generac point of V over K. Let $D = \{D_1, \ldots, D_m\}$ be a P.D. system of K. There exist quantities $z_1, \ldots, z_d \in L = K(x)$ which are independent over K such that $L/K(z_1, \ldots, z_d)$ is separably algebraic (cf. Weil, p. 16 and p. 68). Let $\{z_i^{(\nu)};\ 1 \leq i \leq d,\ \nu \in I\}$ be a set of independent quantities over K such that $z_i = z_i^{(0)}$, $(0) = (0, \ldots, 0)$. Put $M_0 = K(z_1, \ldots, z_d, \ldots, z_1^{p-1}, \ldots, z_d^{p-1})$, $(p-1) = (p-1, \ldots, p-1)$. We extend derivations $D_1, \ldots, D_m$ of K to derivations $\hat{D}_1, \ldots, \hat{D}_m$ of M_0 so that $\hat{D}_j z_i^{(\nu_1 \cdots \nu_m)} = z_i^{(\nu_1 \cdots \nu_j+1 \cdots \nu_m)}$ or $= 0$ according as $\nu_j < p-1$ or

$\nu_j = p-1$. Then $\hat{D} = \{\hat{D}_1, \ldots, \hat{D}_m\}$ is a P.D. system of M_0. Since $M = M_0(x)$ is separably algebraic over M_0, the P.D. system $\hat{D}$ of M_0 is uniquely extended to a P.D. system of M. We will use the same notations for the extended P.D. system. Put $x_i^{(\nu)} = \hat{D}^\nu x_i$, $1 \le i \le n$, $(\nu) \in I$, $(x^{(\nu)}) = (x_1^{(\nu)}, \ldots, x_n^{(\nu)})$ and $(x)_D = (\hat{x}) = (x, \ldots, x^{(\nu)}, \ldots, x^{(p-1)})$ where $x^{(\nu)}$ are arranged by the given ordering of I. Since $\hat{D}^\nu z_i = z_i^{(\nu)}$ and $K(x) \ni z_1, \ldots, z_d$ we have $K(\hat{x}) \ni z_i^{(\nu)}$, $1 \le i \le d$, $(\nu) \in I$ and $K(\hat{x}) = M$ because $\hat{D}_j K(\hat{x}) \subset K(\hat{x})$ for $j = 1, \ldots, m$. The extension M/K is a regular extension of K, hence a variety $\hat{V}_D$ is defined as the locus of $(x)_D$ over K. There exists a canonical projection pr_V of $\hat{V}_D$ onto V.

PROPOSITION 1. <u>Suppose that</u> $z_1', \ldots, z_t'$ <u>are p-independent over</u> $F = K(x^p)$. <u>Then we have</u> $t \le d$ <u>and</u> tq <u>quantities</u> $\{z_1', \ldots, z_t', \ldots, z_1'^{(\nu)}, \ldots, z_t'^{(\nu)}, \ldots\}$ <u>are independent over</u> K.

<u>Proof</u>. Since $L/K(z_1, \ldots, z_d)$ is separably algebraic and $z_1, \ldots, z_d$ are independent over K, $z_1, \ldots, z_d$ are p-independent over $F = K(x^p)$ and $L = F(z_1, \ldots, z_d)$. Hence we have $t \le d$. If $t < d$ we add $d-t$ quantities $z_{t+1}' \ldots z_d'$ to $z_1', \ldots, z_t'$ so that $z_1' \ldots z_d'$ are p-independent over K. Then we have $L = F(z')$, $(z') = (z_1' \ldots z_d')$. Hence there is no non-trivial derivation of L over $K(z_1' \ldots z_d')$ and $L/K(z')$ is separably algebraic. Put $M_0' = K((z')_D)$, $(z')_D = (z', \ldots, z'^{(\nu)} \ldots)$. Then M_0 is $\hat{D}$-invariant, hence $M' = M_0'(x)$ is also $\hat{D}$-invariant because M'/M_0' is separably algebraic. Hence $M' = M$ and M_0' is purely transcendental of dimension dq. Q.E.D.

The proof shows that the variety $\hat{V}_D$ is determined uniquely by K, D and V.

Let $X_1, \ldots, X_n, \ldots, X_1^{(\nu)} \ldots X_n^{(\nu)}, \ldots, X_1^{(p-1)} \ldots X_n^{(p-1)}$ be nq indeterminates, and put $R = K[\hat{X}] = [X, \ldots, X^{(\nu)} \ldots]$. We extend the derivations $D_1, \ldots, D_m$ of K to derivations $\tilde{D}_1 \ldots \tilde{D}_m$ of R by the following recipe:

$$\tilde{D}_j X_i^{(\nu_1 \ldots \nu_m)} = \begin{cases} X_i^{(\nu_1 \ldots \nu_j + 1 \ldots \nu_m)} & \nu_j < p-1 \\ 0 & \nu_j = p-1 \end{cases},$$

$$X_i^{(0)} = X_i \quad .$$

Evidently $\tilde{D} = \tilde{D}_1, \ldots, \tilde{D}_m$ is a P.D. system of the domain R. Let $\mathcal{P}$ denote the prime ideal of $K[x_1, \ldots, x_m]$ of all $f(x)$ with $f(x) = 0$, and $\tilde{\mathcal{P}}$

the ideal of R generated by $P, \widetilde{D}^{\nu} P$, $\nu \in I$.

PROPOSITION 2. Suppose that V is non-singular. Then $\widetilde{P}$ is the prime ideal of all $\hat{f}(\hat{X}) \in R$ with $\hat{f}(\hat{x}) = 0$, and the variety $\hat{V}_D$ is non-singular.

Proof. It is easy to see that $\hat{V}_D \cong (\hat{V}_{\{D_1 \ldots D_{m-1}\}})_{D_m}$, hence if we prove our assertion in the case where $m = 1$, then the general case would be proved by using the induction on m. Assuming $m = 1$, D is simply a standard derivation of K and $q = p$. First we drop the assumption of non-singularity of V. Let $\hat{P}$ denote the prime ideal of R of all $\hat{f}(\hat{X})$ such that $\hat{f}(\hat{x}) = 0$. For every $f \in P$ we have

$$f(x) = 0, \quad \hat{D}(f(x)) = (\widetilde{D}f)(x, x^{(1)}) = 0, \ \ldots, \ \hat{D}^{p-1}(f(x))$$
$$= (D^{p-1}f)\ (x, x^{(1)} \ldots x^{(p-1)}) = 0$$

hence $(\hat{x})$ is a zero of the ideal $\widetilde{P}$. Let $\{f_1(X), \ldots, f_s(X)\}$ be a base of P. Then $\{f_i, \widetilde{D} f_i, \ldots, \widetilde{D}^{p-1} f_i;\ 1 \le i \le d\}$ are a base of $\widetilde{P}$. Let J denote the Jacobian matrix

$$\begin{pmatrix} \dfrac{\partial f_1}{\partial X_1} & \cdots & \dfrac{\partial f_1}{\partial X_n} \\ \vdots & & \\ \dfrac{\partial f_s}{\partial X_1} & \cdots & \dfrac{\partial f_s}{\partial X_n} \end{pmatrix}$$

and $\{\Delta_1(X), \ldots, \Delta_t(X)\}$ be the set of all $(n-d) \times (n-d)$ minors of J such that $\Delta_h(x) \neq 0$. By suitable reordering of $X_1 \ldots X_n$ and $f_1 \ldots f_s$, we may assume that

$$\Delta(X) = \Delta_1(X) = \begin{vmatrix} \dfrac{\partial f_1}{\partial X_{d+1}} & \cdots & \dfrac{\partial f_1}{\partial X_n} \\ \dfrac{\partial f_{n-d}}{\partial X_{d+1}} & \cdots & \dfrac{\partial f_{n-d}}{\partial X_n} \end{vmatrix} .$$

Then $L/K(x_1, \ldots, x_d)$ are separably algebraic and by Proposition 1 the dp quantities $\{x_i^{(\nu)};\ 1 \le i \le d,\ \nu \in I\}$ are independent over K. For $1 \le \nu \le p-1$ we have

$$\widetilde{D}^{\nu} f_i = \Phi_i^{\nu} + \Sigma_{\lambda} \frac{\partial f_i}{\partial X_{\lambda}} X_{\lambda}^{(\nu)}$$

where $\Phi_i^{\nu} \in K[X, X^{(1)} \ldots X^{(\nu-1)}]$. Hence for $\lambda = d+1, \ldots, n$ we have

$$\Delta(X)X_{\lambda}^{(\nu)} + \Sigma_{\mu=1}^{d} \Delta_{\lambda,\mu} X_{\mu}^{(\nu)} + \sum_{i=1}^{n-d} \Delta'_{\lambda,i} \Phi_i^{\nu} = \sum_{i=1}^{n-d} \Delta'_{\lambda,i} \tilde{D}^{\nu} f_i \in \tilde{P}$$

where $\Delta_{\lambda,\mu}$ is obtained by replacing the λ-th column (the column corresponding to X_{λ}) by the first $n-d$ elements of the μ-th column of J and $\Delta'_{\lambda,i}$ are (λ,i)-cofactors of $\Delta(X)$. Hence for $f(\hat{X}) \in R$ there exists a positive integer N such that

$$\Delta(X)^N f(\hat{X}) \equiv g(X, X_1^{(1)} \dots X_d^{(1)} \dots X_1^{(p-1)} \dots X_d^{(p-1)}) \bmod \tilde{P}.$$

If $f \in \hat{P}$, then $g(x, x_1^{(1)} \dots x_d^{(1)}, \dots, x_1^{(p-1)} \dots x_d^{(p-1)}) = 0$ and

$g \in PR \subset \tilde{P}$ because $x_1^{(1)}, \dots, x_d^{(1)}, \dots, x_1^{(p-1)} \dots x_d^{(p-1)}$ are independent

over L. Therefore if N is sufficiently large, we have $\Delta(X)^N \hat{P} \subset \tilde{P}$. Conversely, $\hat{P} \supset \tilde{P}$ because $(\hat{x})$ is a zero of $\tilde{P}$. Hence we have $R[\Delta(X)^{-1}] \tilde{P} \cap R = \hat{P}$, and $\hat{P}$ is an isolated component of $\tilde{P}$. Similarly, we have $\Delta_h(X)^N \hat{P} \subset \tilde{P}$ for $h = 2, \dots, t$ and a sufficiently large integer N. Suppose that V is non-singular. Then for every point (x') of V there exists a $\Delta_h(X)$ such that $\Delta_h(x') \neq 0$. Hence there exist $P(X) \in P$ and $H_1(X), \dots, H_t(X) \in K[X]$ such that $P(X) + \Sigma H_h(X) \Delta_h(X)^N = 1$, and $\hat{P} = (P + \Sigma H_h \Delta_h^N) \hat{P} \subset \tilde{P}$. Hence $\hat{P}$ is equal to $\tilde{P}$. The Jacobian $\tilde{J}$ of the

base $\{f_1, \dots, f_s, \tilde{D} f_1, \dots, \tilde{D}^{p-1} f_1, \dots\}$ is of the form

$$\tilde{J} = \begin{pmatrix} J & & & 0 \\ & J & & \\ & & \ddots & \\ & * & & J \end{pmatrix}$$

with respect to $(X_1, \dots, X_n, X_1^{(1)}, \dots, X_n^{(1)}, \dots, X_1^{(p-1)}, \dots, X_n^{(p-1)})$. Therefore if V is non-singular, then the rank of J at every $(x') \in V$ is equal to $n-d$ and the rank of $\tilde{J}$ at every $(x', \dots) \in \hat{V}_D$ is equal to $p(n-d)$. Hence $\hat{V}_D$ is non-singular. Q.E.D.

PROPOSITION 3. _Suppose that_ V _is non-singular. Let_ $F \supset K$ _be a field and_ $D' = \{D'_1, \dots, D'_m\}$ _a_ P.D. _system of_ F _which is an extension of_ D _to_ F. _If_ $(y) = (y_1, \dots, y_n)$ _is an_ F-_rational point of_ V, _then_ $(y)_D = (y, \dots, y^{(\nu)}, \dots)$, $(y^{(\nu)}) = (D'^{\nu} y)$ _is an_ F-_rational point of_ $\hat{V}_D$. _The morphism_ φ _of_ $K[\hat{x}]$ _into_ F _associated to the specialization_ $(x)_D \to (y)_D$ _satisfies the following commutative diagram:_

$$\begin{array}{ccc} K[(x)_D] & \xrightarrow{\varphi} & F \\ \downarrow D_j & & \downarrow D'_j \\ K[(x)_D] & \xrightarrow{\varphi} & F \end{array}$$

<u>Proof</u>. For every $f \in P$ we have $(\widetilde{D}^\nu f)(y, y^{(1)}, \ldots, y^{(\nu)}) = D'^\nu f(y) = 0$, hence by Proposition 2, $(y)_{D'}$ is a point of $\hat{V}_D$. The second assertion is evident from

$$\varphi \circ \hat{D}_j(x_i^{(\nu)}) = D'_j\ (y_i^{(\nu)})$$

for all $i = 1, \ldots, n$ and $(\nu) \in I$. Q.E.D.

Put $E = K(\hat{x})^{\hat{D}}$. Then E and K are linearly disjoint over k and $E \cdot K = K(\hat{x})$. Put $(u_\mu) = (u_{1\mu}, \ldots, u_{n\mu}) = (\hat{D}^{p-1}\ \omega_\mu^*\ x)$, $(\widetilde{u}) = (u_1, \ldots, u_q) = T((\hat{x}))$. Since $(\hat{x}) = (\ldots, \Sigma_\mu \omega_\mu^{(\nu)}\ u_\mu, \ldots)$ we have $E = k(\widetilde{u})$ and E/k is a regular extension of dimension dq. Let W denote the locus of $(\widetilde{u})$ over k. Then W is a variety defined over k and a birational biregular transformation T defined over K is determined by $T((\hat{x})) = (\widetilde{u})$. Using the notations of Proposition 3 we have $T((y)_{D'}) = (\widetilde{v}) \in E_{F_0}$, $F_0 = F^{D'}$ where $(\widetilde{v}) = (v_1, \ldots, v_q)$ is defined by $(v_\mu) = (\Sigma_\nu (-1)^{|\nu|} \omega_\mu^{*(\nu)}\ y^{(p-1-\nu)})$. By Proposition 3, $(\widetilde{u}) \to (\widetilde{v})$ is a specialization over k and $(\widetilde{v}) \in W_{F_0}$. The point $(y) \in V_F$ is recovered by the relation $(y) = (\Sigma \omega_\mu\ v_\mu)$.

PROPOSITION 4. <u>Suppose that</u> $k = K^D$ <u>is regularly closed and</u> V <u>is nonsingular</u>. <u>Then the set</u> V_K <u>of all</u> K-<u>rational points on</u> V <u>is Zariski dense in</u> V <u>and the set</u> $\{(\xi)_D;\ (\xi) \in V_K\}$ <u>is Zariski dense in</u> $\hat{V}_D$.

<u>Proof</u>. For every $(\xi) \in V_K$, put $R_{K/k}(\xi) = T((\xi)_D)$. Then $R_{K/k}(\xi)$ is a k-rational point $(c_1, \ldots, c_q)$ of W and $(\xi) = (\Sigma \omega_q\ c_q)$, $(\xi)_D = (\ldots \Sigma\ \omega_q^{(\nu)}\ c_q \ldots)$. Conversely, if $(c_1, \ldots, c_q) \in W_k$, then $(\xi) = (\Sigma\ \omega_\mu\ c_\mu)$ is a point of V_K and $(\xi)_D = (\ldots \Sigma\ \omega_q^{(\nu)}\ c_q \ldots)$ is a point of $\hat{V}_D$. Evidently we have

$$V_K \ni (\xi) \to (\xi)_D \in \hat{V}_D \to T((\xi)_D) = (\widetilde{c}) \to (\Sigma\ \omega_q\ c_q) = (\xi),$$
$$W_k \ni (\widetilde{c}) \to (\Sigma \omega_\mu c_\mu) = (\xi) \to (\xi)_D \to T((\xi)_D) = (\widetilde{c}).$$

Hence $R_{K/k}$ is a bijection of V_K onto W_k. If k is regularly closed, then W_k is Zariski-dense, the set $\{(\xi)_D\colon (\xi) \in V_K\}$ is Zariski-dense in $\hat{V}_D$ because the set is equal to $T^{-1}(W_k)$, and finally V_K is Zariski dense in V because of $V_K = pr_V(T^{-1}(W_k))$. Q.E.D.

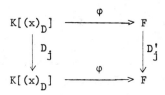

Proof of Theorem 1. Suppose that a field k is regularly closed and K/k is a purely inseparable extension of degree p^m. If $m = 1$ then there exists a standard derivation D of K such that $K^D = k$. Let V be a variety defined over K and $(x) = (x_1 \ldots x_n)$ a generic point of V over K. Then there exists a polynomial $f(X_1 \ldots X_n) \in K[X_1 \ldots X_n]$ such that if $(x') \in V$ and $f(x') \neq 0$ then (x') is simple. Let V' denote the locus of $(x_1, \ldots, x_n, f(x)^{-1})$ over K. Then V' is non-singular and V'_K is Zariski-dense in V' by Proposition 4. Since the projection of V' onto V is regular, V_K is Zariski dense in V. If $m > 1$, there exists a field K' such that $k \subset K' \subset K$ and $[K:K'] = p$. Using the induction on m, we conclude that K is regularly closed. If K/k is a purely inseparable extension of infinite degree, then every variety defined over K is defined over K', $k \subset K' \subset K$ and $[K':k] < \infty$; hence K is regularly closed. Q.E.D.

PROPOSITION 5. *Suppose that K is regularly closed. Let $D = \{D_1, \ldots, D_m\}$ be a P.D. system of K and V a non-singular variety defined over K. Then the set $\{(\xi)_D;\ (\xi) \in V_K\}$ is Zariski dense in $\hat{V}_D$.*

Proof. Since K is regularly closed, $K^p \cong K$ is also regularly closed. Hence $k = K^D$, $K \supset k \supset K^p$ is also regularly closed by Theorem 1. Our assertion follows from Proposition 4. Q.E.D.

The functor $R_{K/k}:V \to W$ is the Weil restriction functor in the special case where K/k is purely inseparable of exponent one. This explicit construction provides a suitable machinery to prove the second theorem.

Let K be a field, V a variety defined over K and $(x) = (x_1, \ldots, x_n)$ a generic point of V over K. Put $L = K(x)$ and $F = KL^p = K(x^p)$. Suppose that $z_1, \ldots, z_m \in K[x]$ are p-independent in L, $z_{\ell+1}, \ldots, z_m$ are p-independent over F, and $z_1, \ldots, z_\ell \in F(z_{\ell+1}, \ldots, z_m)$. We have

$$z_i = f_i(z_{\ell+1}, \ldots, z_m), \quad f_i \in F[Z_{\ell+1}, \ldots, Z_m] .$$

Since L/K is seperably generated, $L^p \cap K = K^p$ and K and L^p are linearly disjoint over K^p. Each derivation D of K is extended uniquely to a derivation of F/L^p which will be denoted by the same letter D. We also have $[L:F] = p^d$ where d is the dimension of V. We assume that $d > 0$. For $f(Z) \in F[Z_{\ell+1} \ldots Z_m]$, $f(Z) = \Sigma\, a_\mu Z^\mu$, put $f^D(Z) = \Sigma\, D(a_\mu)Z^\mu$.

PROPOSITION 6. *There exists a P.D. system $\{D_1, \ldots, D_\ell\}$ of K such that*

$$|f_i^{D_j}(z_{\ell+1}, \ldots, z_m)| \neq 0 .$$

Proof. Let S be a p-base of $L/F(z_{\ell+1}, \ldots, z_m)$ and C a p-base of K. Then C is a p-base of F/L^p and $C \cup \{z_{\ell+1}, \ldots, z_m\} \cup S$ is a p-base of L. On the other hand, $\{z_1, \ldots, z_m\} \cup S$ is a p-independent set, hence by the exchange principle of independency, there exist $\gamma_1, \ldots, \gamma_\ell \in C$ such that $(C - \{\gamma_1, \ldots, \gamma_\ell\}) \cup \{z_1, \ldots, z_m\} \cup S$ is a p-base of L. Put $F_0 = L^p((C - \{\gamma_1, \ldots, \gamma_\ell\}) \cup \{z_{\ell+1}, \ldots, z_m\} \cup S)$. We have $L = F_0(\gamma_1, \ldots, \gamma_\ell) = F_0(z_1, \ldots, z_\ell)$. Let $\tilde{D}_1, \ldots, \tilde{D}_\ell$ denote the partial derivations

$\frac{\partial}{\partial\gamma_1}, \ldots, \frac{\partial}{\partial\gamma_\ell}$ of L/F_0. Then $D_1 = \tilde{D}_1|_K, \ldots, D_\ell = \tilde{D}_\ell|_K$ are derivations of K over $K^p(C - \{\gamma_1, \ldots, \gamma_\ell\})$ and $\{D_1, \ldots, D_\ell\}$ is a P.D. system of K. We also have $\tilde{D}_j f(z_{\ell+1}, \ldots, z_m) = f^{D_j}(z_{\ell+1}, \ldots, z_m)$ for $f(Z) \in F[Z_{\ell+1}, \ldots, Z_m]$. Since $\{z_1, \ldots, z_\ell\}$ are p-bases of L/F_0, the Jacobian $|\tilde{D}_i f_j| = |f_j^{D_i}(Z_{\ell+1}, \ldots, Z_m)|$ is not equal to 0. Q.E.D.

Suppose now that $[K:K^p] \geq p^m$. Then there exist $\gamma_{\ell+1}, \ldots, \gamma_m \in C$ such that $\{\gamma_1, \ldots, \gamma_m\}$ are p-independent. Let $D = \{D_1, \ldots, D_m\}$ be the P.D. system of K such that $D_i\gamma_j = \delta_{ij}$ and $D_i\gamma = 0$ for $\gamma \neq \gamma_1, \ldots, \gamma_m$.

PROPOSITION 7. *The Jacobian* $J = |\hat{D}_i z_j|$ $1 \leq i, j \leq m$ *is not equal to* 0.

Proof. We have

$$J = \begin{vmatrix} f_1^{D_1}(z) + \Sigma \frac{\partial f_1}{\partial z_\lambda} \hat{D}_1 z_\lambda & \ldots f_\ell^{D_1}(z) + \Sigma \frac{\partial f_\ell}{\partial z_\lambda} \hat{D}_1 z_\lambda, & \hat{D}_1 z_{\ell+1} & \cdots & \hat{D}_1 z_m \\ \vdots & & & & \\ f_1^{D_m}(z) + \Sigma \frac{\partial f_1}{\partial z_\lambda} \hat{D}_m z & \ldots f_\ell^{D_m} + \Sigma \frac{\partial f_\ell}{\partial z_\lambda} \hat{D}_m z, & \hat{D}_m z_{\ell+1} & \cdots & \hat{D}_m z_m \end{vmatrix}$$

$$= \begin{vmatrix} f_1^{D_1}(z) & \ldots & f_\ell^{D_1}(z) & \hat{D}_1 z_{\ell+1} & \cdots & \hat{D}_1 z_m \\ \vdots & & & & & \\ f_1^{D_m}(z) & \ldots & f_\ell^{D_m}(z) & \hat{D}_m z_{\ell+1} & \cdots & \hat{D}_m z_m \end{vmatrix}$$

By Proposition 1, the $m(m-\ell)$ quantities $\hat{D}_\mu z_\lambda$, $1 \leq \mu \leq m$, $\ell+1 \leq \lambda \leq m$ are independent over L. By Proposition 6, we have $J \neq 0$.

Proof of Theorem 2. As in the proof of Theorem 1, we may assume that V is non-singular. By Proposition 3, $(x)_{\hat{D}} \to (\xi)_D$ is a specialization over K for every $(\xi) \in V_K$. Let $\hat{\varphi}_{(\xi)}$ denote the morphism of $K[\hat{x}]$ onto K associated to the specialization. By Proposition 5, the set $\{(\xi)_D; (\xi) \in V_K\}$ is Zariski dense in $\hat{V}_D$ hence there exist infinitely many $(\xi) \in V_K$ such that $\hat{\varphi}_{(\xi)}(J) \neq 0$ because $J \neq 0$ by Proposition 7. By Proposition 3 we have

$$\hat{\varphi}_{(\xi)}(J) = |D_i(\hat{\varphi}_{(\xi)}(z_j))|_{1 \le i,j \le m} \neq 0 .$$

Then $\hat{\varphi}_{(\xi)}(z_1), \ldots, \hat{\varphi}_{(\xi)}(z_m)$ are p-independent in K. Evidently $\hat{\varphi}_{(\xi)}(z) = \varphi_{(\xi)}(z)$ for $z \in K[x]$. Q.E.D.

REFERENCES

CHERLIN, G. VAN DEN DRIES, L., AND MACINTYRE, A., The elementary theory of regularly closed fields. To appear.

JACOBSON, N., Lectures in Abstract Algebra, Vol. III. van Nostrand, New York, 1964, and Springer-Verlag, New York, 1975.

WEIL, A., Foundations of Algebraic Geometry. American Math. Soc. Colloq. Publs. Vol. XXIX, New York, 1946, 2d ed. 1962.

Department of Mathematics, Yale University
Box 2155 Yale Station
New Haven, Connecticut 06520

Contemporary Mathematics
Volume 13, 1982

SOME GEOMETRIC APPLICATIONS OF A DIFFERENTIAL EQUATION IN CHARACTERISTIC $p > 0$ TO THE THEORY OF ALGEBRAIC SURFACES

by
Piotr Blass
(with the cooperation of James Sturnfield
and Jeffrey Lang)

INTRODUCTION

The purpose of this paper is to outline a program of work in the theory of the Zariski surfaces (see [BP1], [BP2], [LJ] and the definition below). We show how a method which was inspired by the work of Jacobson in 1937 and then continued in the 1950's by among others Barsotti, Hochschild, Cartier, and Samuel plays a decisive role in the geometry of Zariski surfaces.

Our approach is to treat a certain formula inspired by Jacobson [JN1] and proven by Hochschild [Ho] and Barsotti [BA] as a differential equation over a field of positive characteristic. We call this equation (JHBS)-see Proposition 1 below. It was amply illustrated by the work of my former student at Purdue, Jeffrey Lang [LJ] that this differential equation can in many cases be effectively solved. We show here how this yields some very subtle numerical invariants of Zariski surfaces such as the Artin invariant (provided that there is no torsion in Pic or that we know the torsion part of Pic). In a future paper we hope to also apply this differential equation to the Artin period map and to the Chern class map in crystalline cohomology.

It is a pleasure to dedicate this paper to Nathan Jacobson. Since I am more a geometer than an algebraist, I had recently to discover Jacobson's deep work for myself through geometry.

Some of the problems discussed in this paper belong properly in the framework of purely inseparable descent of Cartier and Grothendieck and the related notion of p-curvature introduced by Grothendieck and Katz. Jacobson's fundamental work on restricted lie algebras of derivations laid the foundations for their work.

On a more technical level, I wish to mention that the differential equation that we study is mentioned as an exercise in Jacobson's textbook [JN2] and also that it seems likely that the methods of semi-linear algebra [JN3] may be very useful in solving that differential equation. This paper does not

0271-4132/82/0546/$04.25

give complete proofs in some cases. We hope that the well-intentioned reader will have no difficulty in filling in the gaps. In any case, we hope to do so in a future publication. Our main purpose is to outline an emerging theory. The reader who wants to learn more about Zariski surfaces is referred to my thesis [BP1], [BP2] and Jeff Lang's forthcoming thesis [LJ].

The study of Zariski surfaces carried out in §1 yields the following result in pure algebra: There are infinitely many non-isomorphic fields between $k(X_1 \ldots X_n)$ and $k(X_1^{1/p}, \ldots, X_n^{1/p})$. Such a result would be difficult if not impossible to prove by field theory alone. Thus the geometry of ZS's provides a useful complement to the "Galois theoretic" methods pioneered by Jacobson for the study of fields in characteristic $p > 0$.

Zariski surfaces form admittedly a fairly narrow class of algebraic surfaces. Their most important property is that they can be joined with the plane $\mathbb{P}^2$ by a dominant purely inseparable map. This naturally leads us to introduce a new concept of equivalence for surfaces in characteristic $p > 0$ which we call here β-equivalence. We hope that this concept will prove useful in a much broader context than Zariski surfaces. As some evidence that this indeed may be the fact, we summarize here our recent solution [BP3] of the problem of unrationality of Enriques' surfaces in characteristic two, which is best phrased in terms of β-equivalence.

Finally, I cannot resist the temptation, and I announce a number of open problems and conjectures in the theory of algebraic surfaces. I hope that they will provide an added stimulus for people who wish to work in this beautiful area of mathematics.

ACKNOWLEDGEMENTS

Several theorems in this paper were proven jointly with James Sturnfield. I thank him for his cooperation. I also rely heavily on the extensive computations carried out by Jeffrey Lang in his 1981 Purdue thesis. I am also indebted to M. Artin, Abhyankar, Lipman, Heinzer, Haboush and A. Fauntleroy for some very useful discussions of these topics. I have tried to acknowledge joint work in the text. The responsibility for all errors is, of course, mine. I wish to thank K. Lunetta and C. Underwood for typing the preliminary version.

§0. PRELIMINARIES AND NOTATION

k is an algebraically closed field of characteristic $p > 0$.

X, a smooth projective surface over k, is called a Zariski surface (or ZS) if there exists a purely inseparable rational dominant map of degree p, $\pi: X \to \mathbb{P}^2$. (Equivalently, there exists a purely inseparable extension of degree

p, $k(X) \subsetneq k(P^2) = k(t_1,t_2)$ where the t_i's are indeterminates.)

Next we outline a construction procedure for ZS's which is fundamental for this paper.

Let $f_{pe}(x_0,x,y)$ be a form of degree pe which is not a p-th power of a form of degree e. We construct a two dimensional variety $\overline{F}$ by glueing together the three, possibly singular, surfaces,

$$F_0:\ z^p = f_{pe}(1,x,y)$$
$$F_1:\ z_1^p = f_{pe}(x_0,1,\overline{y})$$
$$F_2:\ z_2^p = f_{pe}(\overline{x}_0,\overline{x},1)$$

the glueing of F_0 and F_1 is given by $x \to 1/x_0$, $y \to \overline{y}/x_0$, $z \to z_1/x_0^e$, and similarly for the other cases.

We will denote by ρ the canonical map $\rho : \overline{F} \to P^2$ and by $\pi:\widetilde{F} \to \overline{F}$ a minimal desingularization of $\overline{F}$. Clearly $\widetilde{F}$ is a Zariski surface.

If $\overline{F}$ is a normal surface, the map $\rho: \overline{F} \to P^2$ is simply the normalization of P^2 in the field $k(F_0) = k(F_1) = k(F_2)$. We will refer to $\overline{F}$ and $\widetilde{F}$ as surfaces defined by $z^p = f_{pe}(x_0,x,y)$ in symbols $\overline{F},\widetilde{F}$: $z^p = f_{pe}(x_0,x,y)$.

Remark: It is more natural to think of $\overline{F}$ as embedded in the weighted projective space $P(1,1,1,e)$. We do not exploit this point of view in the present paper. We hope to do so in a subsequent one.

The singularities of $\overline{F}$ may be very difficult to understand and resolve in general. Therefore, in this paper we will mainly discuss the "generic case" of this construction where the singularities of $\overline{F}$ are as mild as possible.

Definition. A form $f_{pe}(x_0,x,y)$ is called generic, or of type (B*), (using notation analogous to J. Lang [LJ]) if the corresponding surface $\overline{F}$: $z^p = f_{pe}(x_0,x,y)$ has only finitely many singular points all of which are double points locally given by $z^p = uv$ + higher order terms. The corresponding smooth surface $\widetilde{F}$: $z^p = f_{pe}(x_0,x,y)$ will be referred to as a generic ZS of degree (p,pe) (or a ZS of type B* and of degree (p,pe)).

Let us denote by A_p^{pe} the affine space of all the forms of degree pe. Let GA_p^{pe} be the subset of generic forms (of type B*). It can be shown that GA_p^{pe} is a Zariski open dense subset of A_p^{pe}. We omit the proof. We will also need the corresponding projective spaces: $P_p^{pe} = A_e^{pe} - \{0\}/k^*$ and $GP_p^{pe} = GA_e^{pe}/k^*$. Let X be a ZS. We will denote by b_i the étale betti numbers of X and by ρ the base number (see [Mi], p. 216). Since X is unirational, we have $\rho = b_2$ and it can be shown that the discriminant of the intersection form on $N(X) = \text{Pic } X/\text{Pic}^n X$ is an even power of the prime p. We denote it by $p^{2\sigma_0}$ and σ_0 is called the M. Artin invariant of X.

Finally we introduce some notations from Samuel [SP] and Lang [LJ].

Let $F : z^p = f(x,y)$ be an affine, normal surface, the ring $k[F] \approx k[x^p, y^p, f] \subset k[x,y]$. Consider the derivation of $k[x,y]$ given by $Dw = \frac{\partial f}{\partial x}\frac{\partial w}{\partial y} - \frac{\partial f}{\partial y}\frac{\partial w}{\partial x}$. In our case, the units of $k[F]$ are just elements of k. We have the following proposition due to Samuel.

PROPOSITION 1. *$C\ell\ k[F] \approx$ the additive group of polynomial solutions of the differential equation*

(JBHS) $$D^{p-1} t - at = -t^p$$

where a is defined by $D^p = aD$.

§1. THE GEOMETRY OF GENERIC ZARISKI SURFACES

In this section, we compute the usual numerical invariants of a generic ZS of degree (p,pe). This is joint work with Jim Sturnfield. We then apply this knowledge to field theory. Then we introduce a basic exact sequence involving $\text{Pic}\ \tilde{F}$. Finally, we link it with the JBHS differential equation using Samuel [SP] and we quote some results of Jeffrey Lang. We also give a formula for the Artin invariant of a generic Zariski surface of degree (p,pe) in the case that there is no torsion in Pic. We conclude with some examples of the theory.

Let $f_{pe}(x_0,x,y) \in GA_e^{pe}$ be a generic form (see notations). Let $\tilde{F} \overset{\pi}{\to} \overline{F} \overset{\rho}{\to} P^2$ be the corresponding surfaces. We denote by $K_{\tilde{F}}$ the canonical divisor and by $\tilde{\ell}_{\tilde{F}} = (\rho \circ \pi)^* \mathcal{O}_{P^2}(1)$ as well as the corresponding divisor.

Lemma 1. (Abhyankar).

$$K_{\tilde{F}} = (pe - e - 3)\tilde{\ell}_{\tilde{F}} .$$

Proof. The proof consists in computing the divisor of a differential using the explicit equations for the F_0, F_1, F_2 and was carried out by Abhyankar. We omit the details.

Lemma 2. (James Sturnfield). *$\overline{F}$ has $(pe)^2 - 3pe + 3$ singularities all of the form $z^p = uv$ + higher order terms.*

Idea of Proof. Compute the intersection number of the curves

$\frac{\partial f(1,x,y)}{\partial x} = 0$ and $\frac{\partial f_{pe}(1,x,y)}{\partial y} = 0$ at infinity.

THEOREM 3. (P. Blass - J, Sturnfield).

i) $p_g(\widetilde{F}) = p_a(\widetilde{F})$ = number of non-negative integral solutions of the inequality $\alpha + \beta + e\gamma \le pe - e - 3$ = number of integer points in the interior of the tetrahedron spanned by $(0,0,0)$, $(pe,0,0)$, $(0,pe,0)$ and $(0,0,p) = \frac{1}{12}(p-1)[e^2 p(2p-1)-9pe+12]$

ii) $b_1(\widetilde{F}) = b_3(\widetilde{F}) = 0$

iii) $b_2(\widetilde{F}) = \rho(\widetilde{F}) = [(pe)^2 - 3pe + 3](p-1) + 1$

iv) $(\widetilde{\ell}_{\widetilde{F}})^2 = p$ and $K^2_{\widetilde{F}} = (pe - e - 3)^2 p$.

The proof is straightforward since we remember that each singularity of $\overline{F}$ contributes $p - 1$ exceptional curves on $\widetilde{F}$ except perhaps for i), which follows immediately from the technical Lemma 8 which we put at the end of this section.

Corollary. For every $n \ge 2$ there exist infinitely many non-isomorphic fields between $k(X_1,X_2, \ldots, X_n)$ and $k(X_1^{1/p}, X_2^{1/p}, \ldots, X_n^{1/p})$ where the X_i 's are indeterminates.

Proof. By the theorem, there exists a sequence of generic ZS, $\widetilde{F}_1, \widetilde{F}_2, \ldots, \widetilde{F}_n, \ldots$ with $p_g(\widetilde{F}_n) \to +\infty$ with n. Obviously, we may assume $k(X_1,X_2) \subsetneq k(\widetilde{F}_n) \subsetneq k(X_1^{1/p}, X_2^{1/p})$. This proves the corollary for $n = 2$ since p_g is a birational invariant. To prove the general case, consider the varieties $V_i \simeq \widetilde{F}_i \times A^{n-2}$. Obviously, $h^{2,0}(V_i) = \dim H^0(V_i, \Omega^2_{V_i})$ satisfies $h^{2,0}(V_i) \to +\infty$ as $i \to \infty$ and the corollary follows from the birational invariance of the numbers $h^{2,0}$ see [Ha] p. 190, exercise 8.7, q.e.d.

In order to study the Artin invariant of $\widetilde{F}$ we first have to establish an exact sequence.

THEOREM 4. In the notations of Theorem 3, let A be the free abelian group generated by the curve $\widetilde{\ell}_{\widetilde{F}}$ and by the exceptional curves for the desingularization $\pi: \widetilde{F} \to \overline{F}$; then the discriminant of the intersection form on A is equal to $p^{(pe)^2 - 3pe+4}$. Moreover, if all the singularities of $\overline{F}$ are contained in F_0, a condition which can always be realized by a linear change of the coordinates x_0,x,y, then we have the exact sequence

$$0 \to A \to \operatorname{Pic} \widetilde{F} \to C\ell\, k[F_0] \to 0 \tag{4.1}$$

Outline of Proof. The intersection matrix of the generators of A is given by $(pe)^2 - 3pe + 3$ blocks of the form

$$\underbrace{\begin{bmatrix} -2 & 1 & & & \\ 1 & -2 & \ddots & & \\ & \ddots & \ddots & \ddots & \\ & & \ddots & -2 & 1 \\ & & & 1 & -2 \end{bmatrix}}_{p-1} \Bigg\} \, p-1 \qquad \text{with determinant} = p$$

and by one block $[p]$ corresponding to $\tilde{\ell}^2 = p$. Thus the curves generating A are independent in $\operatorname{Pic} \tilde{F}$, $\operatorname{Pic} \tilde{F} = C\ell\, \tilde{F}$ and there is an obvious map $C\ell\, \tilde{F} \twoheadrightarrow C\ell\, k[F_i]$. We leave the remaining details to the reader.

Let us now assume that all the singularities of $\overline{F}$ are in F_0; let us write $f(x,y)$ for $f_{pe}(1,x,y)$. Samuel's theory now applies to $F_0 : z^p = f(x,y)$ and consequently $C\ell\, k[F_0]$ is isomorphic to the group of polynomial solutions of the (JBHS) equation

$$\text{(JBHS)} \qquad D^{p-1}t - at = -t^p \quad \text{corresponding to}$$
$$z^p = f(x,y) \quad \text{(see notations, Proposition 1)}$$

THEOREM 5. (J, Lang). $C\ell\, k[F_0]$ *is a finite group isomorphic to the direct sum of* ν *copies of* $\mathbb{Z}/p\mathbb{Z}$. *Moreover, any polynomial* $t(x,y)$ *satisfying the* (JBHS) *equation has degree* $\leq \deg f - 2$ $(\leq pe - 2$ in our case.)

Proof. See Lang [LJ].

Let us now assume in addition that $\operatorname{Pic} \tilde{F}$ has no torsion. We do not know whether or not this is always true for a generic Zariski surface, this remains to be investigated. However, we do know that it is true in certain cases. (See examples below.)

PROPOSITION 6. *If* $\operatorname{Pic} \tilde{F}$ *has no torsion, then* $\operatorname{Pic} \tilde{F} = NS\, \tilde{F} = N(\tilde{F})$ *because the Picard scheme of* $\tilde{F}$ *is reduced and discrete, see* [Mi] p. 215-6, *and we have the following formula for the Artin invariant of* $\tilde{F}$:

$$(6.1) \qquad \sigma_0(\tilde{F}) = \frac{(pe-1)(pe-2)}{2} + 1 - \nu$$

where $(\mathbb{Z}/p\mathbb{Z})^\nu$ *is the group of solutions of the differential equation* (JBHS).

Proof. This follows from the exact sequence 4.1 and basic properties of quadratic forms.

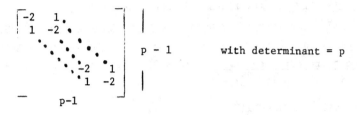

Remark 7. A similar theory can be developed for the surface $F: z^p = f_{p+1}(x,y)$ where $f_{p+1}(x,y)$ is a generic polynomial of degree $p+1$. If $\overline{F}$ is the closure of F in $\mathbb{P}^3$ and $\widetilde{F}$ is a minimal desingularization of $\overline{F}$ and if we assume that $\mathrm{Pic}\,\widetilde{F}$ has no torsion, then

$$(7.1) \qquad \sigma_0(\widetilde{F}) = \frac{p(p+1)}{2} - \nu$$

where $C\ell\; k[F_1] \approx (\mathbb{Z}/p\mathbb{Z})^\nu$ and ν can be computed from the (JBHS) differential equation. The reader may wish to consult [BP-1] Ch. IV for some details of the geometry of such an $\widetilde{F}$.

Some examples.

1) $\widetilde{F} : z^2 = f_6(x_0,x,y)$, $f_6 \in GA_2^6$. In this case $\overline{F}$ has 21 singularities. Thus $b_2(\widetilde{F}) = 22$. $K_{\widetilde{F}} \equiv 0$. Thus $\widetilde{F}$ is a K3 surface and $\mathrm{Pic}\,\widetilde{F}$ has no torsion. J. Lang has shown [LJ] that $1 \leq \nu$ for a generic $\widetilde{F}$ thus $\sigma_0 = 10 + 1 - \nu \leq 10$ as is well known ([S-R]).

2) $\widetilde{F} : z^3 = f_6(x_0,x,y)$, Chap. 3, here J. Lang [LJ] shows that for an open dense subset of GA_2^6 $\nu = 0$ thus $\sigma_0 = 11$, otherwise $\sigma_0 \leq 11$. We remark that $\mathrm{Pic}\,\widetilde{F}$ is automatically torsion free in this case.

3) $F : z^3 = f_4(x,y)$ Chap. 3. Let $\widetilde{F}$ be the smooth projective model of $k(F)$. For a generic choice of the fourth degree polynomial $f_4(x,y)$, $\widetilde{F}$ is a K3 surface. $\sigma_0(\widetilde{F}) = 6 - \nu$ see (7.1) above. J. Lang has shown that $\nu = 0$ for a dense open set of such $f_4(x,y)$, he also shows that $z^3 = x^4 + y^4 - xy$ has $\nu = 5$, thus $\sigma_0 = 1$.

We end with a technical lemma.

Lemma 8. (P. Blass - J. Sturnfield). Let $\widetilde{F} \xrightarrow{\pi} \mathbb{P}^2$ be a generic ZS of degree (p,pe), $e > 1$, $\widetilde{\ell} = \pi^* \mathcal{O}_{\mathbb{P}^2}(1)$. Then $\dim H^0(\widetilde{\ell}) = 3$ if $e > 1$ and $\dim H^0((pe-e-3)\widetilde{\ell}) = p_g(\widetilde{F}) = \#$ of distinct monomials $x^\alpha y^\beta z^\gamma$ with $\alpha + \beta + \gamma e \leq pe - e - 3$.

Proof. By a suitable change of coordinates, we may assume that $F : z^p = f_{pe}(1,x,y)$ contains all the singularities and $\widetilde{\ell}$ is the line at infinity. Then in order to find $H^0(\widetilde{\ell})$ we only need to find polynomials $p(x,y,z) \in k[F_0]$ which have a pole of order ≤ 1 on $\widetilde{\ell}$. Clearly, $1,x,y$ have that property. However, the function z has under our glueing a pole of order e along $\widetilde{\ell}$. Thus if $e \geq 2$ $z \notin H^0(\widetilde{\ell})$. Also $p(x,y,z)$ has a unique representation $p = \Sigma\, c_{\alpha\beta\gamma}\, x^\alpha y^\beta z^\gamma$ with $0 \leq \gamma \leq p-1$. In order to compute the order of p along $\widetilde{\ell}$, we write the rational function p on the chart

$F_1: z_1^p = f_{pe}(x_0,1,\bar{y})$ as $p = \Sigma c_{\alpha\beta\gamma} x_0^{-\alpha-\beta-\gamma e} \bar{y}^{\beta} z_1^{\gamma}$. If $p \in H^0(\tilde{\ell})$ then $x_0 p$ is regular on F_i since x_0 has a simple zero along $\tilde{\ell}$. Thus $x_0 p = \Sigma c_{\alpha\beta\gamma} x_0^{-\alpha-\beta-\gamma e+1} \bar{y}^{\beta} z_1^{\gamma} \in k[F_1]$. This means that if $c_{\alpha\beta\gamma} \neq 0$ then $\alpha + \beta + \gamma e - 1 \leq 0$ as can be seen by collecting all terms with a fixed γ and noting that monomials $x_0^m y^n$, where $m, n \in Z$ are independent in $k[F_1]$. If $e \geq 2$ the only solutions (α,β,γ) are $(0,0,0)$, $(1,0,0)$ and $(0,1,0)$. Therefore, p is a combination over k of $1,x,y$. Similarly, multiplying by x_0^{pe-e-3} we conclude that $H^0((pe - e - 3)\tilde{\ell}_{\tilde{F}})$ is spanned by the desired monomials in x_0, x, y. q.e.d.

2. INTRODUCTION TO MODULI AND THE (CONJECTURAL) ARTIN STRATIFICATION BY σ_0 OR ν.

Throughout this section we will assume for technical reasons that $e \geq 2$ and that $pe - e - 3 > 0$ and also $e + 3 \not\equiv 0(p)$. I hope that that last condition will be eventually removed; it is used only in the proof of Lemma 1 below.

In this section we tackle the following problem: Let $\tilde{F}$, $\tilde{G}$ be generic ZS defined by $z^p = f_{pe}(x_0,x,y)$, and $z^p = g_{pe}(x_0,x,y)$ respectively. If $\tilde{F}$ is isomorphic to $\tilde{G}$ as surfaces over k, what can be said about the relationship of the forms $f_{pe}(x_0,x,y)$ and $g_{pe}(x_0,x,y)$? It turns out that the answer is quite simple. $\tilde{F} \approx \tilde{G}$ if and only if f_{pe} can be obtained from g_{pe} by a linear change of coordinates followed by adding the p-th power of a form of degree e. See Theorem 9 below.

This leads us naturally to the following construction of a moduli space for generic Zariski surfaces of degree (p,pe).*

Consider the action of $GL(3)$ on $A_p^{pe} \approx$ the space of homogeneous forms in x_0, x, y of degree pe. Let $B_p^{pe} \subseteq A_p^{pe}$ be the subspace generated by the monomials $x_0^{\alpha} y^{\beta} x^{\gamma}$ with $\alpha + \beta + \gamma = pe$ and $p|\alpha$, $p|\beta$, $p|\gamma$. Clearly B_p^{pe} is stable under the action of $GL(3)$. There is, therefore, an induced action on $C_p^{pe} = A_p^{pe}/B_p^{pe}$. Now the set of generic forms or forms of type B* (see notations) $GA_p^{pe} \subseteq A_p^{pe}$ is stable under the action of $GL(3)$ and it is easy to see that the set of equivalence classes of elements of GA_p^{pe} forms an open and dense subset of C_p^{pe}.

We will denote this subset by GC_p^{pe}. Now $GL(3)$ acts on GC_p^{pe} and consequently $PGL(3)$ acts on GC_p^{pe}/k^*. In fact $PGL(3)$ acts on the projective

---

*We are using the word moduli somewhat loosely here. I still have not investigated the question whether or not the space that we obtain is in fact a coarse or a fine moduli space in the sense of Mumford [Mu].

space $C_p^{pe} - \{0\}/k^*$. We will show that every point of GC_p^{pe} is stable in the sense of geometric invariant theory and this will show that the universal categorical quotient $GC_p^{pe}/PGL(3)$ exists. As the next proposition will show, this quotient denoted GM_p^{pe} is a good candidate for a coarse moduli space for Zariski surfaces of type B* of degree (p,pe).

We use the same notions and notation as J. Shah [Sh] who in turn follows Mumford [Mu].

By a suitable choice of coordinates, any given one-parameter subgroup λ of PGL_3 is diagonalized and has the form

$$\begin{bmatrix} t^{r_0} & 0 & 0 \\ 0 & t^{r_1} & 0 \\ 0 & 0 & t^{r_2} \end{bmatrix}$$

such that $\Sigma r_i = 0$ and $r_0 \geq r_1 \geq r_2$, $r_0 > 0$.

We wish to show that the class of any form $f_{pe}(x_0,x,y) \in GA_p^{pe}$ is stable. We consider three cases:

<u>Case 1</u>. The term $x_0^{pe-1}x$ or $x_0^{pe-1}y$ is present in $f_{pe}(x_0,x,y)$. Then $\mu(f,\lambda) \geq (pe-1)r_0 + r_1 = (pe-1)r_0 + r_1 + r_2 - r_2 = (pe-2)r_0 - r_2 > 0$ or $(pe-1)r_0 + r_2 = (pe-2)r_0 + r_0 + r_1 + r_2 - r_1 = (pe-2)r_0 - r_1 > 0$.

<u>Case 2</u>. $x_0^{pe-2}xy$ is present in f. $\mu \geq (pe-2)r_0 + r_1 + r_2 = (pe-3)r_0 + r_0 + r_1 + r_2 = (pe-3)r_0 > 0$.

<u>Case 3.</u> The terms mentioned in Cases 1 and 2 above are missing, but then both terms $x_0^{pe-2}x^2$ and $x_0^{pe-2}y^2$ appear. Suppose $(pe-2)r_0 + 2r_1 \leq 0$ and $(pe-2)r_0 + 2r_2 \leq 0$. Then $(pe-2)r_0 + r_1 + r_2 \leq 0$ so that $(pe-3)r_0 \leq 0$, a contradiction. Thus $\mu([f],\lambda) > 0$ which shows stability by Mumford's theorem. We conclude that under the action of PGL(3) on C_p^{pe} the open set GC_p^{pe} is stable and it consists of stable points in the sense of Mumford. We conclude that the universal categorical quotient $GC_p^{pe}/PGL(3)$ exists and we denote it GM_p^{pe}. This is our 'moduli space' for generic ZS's of degree (p,pe). As we said in the beginning of the section, this terminology is partially justified by the following theorem:

<u>THEOREM 9</u>. <u>Assume that</u> $pe - e - 3 \geq 1$ <u>and that</u> $p \nmid e + 3$. <u>Let</u> $\widetilde{F} : z^p = f_{pe}(x_0,x,y)$ <u>and</u> $\widetilde{G} : z^p = g_{pe}(x_0,x,y)$ <u>be two generic Zariski surfaces defined by generic forms</u> f_{pe}, g_{pe} [of type B*, see notation]. <u>Then</u> $\widetilde{F}$ <u>is isomorphic to</u> $\widetilde{G}$ <u>as</u> k-<u>schemes if and only if</u> f_{pe} <u>and</u> g_{pe} <u>define the same</u> element of GM_p^{pe} (i.e. iff f_{pe} and g_{pe} can be obtained from each other by

a linear change of coordinates followed by adding the p-th power of a form of degree e).

Summary of the proof of the "only if" part: Let $\alpha: \tilde{F} \to \tilde{G}$ be a k-isomorphism. By Lemma 1 $(pe-e-3)(\alpha * \mathcal{L}_{\tilde{G}} - \mathcal{L}_{\tilde{F}}) = 0$. Since $p \mid e+3$ and Pic $\tilde{F}$ has no torsion prime to p therefore $\alpha^* \tilde{\mathcal{L}}_{\tilde{G}} = \tilde{\mathcal{L}}_{\tilde{F}}$. Now $\dim H^0(\tilde{\mathcal{L}}_{\tilde{F}}) = 3$ by Lemma 8 and this leads to a commutative diagram with $\bar{\alpha}$, γ isomorphism and γ linear. (Compare [Ha], II, 7).

$$\begin{array}{ccccc}
\tilde{F} & \xrightarrow{\pi_{\tilde{F}}} & \bar{F} & \xrightarrow{\rho_{\bar{F}}} & P^2 \\
\wr\downarrow \alpha & & \wr\downarrow \bar{\alpha} & & \wr\downarrow \gamma \\
\tilde{G} & \xrightarrow{\pi_{\tilde{G}}} & \bar{G} & \xrightarrow{\rho_{\bar{G}}} & P^2
\end{array}$$

The remainder of the proof can now be reduced to the following lemma whose proof we leave as an exercise in algebra.

Lemma (P. Blass, W. Heinzer). Let $F: z^p = f(x,y)$ and $G: z_1^p = g(x,y)$ be normal surfaces in A^3 such that the rings $k[F]$ and $k[G]$ are isomorphic over $k[x,y]$. Then $f - cg = h^p$ for some $h \in k[x,y]$ and $c \in k$.

Conjectural stratification by σ_0 (or ν). We state here two conjectures which were crystallized in the author's conversations with M. Artin and Jeffrey Lang. The idea is the following. As the form $f_{pe}(x_0,x,y)$ varies in GA_p^{pe} the number of solutions p^ν of the differential equation (JBHS) will also change. This will result in a change of σ_0 (see formula 4.1 where we assume that Pic $\tilde{F}$ is torsion free). By analogy with Artin's study of supersingular K3 surfaces, we are led to the following conjectures. In what follows, we will assume that Pic $\tilde{F}$ is torsion free for all the surfaces in GM_p^{pe} although this is probably not essential.

CONJECTURE 1. Let $M_\tau \subset GM_p^{pe}$ be the set of surfaces with the Artin invariant $\sigma_0 \le \tau$ (or equivalently with $\nu \ge \frac{(pe-1)(pe-2)}{2} + 1 - \tau$). Then we have a filtration $M_{(\sigma_0)\min} \subset \dots \subset M_{\tau-1} \subset M_\tau \subset \dots \subset M_{(\sigma_0)\max} \subset GM_p^{pe}$, where $(\sigma_0)\max$, $(\sigma_0)\min$ are the largest and the smallest possible values of the Artin invariant and we conjecture that $M_{\tau-1}$ is closed in M_τ and that $M(\sigma_0)\max$ is dense in GM_p^{pe}. Of course, we could also state this conjecture in terms of ν (as long as Pic has no torsion).

CONJECTURE 2 (M. Artin). Codimension of $M_{\tau-1}$ in M_τ is $\le p_g(\tilde{F})$.

This last conjecture resembles Lefschetz's theorem in char. zero which says that a cycle has to satisfy p_g conditions in order to be algebraic (vanishing of all the periods of abelian integrals).

§3. β-EQUIVALENCE.

Let S be the set of all (isomorphism classes) of smooth projective surfaces over $k = \bar{k}$, char. $k = p > 0$. We wish to introduce on S the weakest equivalence relation that identifies two surfaces X and Y provided that there exists a purely inseparable dominant rational map $\pi : X \to Y$. It is easy to see that the right definition is as follows.

Definition. X is said to be β-equivalent to Y where $X, Y \in S$ iff there exists a k-inclusion of fields $k(X)^{p^n} \hookrightarrow k(Y)$ such that $k(Y)$ becomes a finite purely inseparable extension of $k(X)^{p^n}$ for some $n \geq 0$.

It is straightforward to check that this is an equivalence relation. We write $X \sim_\beta Y$ and we denote by $[X]_\beta$ the equivalence class of X.

A number of étale cohomology invariants are preserved under β-equivalence:

Proposition 10. If $X \sim_\beta Y$, then

1) $b_1(X) = b_1(Y)$, i.e. $\dim \mathrm{Alb}(X) = \dim \mathrm{Alb}(Y)$.

2) $\lambda(X) = b_2(X) - \rho(X) = b_2(Y) - \rho(Y) = \lambda(Y)$.

3)* $\pi_1^{et}(X) \simeq \pi_1^{et}(Y)$ (conjecture).

Proof. Omitted. (I have checked 3) carefully only in the simply connected case.) As an application of β-equivalence, we sketch here the proof of the following theorem (details will appear elsewhere; see [BP3]).

THEOREM 11. *A classical (or a supersingular) Enriques surface X in characteristic two is always β-equivalent to $\mathbb{P}^2$, thus certainly unirational.*

Summary of proof: We have a diagram $X_1 \xrightarrow{u} \tilde{X} \xrightarrow{v} X$ where u is the desingularization of the purely inseparable cover $\tilde{X}$ of X constructed in [B-M] p. 220. Since $X_1 \sim_\beta X$, Proposition 10 and [B-M] imply that $b_1(X_1) = b_1(X) = 0$ and $\lambda(X_1) = \lambda(X) = 0$. In [BP3] we use duality to show that X_1 is ruled or K3. Thus either X_1 is rational and we are done or it is K3 and supersingular. Then by [R-Š], $X_1 \sim_\beta P^2$ thus $X \sim_\beta P^2$. q.e.d.

Remark: We propose to call any surface β-equivalent to $\mathbb{P}^2$ a generalized Zariski surface or GZS for short. The class of GZS's is closed under specialization. [(BP-ML)].

§4. OPEN PROBLEMS.

1. Find a surface $X \in [P^2]_\beta$ which is not a Zariski Surface.

Remark: For example, a generic surface $z^{25} = f_{26}(x,y)$ is a reasonable candidate in char. 5, but I do not know how to prove that it is not a ZS.

2. Does there exist a surface X of general type and such that if $Y \sim_\beta X$ then Y must also be of general type?

3. Zariski's problem: For $p \geq 5$ find a ZS with $p_g = 0$ but non-rational.

Remark: This was solved by P. Blass in char. two in 1977 (see [BP1]) and by W. E. Lang in char. three in 1978 (see [LW]).

4. Problem due to Piotr Blass and W. E. Lang: Let X be a Zariski Surface with $p_g = 0$ and $\mathrm{Pic}^\tau X$ trivial. Does X have to be rational?

5. Is every simply-connected and unirational surface β-equivalent to P^2 ?

6. X is a ZS with $p_g = 0$ and Pic X torsion free. Is X rational?

7. Is it possible to construct a sequence of surfaces $X_1, X_2, \ldots, X_n, \ldots$ over k with $b_2(X_i) - \rho(X_i) \to +\infty$ and such that every smooth surface over k is dominated by one of the surfaces X_i (via a rational map).

Finally, three problems in characteristic zero.

8. Find a surface X which is not birationally equivalent to a normal surface in P^3.

9. Let $X \subset P^3$ be a normal surface with only one isolated singularity $g \in X$. Let $\tilde{X} \to X$ be a desingularization. Suppose that $H^1(\tilde{X}, \mathcal{O}_{\tilde{X}}) \neq 0$. Is $\tilde{X}$ necessarily ruled?

10. Mumford's question: Let $F \subseteq P^3$ be a surface of degree five. Let $\tilde{F} \xrightarrow{\pi} F$ be a desingularization. Suppose that $H^1(\tilde{F}, \mathcal{O}_{\tilde{F}}) \neq 0$. Does $\tilde{F}$ have to be a (birationally) ruled surface?

REFERENCES

[AR] M. Artin, Supersingular K3 surfaces, Ann. École Norm. Sup. 7(1974), 543-568.

[BA] I. Barsotti, Repartitious in Abelian Varieties, Ill. J. Math. 2(1958), 43-69. (See especially pp. 58, 59).

[BP1] P. Blass, Zariski Surfaces, U. of Michigan Thesis (1977), to appear in Dissertationes Mathematicae.

[BP2] P. Blass, Zariski Surfaces, C. R., Acad. of Science Canada II (1980).

[BP3] P. Blass, Unirationality of Enriques Surfaces in Characteristic two, to appear.

[BP-ML] P. Blass and M. Levine, Families of Zariski Surfaces, to appear.

[BM] E. Bombieri and D. Mumford, Enriques Classification of Surfaces in Characteristic $p > 0$, III. Invent. Math. 35(1976), 197-232.

[Ha] R. Hartshorne, Algebraic Geometry, Graduate texts in mathematics, Springer-Verlag, 1977.

[Ho] G. Hochschild, Simple algebras with purely inseparable splitting fields of exponent one, Trans. Amer. Math. Soc. 79(1955), 477-489.

[JN1] N. Jacobson, Abstract derivations and Lie Algebras, Trans. Amer. Math. Soc. 42(1937), 206-224.

[JN2] N. Jacobson, Basic Algebra II, Freeman, 1980.

[JN3] N. Jacobson, The Theory of Rings, Math. Surveys, Amer. Math. Soc. II (1943).

[LJ] J. Lang, Purdue Thesis (1981).

[LW] W. E. Lang, Quasi-elliptic surfaces in characteristic three, Ann. École Norm. Sup. 12(1979), 473-500.

[Mi] J. S. Milne, Étale cohomology, Princeton University Press (1979).

[Mu] D. Mumford, Geometric Invariant Theory, Academic Press, 1965.

[SH] J. Shah, A complete moduli space for K3 surfaces of degree 2, Ann. of Math. 112 (1980), 480-510 (see especially pp. 489-490).

[SP] P. Samuel, Lectures on UFD's, Tata Institute of Fundamental Research, Bombay, 1964.

[SR] I. Šafarevič and A. N. Rudakov, Supersingular K3 surfaces over Fields of Characteristic Two, Izv. Akad. Nauk USSR 13 (1979), 147-165.

UNIVERSITY OF PENNSYLVANIA
and
THE INSTITUTE FOR ADVANCED STUDY
PRINCETON, N. J. 08540

Contemporary Mathematics
Volume 13, 1982

ON THE INVERSE PROBLEM IN DIFFERENCE GALOIS THEORY

by Ronald P. Infante

Throughout this paper F is an inversive difference field of characteristic zero with field of constants C, and G is a connected C-group whose universal field contains the underlying field of F. By "the inverse problem in the Galois theory of difference fields" we mean the question: Does there exist a strongly normal extension of F, M, with $\text{Gal}(M,F) = G$? Recall that M is called a strongly normal extension of F if the following conditions obtain:

(i) F is algebraically closed in M,

(ii) the field of constants M, C is the field of constants of F,

(iii) for every difference isomorphism σ of M/F into an overfield of M we have $M\langle\sigma M\rangle = M\langle C_\sigma\rangle = \sigma M\langle C_\sigma\rangle$ where C_σ is the field of constants of $M\langle\sigma M\rangle$.

The theory of strongly normal extensions suggests that we consider fields $M = F\langle\alpha\rangle = F\cdot C(\alpha)$ where α is an element of G which is "generic" over F. Assume such an M is strongly normal over F and C is algebraically closed. Write $\alpha_1 = \alpha\beta$ where β is an element of G which is rational over M. For each c in G which is rational over C define a new difference structure on M by transforming α to $c^{-1}\alpha$. The assignment $\alpha \mapsto \alpha\beta$, defines a "new" difference isomorphism of M/F. By an argument given below, β must be rational over the constants of this new difference structure. Since this is true for each c, we have the fact that β is fixed by every old difference automorphism of M/F so β is rational over F. Therefore, for our discussion of the inverse problem, we restrict our attention to fields $M = F\langle\alpha\rangle$ where $\alpha_1 = \sigma\beta$ with β rational over F.

In general, the notation and terminology of Cohn's book [1] is followed for difference algebra while Chapter V of Kolchin's book [4] is our reference for algebraic groups.

0271-4132/82/0547/$02.00

§1. If $\mathcal{U}$ is the universal field for G, we may assume that $\mathcal{U}$ has infinite transcendence degree over the underlying field of F. Choose an infinite subset of $\mathcal{U}$, S, which is algebraically independent over the underlying field of F. Regarding the elements of S as constants, form the difference field P(S) where P is the field of periodic elements of F. Set P* equal to an algebraic closure of P(S). We may take the underlying field of P* to be a subfield of $\mathcal{U}$ and hence, we will identify C*, the constants of P*, with a sub-field of $\mathcal{U}$. Set $G' = G_{C*}$.

Choose an element α of G with F algebraically closed in $F(\alpha)$. If β is an element of G_F, define a difference structure on $F(\alpha)$ by setting $\alpha_1 = \alpha\beta$ and write $M = F<\alpha>$.

PROPOSITION 1: _If_ $C_M = C$ _then_ M _is a strongly normal extension of_ F _and_ $Gal(M,F) = G$.

Proof: If U is an overfield of M and σ a difference isomorphism of M/F into U, then we may assume that the underlying field of U is contained in $\mathcal{U}$. Set $\bar{\alpha} = \sigma\alpha$ and $h = \alpha\bar{\alpha}^{-1}$. Note that $\bar{\alpha}_1 = \bar{\alpha}\beta$ since $\alpha_1 = \alpha\beta$ and β is rational over F. Consider $h_1 = \alpha_1\bar{\alpha}_1^{-1} = \alpha\beta(\bar{\alpha}\beta)^{-1} = \alpha\bar{\alpha}^{-1} = h$. Thus h is rational over C and from this it follows that M is strongly normal over F.

To show that Gal(M,F) is C-isomorphic to G, we take $U = M<P*>$. We have seen that if σ is an isomorphism of M/F into U then $\sigma(\alpha) = h_\sigma^{-1}\alpha$ where $h_\sigma \in G_{C*}$. Conversely, if h is an element of G_{C*}, the the formula $\sigma_h\alpha = h^{-1}\alpha$ defines a difference isomorphism of M/F into U. It is clear that the correspondence $h \to \sigma_{h^{-1}}$ is an isomorphism of G_{C*} onto the group of all isomorphisms of M/F into U. Since $C_{\sigma_h} = C_{h^{-1}} = C_h$, this isomorphism is a C-isomorphism.

COROLLARY: _If_ G _is a linear group and_ C _is algebraically closed then, again assuming_ $C_M = C$, M _is a Picard-Vessiot extension of_ F.

Proof: We may assume $G \subseteq G(k,\mathcal{U})$ for some positive integer k and $M = F<\alpha>$ where α is a k k non-singular matrix (a_{ij}). Let $\{a^{(1)},\ldots,a^{(n)}\}$ be a sub-set of $\{a_{ij}\}$ which is linearly independent over C and satisfies

$C(\alpha) = C(a^{(1)},\ldots,a^{(n)})$; then $M = F\langle\alpha\rangle = F\langle a^{(1)},\ldots,a^{(n)}\rangle$. Let y be a difference indeterminate over M and set:

$$f(y) = \frac{\hat{C}(y, a^{(1)},\ldots,a^{(n)})}{\hat{C}(a^{(1)},\ldots,a^{(n)})}$$

where $\hat{C}(x^{(1)},\ldots,x^{(m)})$ is the Casorati determinant of $x^{(1)},\ldots,x^{(m)}$. A standard argument shows that $f \in F\{y\}$ and since $f(a^{(j)}) = 0$ for each j, we see that M is a Picard-Vessiot extension of F.

§2. Proposition 1 shows that solving the inverse problem requires the choice of β so that the constants of $F\langle\alpha\rangle$ are in C. For instance, if β is rational over constants and of finite order or, more generally, if there is a positive integer n with $\beta\beta,\ldots,\beta_n = e$ then α will be periodic so $F\langle\alpha\rangle$ will have constants not in F. Note also that β cannot be of the form a_1a^{-1} for some a rational over F. For then αa^{-1} is rational over the constants of $F\langle\alpha\rangle$ but is not rational over F. We mention two results due to Charles Franke [2] which explains the one-dimensional affine case.

PROPOSITION 2: (i) *If* F *contains no solution of* $y = y + B$ *and* a *satisfies* $a_1 = a + B$ *then* $C_{F\langle\alpha\rangle} = C$.

(ii) *If* F *contains no non-zero solution of* $y_1 = B^n y$ *for any positive integer* n *and* a *is a solution of* $a_1 = Ba$ *then* $C_{F\langle\alpha\rangle} = C$.

It is worth noting that to prove $C_M = C$, it is sufficient to show that C_M is fixed by every element of $\mathrm{Gal}(M,F)$. For if $\sigma \in \mathrm{Gal}(M,F)$ is generic one shows $M \cap \sigma M = F$ (see [3]) so the fixing of C_M implies $C_M \subset F$.

PROPOSITION 3: *If* G *is simple and* $F = P$ *and* $\beta \in G_F$ *is such that* $\beta\beta_1 \ldots \beta_n \neq e$ *for any positive integer* n, *set* $M = F\langle\alpha\rangle$ *with* $\alpha_1 = \alpha\beta$. *Then* $C_M = C$.

Proof: The argument of proposition 1 shows that we may identify the set of isomorphisms of M/F into an overfield, U, of M with $G_{C'}$, where C' is the field of constants of U. If P_M is the field of periodic elements of M recall that P_M is the algebraic closure of C_M in M. Choose U so that its underlying field is $\mathcal{U}$ and set $H = \{\sigma \in \mathrm{Gal}(M,F) : \sigma(x) = x \ \forall\, x \in P\}$.

Since M is strongly normal over P_M, H is a connected partial C-subgroup of G (see [3]). Since P_M is stable under $\mathrm{Gal}(M,F)$, H is a normal partial C-subgroup of G. If $\bar{H}$ is the completion of H in G then, since $\bar{H}$ is connected and normal in G, $\bar{H} = \{e\}$ or $\bar{H} = G$.

If $\bar{H} = \{e\}$ then $H = \{e\}$ so $P_M = M$. But $M = F\langle\alpha\rangle$ so there is an n with $\alpha_{n+1} = \alpha$. Since $\alpha_{n+1} = \alpha\beta\beta_1 \dots \beta_n$ we have $\beta\beta_1 \dots \beta_n = e$ contrary to hypotheses. Hence $\bar{H} = G$; but $H = \bar{H}_{C*} = G_{C*}$ so P_M is fixed by every element of $\mathrm{Gal}(M,F)$ which implies C_M is fixed by every element of $\mathrm{Gal}(M,F)$ and thus $C_M = C$.

COROLLARY: If $F = C$ is algebraically closed then the inverse problem is solvable over F for any simple C-group G.

REFERENCES

[1] Cohn, R., Difference Algebra, Wiley, New York, 1965.

[2] Franke, C., Picard-Vessiot Theory of Linear Homogeneous Difference Equations. Trans. Amer. Math. Soc. 108 (1963), 491-515.

[3] Infante, R., Strong Normality and Normality for Difference Fields, Aequations, Math. 20 (1980), 159-165.

[4] Kolchin, E., Differential Algebra and Algebraic Groups, Academic Press, New York, 1972.

Department of Mathematics
Seton Hall University
South Orange, New Jersey 07079

Contemporary Mathematics
Volume 13, 1982

QUADRATIC HOPF ALGEBRAS AND GALOIS EXTENSIONS

H. F. Kreimer

Let R be a commutative ring, and let J be a Hopf algebra which is a finitely generated projective R-module. Generalizing Chase and Sweedler [3], T. Early and the author [5] have defined the concept of a J-Galois algebra S for S a not necessarily associative R-algebra. Such an algebra S has normal basis if $S \cong J$ as J^*-modules, J^* = the linear dual of J. If J^* is commutative, then the set of isomorphism classes of J-Galois algebras S forms an abelian group T(J), and the Picard invariant is a homomorphism from T(J) onto the Harrison cohomology group $H^1(J^*,P)$. In this context a question considered by Childs and Magid becomes that of asking if the Picard invariant is onto $H^1(J^*,P)$ when restricted to the subgroup A(J) of T(J) consisting of associative J-Galois algebras.

In this paper we show that the Picard invariant is onto if J is a Hopf algebra which is a free R-module of rank 2, thus generalizing the case $J^* = RG$, G a group of order 2, treated in [4].

In order to obtain this result, we first study the structure of such J, and show that J must be commutative and cocommutative, generated by 1 and x where $x^2 = qx$, $\Delta(x) = x \otimes 1 + 1 \otimes x + p(x \otimes x)$, where $pq = -2$. It follows that if 2 is a unit of R, then J = RG, G a group of order 2. We then examine the structure of a not necessarily associative J-Galois algebra S and describe a method of altering the multiplication of S to yield an associative J-Galois algebra. This shows that the Picard invariant is onto. In addition, we show that the group T(J) is isomorphic to $A(J) \times G(p)^2$, where $G(p) = \{a \text{ in } R \mid 1+ap \text{ is in } U(R)\}$.

In the final section of this paper we show that with J generated as above by 1, x, and with p, q as above with $pq = -2$, then the subgroup N(J) of A(J) consisting of J-Galois algebras with normal basis is isomorphic to $G(p^2)/G(p)$. If $p = 2u$, u a unit of R, we recover a result of C. Small [6].

1. QUADRATIC HOPF ALGEBRAS

Let R be a commutative ring with identity element 1, let J be a Hopf algebra over R with antipode S, and use the unit map to identify R with

0271-4132/82/0548/$03.25

a subring of J. Let $\Delta: J \to J \otimes J$ denote the comultiplication or diagonal map, and let $\varepsilon: J \to R$ denote the counit or augmentation map. Throughout the paper assume that J is a free R-module of rank 2.

1.1. LEMMA: <u>An R-module M is free of rank 1 if and only if $R \oplus M$ is a free R-module of rank 2.</u>

<u>Proof</u>: Clearly, if M is a free R-module of rank 1, then $R \oplus M$ is a free R-module of rank 2. Conversely, suppose $R \oplus M$ is free of rank 2. Then M is a finitely generated, projective module which must have rank one [2, Chap.II, §5, No. 3, Def. 2]. Moreover M is isomorphic to the exterior power $\Lambda^2(R \oplus M)$, which is a free R-module of rank 1 [1, Chap. III; and 2, Chap. II, §5, No. 3].

Since J is isomorphic to $R \oplus \operatorname{Ker} \varepsilon$ as an R-module, $\operatorname{Ker} \varepsilon$ is a free R-module of rank 1. If x is a free generator for $\operatorname{Ker} \varepsilon$, then 1 and x form a basis for J over R. Since $\varepsilon(x^2) = (\varepsilon(x))^2 = 0$, $x^2 = qx$ for some element q of R. An application of ε to either factor of $J \otimes J$ will map Δx onto x, and therefore Δx must have the form $1 \otimes x + x \otimes 1 + px \otimes x$ for some element p of R. Furthermore $\varepsilon(S(x)) = \varepsilon(x) = 0$, so $S(x) = kx$ for some k in R. But then $0 = \varepsilon(x) = S(1)x + S(x)\cdot 1 + pS(x)\cdot x = x + k(1 + pq)x$ and $k(1 + pq) = -1$. Also $q(1 \otimes x + x \otimes 1 + px \otimes x) = \Delta(qx) = \Delta(x^2) = (\Delta x)^2 = (1 \otimes x + x \otimes 1 + px \otimes x)^2 = q(1 \otimes x + x \otimes 1) + (2 + 4pq + p^2q^2)x \otimes x$, and consequently $p^2q^2 + 4pq + 2 = pq$ or $(pq + 1)(pq + 2) = 0$. Since $k(1 + pq) = -1$, it now follows that $pq + 2 = 0$ and $k = 1$.

Now let J^* be the dual of the R-module J, and let ϕ denote the element of J^* such that $\phi(1) = 0$ and $\phi(x) = 1$. Then ε and ϕ form a basis for J^* over R which is dual to the basis 1, x for J. Note that J^* is a Hopf algebra over R, ε is the identity element of the algebra J^*, ϕ generates the kernel of the counit map of J^*, $\phi^2 = p\phi$, and the image of ϕ under the comultiplication map for J^* is $\varepsilon \otimes \phi + \phi \otimes \varepsilon + q\phi \otimes \phi$.

1.2 THEOREM: <u>Let J be a Hopf algebra over R, which is a free R-module of rank 2. There exists an element x of J such that $\varepsilon(x) = 0$ and 1,x form a basis for J. Moreover, if 1,x is such a basis for J and ε,ϕ is the dual basis for J^*, then:</u>

(1) $x^2 = qx$ <u>and</u> $\Delta x = 1 \otimes x + x \otimes 1 + px \otimes x$, <u>where</u> p <u>and</u> q <u>are elements of</u> R <u>such that</u> $pq + 2 = 0$;

(2) <u>the space of integrals of</u> J^* <u>is freely generated as an R-module by</u> $p\varepsilon - \phi$.

<u>Proof</u>: There remains only statement (2) to be proved. An element $a_1\varepsilon + a_2\phi$ of J^* is in the space of integrals if and only if $0 = \phi(1)\cdot(a_1\varepsilon + a_2\phi) = \phi\cdot(a_1\varepsilon + a_2\phi) = (a_1 + pa_2)\phi$. But if $(a_1 + pa_2)\phi = 0$, then $a_1 = -pa_2$ and $a_1\varepsilon + a_2\phi = (-a_2)(p\varepsilon - \phi)$. Also if $0 = a(p\varepsilon - \phi) = ap\varepsilon + (-a)\phi$, then $a = 0$.

An element y of J is also a generator for $\mathrm{Ker}\,\varepsilon$ if and only if $y = ux$ for some unit u in R. But if $y = ux$, then $y^2 = qu^2x = q'y$, where $q' = qu$, and $\Delta y = u(1 \otimes x + x \otimes 1 + px \otimes x) = 1 \otimes y + y \otimes 1 + p'y \otimes y$, where $p' = pu^{-1}$. Also it is readily verified that elements p and q of R such that $pq + 2 = 0$ may be used to construct a Hopf algebra with basis consisting of the identity element 1 of the algebra and an element x such that $x^2 = qx$, $\varepsilon(x) = 0$, $\Delta x = 1 \otimes x + x \otimes 1 + px \otimes x$, and $S(x) = x$.

If z is a grouplike element of J, then $\varepsilon(z) = 1$ and z must be of the form $1 + ax$. But also $(1 + ax) \otimes (1 + ax) = \Delta(1 + ax) = 1 \otimes 1 + a(1 \otimes x + x \otimes 1 + px \otimes x)$, and so $a^2 = pa$. In particular, 1 and $1 + px$ are grouplike elements. Elements 1 and $1 + ax$ form a basis for J over R if and only if the determinant $\begin{vmatrix} 1 & o \\ 1 & a \end{vmatrix} = a$ is a unit in R; and if a is a unit such that $a^2 = pa$, then $a = p$. Thus J is the group algebra of the group of order 2 if and only if p is a unit in R. Since $pq + 2 = 0$, J must be this group algebra whenever 2 is a unit in R. If R is a field of characteristic 2, then there are the following three isomorphism classes of 2-dimensional Hopf algebras. If $p \neq o$, then J is the group algebra of the group of order 2. If $p = o$ but $q \neq o$, then J is the dual of this group algebra. Finally, if $p = o = q$, then J is the enveloping algebra of the 1-dimensional, 2-restricted Lie algebra over R.

2. GALOIS ALGEBRAS

Although ultimately associative R-algebras may be the objects of interest, it is useful to examine more general algebras at first. The term R-algebra will designate an ordered pair (B,μ) consisting of an R-module B and a multiplication map $\mu: B \otimes B \to B$ which is a homomorphism of R-modules. For elements b,c of B, $\mu(b \otimes c)$ will usually be denoted simply by $b\cdot c$ or bc. Also the identity automorphism of an R-module X will be denoted by the symbol X also. The following definitions are repeated from [5].

2.1 Definition. An R-algebra (B,μ) is called J-Galois if:

(1) B is a faithful, finitely generated, projective R-module;

(2) there exists an R-algebra homomorphism $\alpha: B \to B \otimes J$ such that $(\alpha \otimes J)\cdot\alpha = (B \otimes \Delta)\cdot\alpha$ and $(B \otimes \varepsilon)\cdot\alpha = B$;

(3) $(\mu \otimes J)\cdot(B \otimes \alpha)$ is an isomorphism of $B \otimes B$ onto $B \otimes J$.

To any R-module homomorphism $\alpha: B \to B \otimes J$ there corresponds an R-module homomorphism of J^* into $\mathrm{Hom}(B,B)$ under the canonical isomorphisms $\mathrm{Hom}(B,B \otimes J) \cong \mathrm{Hom}(B,\mathrm{Hom}(J^*,B)) \cong \mathrm{Hom}(B \otimes J^*,B) \cong \mathrm{Hom}(J^*,\mathrm{Hom}(B,B))$. If α satisfies condition (2) of the preceding definition, then the corresponding homomorphism of J^* into $\mathrm{Hom}(B,B)$ is an R-algebra homomorphism by which B becomes a left J^*-module and $\phi(bc) = \phi(b)\cdot c + b\cdot\phi(c) + q\phi(b)\cdot\phi(c)$ for elements b,c of B. The algebra J is J-Galois with respect to the comultiplication map $\Delta: J \to J \otimes J$ [5, Prop. 1.2].

2.2 Definition. A J-Galois algebra (B,μ) is said to have a normal basis if B and J are isomorphic as left J^*-modules. Two J-Galois algebras are said to be isomorphic if there exists an R-algebra isomorphism between them which is also an isomorphism of left J^*-modules.

Now let (B,μ) be a J-Galois algebra which has a normal basis. Then B is a free R-module of rank 2, and there exists a basis b_1, b_2 for B (corresponding to the basis $1,x$ for J) such that $\phi(b_1) = 0$ and $\phi(b_2) = b_1 + pb_2$. Moreover there exists a unit $u = a_o\varepsilon \otimes \varepsilon + a_1\varepsilon \otimes \phi + a_2\phi \otimes \varepsilon + a_3\phi \otimes \phi$ in the algebra $J^* \otimes J^*$ such that (B,μ) is isomorphic to the algebra $(J,\hat{u})$ which is constructed in [5, Prop. 1.6 and the paragraph preceding Prop. 1.6]. The following rules for multiplication in B are obtained by tracing the construction of $(J,\hat{u})$ and using the relation $pq + 2 = 0$: $b_1b_1 = a_ob_1$, $b_1b_2 = a_1b_1 + (a_o + pa_1)b_2$, $b_2b_1 = a_2b_1 + (a_o + pa_2)b_2$, and $b_2b_2 = a_3b_1 + (qa_o + a_1 + pqa_1 + a_2 + pqa_2 + 2pa_3 + p^2qa_3)b_2 = a_3b_1 + (qa_o - a_1 - a_2)b_2$. Finally note that under the regular representation of the algebra $J^* \otimes J^*$ there corresponds to any element $u = a_o\varepsilon \otimes \varepsilon + a_1\varepsilon \otimes \phi + a_2\phi \otimes \varepsilon + a_3\phi \otimes \phi$ of $J^* \otimes J^*$ the following matrix over R.

$$(2.3) \qquad \begin{bmatrix} a_o & a_1 & a_2 & a_3 \\ 0 & a_o + pa_1 & 0 & a_2 + pa_3 \\ 0 & 0 & a_o + pa_2 & a_1 + pa_3 \\ 0 & 0 & 0 & a_o + pa_1 + pa_2 + p^2a_3 \end{bmatrix}$$

Therefore u is a unit in $J^* \otimes J^*$ if and only if $a_o, a_o + pa_1, a_o + pa_2$, and $a_o + pa_1 + pa_2 + p^2a_3$ are units in R.

2.4 THEOREM: Let (B,μ) be a J-Galois algebra which has a normal basis.

(1) $\operatorname{Ker}\phi$ and $\operatorname{Ker}(p\varepsilon - \phi)$ are free R-modules of rank 1.

(2) The map $p\varepsilon - \phi$ is an R-module homomorphism of B onto $\operatorname{Ker}\phi$.

(3) $(\operatorname{Ker}\phi)^2 = \operatorname{Ker}\phi = (\operatorname{Ker}(p\varepsilon - \phi))^2$.

Proof: For any element $a_1b_1 + a_2b_2$ of B, $\phi(a_1b_1 + a_2b_2) = a_2b_1 + pa_2b_2$ and $(p\varepsilon - \phi)(a_1b_1 + a_2b_2) = (pa_1 - a_2)b_1$. It follows readily that b_1 generates $\operatorname{Ker}\phi$ and $b_1 + pb_2$ generates $\operatorname{Ker}(p\varepsilon - \phi)$. Also $(p\varepsilon - \phi)(-qb_1 + b_2) = (-pq - 1)b_1 = b_1$. Since $b_1b_1 = a_ob_1$ and $(b_1 + pb_2)(b_1 + pb_2) = b_1b_1 + pb_1b_2 + pb_2b_1 + p^2b_2b_2 = (a_o + pa_1 + pa_2 + p^2a_3)b_1$, where a_o and $a_o + pa_1 + pa_2 + p^2a_3$ are units in R, $(\operatorname{Ker}\phi)^2 = \operatorname{Ker}\phi = (\operatorname{Ker}(p\varepsilon - \phi))^2$.

2.5 Corollary: Let (B,μ) be a J-Galois algebra.

(1) $\operatorname{Ker}(p\varepsilon - \phi)$ is a projective R-module of rank one, and the restriction of μ is an isomorphism of $\operatorname{Ker}(p\varepsilon - \phi) \otimes \operatorname{Ker}(p\varepsilon - \phi)$ onto $\operatorname{Ker}\phi$.

(2) The map $p\varepsilon - \phi$ is a homomorphism of B onto $\operatorname{Ker}\phi$.

(3) <u>There is a unique element</u> e <u>of</u> B <u>such that</u> $e^2 = e$ <u>and</u> e <u>is a free generator of the</u> R-<u>module</u> $\operatorname{Ker} \phi$.

(4) <u>There exist elements</u> c <u>of</u> B <u>and</u> v <u>of</u> R <u>such that the</u> R-<u>module</u> B <u>is the direct sum of</u> $\operatorname{Ker}(p\varepsilon - \phi)$ <u>and a submodule freely generated by</u> $c, \phi(c) = pc - e, ec = c = ce$, <u>and</u> $c^2 + qc + ve = 0$. <u>Moreover</u> $pc - e$ <u>is an element of</u> $\operatorname{Ker}(p\varepsilon - \phi)$ <u>and</u> $(pc - e)^2 = (1 - p^2 v)e$.

<u>Proof</u>: Given any prime ideal of R, there is an element a in the complement of that prime ideal such that for the localizations with respect to a, (B_a, μ_a) is a J_a-Galois algebra which has a normal basis [5, Prop. 2.1 and the paragraph following Prop. 2.1]. Then $\operatorname{Ker} \phi_a$ and $\operatorname{Ker}(p\varepsilon - \phi)_a$ are free R_a-modules of rank 1, $(p\varepsilon - \phi)_a$ is an R_a-module homomorphism of B_a onto $\operatorname{Ker} \phi_a$, and isomorphisms of $\operatorname{Ker} \phi_a \otimes \operatorname{Ker} \phi_a$ onto $\operatorname{Ker} \phi_a$ and $\operatorname{Ker}(p\varepsilon - \phi)_a \otimes \operatorname{Ker}(p\varepsilon - \phi)_a$ onto $\operatorname{Ker} \phi_a$ are obtained by restricting μ_a. By [2, Chap. II, 5, Theorem 2], $\operatorname{Ker} \phi$ and $\operatorname{Ker}(p\varepsilon - \phi)$ are rank one, projective R-modules. By [2, Chap. II, §3, Theorem 1], $p\varepsilon - \phi$ is an R-module homomorphism of B onto $\operatorname{Ker} \phi$ and isomorphisms of $\operatorname{Ker} \phi \otimes \operatorname{Ker} \phi$ onto $\operatorname{Ker} \phi$ and $\operatorname{Ker}(p\varepsilon - \phi) \otimes \operatorname{Ker}(p\varepsilon - \phi)$ onto $\operatorname{Ker} \phi$ are obtained by restricting μ. Since $\operatorname{Ker} \phi$ is a rank one, projective R-module and $\operatorname{Ker} \phi \otimes \operatorname{Ker} \phi$ is isomorphic to $\operatorname{Ker} \phi$, $\operatorname{Ker} \phi$ must be isomorphic to R as an R-module. If b generates the R-module $\operatorname{Ker} \phi$, then so must b^2. Therefore $b^2 = ab$ for some unit a in R, $e = a^{-1}b$ generates $\operatorname{Ker} \phi$, and $e^2 = a^{-2}b^2 = a^{-1}b = e$. If f is a generator of the R-module $\operatorname{Ker} \phi$, then $f = ae$ for some unit a in R. If also $f^2 = f$, then $ae = (ae)^2 = a^2 e$ and hence $a = a^2$ and $a = 1$. Note that, since $\phi(e) = 0$, $\phi(eb) = e \cdot \phi(b)$ and $\phi(be) = \phi(b) \cdot e$ for any element b of R.

Since $\operatorname{Ker} \phi$ is freely generated by e, any element c of B for which $pc - \phi(c) = e$ generates a free submodule of B such that B is the direct sum of this submodule and $\operatorname{Ker}(p\varepsilon - \phi)$. If $pc - \phi(c) = e$, then $pece - \phi(ece) = e(p - \phi(c))e = e$. Therefore the element c may be chosen so that $ec = c = ce$. Then $\phi(c^2 + qc) = c \cdot \phi(c) + \phi(c) \cdot c + q\phi(c) \cdot \phi(c) + q\phi(c) = c(pc - e) + (pc - e)c + q(pc - e)^2 + q(pc - e) = (2c + q(pc - e) + qe)(pc - e) = (2 + pq)c(pc - e) = 0$. Therefore $c^2 + qc = -ve$ for some v in R. Clearly, $(p\varepsilon - \phi)(pc - e) = (p\varepsilon - \phi)\phi(c) = 0$, since $\phi^2 = p\phi$; and $(pc - e)^2 = p^2 c^2 - 2pc + e = p^2 c^2 + p^2 qc + e = p^2(c^2 + qc) + e = -p^2 ve + e = (1 - p^2 v)e$.

2.6 <u>Corollary</u>: <u>Let</u> (B, μ) <u>be a</u> J-<u>Galois algebra and let</u> e <u>be the idempotent element of</u> B <u>which generates</u> $\operatorname{Ker} \phi$. <u>There exists an</u> R-<u>module homomorphism</u> $\nu: B \otimes B \to B$ <u>such that</u> (B, ν) <u>is an associative and commutative</u> J-<u>Galois algebra with identity element</u> e.

<u>Proof</u>: Choose elements c of B and v of R as in statement (4) of the preceding corollary. For any elements z, z' of $\operatorname{Ker}(p\varepsilon - \phi)$, set $\nu(z \otimes z') = \mu(z \otimes z')$, $\nu(z \otimes c) = pvz + \mu(\mu(z \otimes (pc - e)) \otimes c) = \nu(c \otimes z)$, and

$\nu(c \otimes c) = -qc - ve$; and extend ν bilinearly to all of $B \otimes B$. To prove the properties of (B,ν) in the statement of the corollary, it is necessary to verify certain equations. These equations will be valid provided they are valid for every localization of J and (B,ν) with respect to a prime ideal of R. [cf. 2, Chap. II, §3, No. 3, Thm. 1 and Cor. 1 and 2]. Since $\mathrm{Ker}(p\varepsilon - \phi)$ is a finitely generated, projective R-module, it is a free R-module when R is a local ring. Without loss of generality it may be assumed that $\mathrm{Ker}(p\varepsilon - \phi)$ is freely generated by a single element z_o. Since the restriction of μ is an isomorphism of $\mathrm{Ker}(p\varepsilon - \phi) \otimes \mathrm{Ker}(p\varepsilon - \phi)$ onto $\mathrm{Ker}\,\phi$, $\mu(z_o \otimes z_o) = ue$ for some unit u of R. If $pc - e = az_o$, then $a^2ue = \mu((pc - e) \otimes (pc - e)) = (1 - p^2v)e$ and $a^2u = 1 - p^2v$. The rules for the multiplication ν become: $z_o^2 = ue$, $z_oc = pvz_o + auc = cz_o$, and $c^2 = -qc - ve$. Obviously (B,ν) is a commutative algebra; and $ez_o = (pc - az_o)z_o = pcz_o - aue = p^2vz_o + pauc - aue = p^2vz_o + au(pc - e) = p^2vz_o + a^2uz_o = z_o$, while $ec = (pc - az_o)c = pc^2 - az_oc = -pqc - pve - pavz_o - a^2uc = -pqc - pv(e + az_o) - a^2uc = -pqc - p^2vc - a^2uc = -pqc - c = -(pq + 1)c = c$. To prove that (B,ν) is associative, there remains only to verify that $(z_oz_o)c = z_o(z_oc)$ and $(z_oc)c = z_o(cc)$. But $z_o(z_oc) = pvz_o^2 + aucz_o = u(pve + acz_o) = u(pv(pc - az_o) + a(pvz_o + auc)) = u(p^2vc + a^2uc) = uc = (z_oz_o)c$; and $(z_oc)c = pvz_oc + auc^2 = p^2v^2z_o + pauvc - qauc - auve = p^2v^2z_o + auv(pc - e) - qauc = p^2v^2z_o + a^2uvz_o - qauc = vz_o - qauc$, while $z_o(cc) = -qz_oc - vz_o = -pqvz_o - qauc - vz_o = -(pq + 1)vz_o - qauc = vz_o - qauc$.

Now observe that $\phi(z_oc) = pv\phi(z_o) + au\phi(c) = p^2vz_o + a^2uz_o = z_o$, since $\phi(c) = pc - e = az_o$; while $z_o\phi(c) + \phi(z_o)\cdot c + q\phi(z_o)\cdot\phi(c) = aue + pz_oc + pqaue = (pq + 1)aue + p^2vz_o + pauc = au(pc - e) + p^2vz_o = a^2uz_o + p^2vz_o = z_o$. Also $\phi(c^2) = -q\phi(c) = -qaz_o$, while $c\cdot\phi(c) + \phi(c)\cdot c + q\phi(c)\cdot\phi(c) = 2az_oc + qa^2z_o^2 = a(-pqc + qaz_o)z_o = -qaez_o = -qaz_o$. Now consider the homomorphism $\gamma = (\nu \otimes J)(B \otimes \alpha)$ of $B \otimes B$ into $B \otimes J$. If b and b' are elements of B, $\gamma(b \otimes z_o + b' \otimes c) = (bz_o + b'c) \otimes 1 + (pbz_o + ab'z_o) \otimes x$, and γ may be represented by the matrix $\begin{vmatrix} z_o & pz_o \\ c & az_o \end{vmatrix}$. Since the determinant of this matrix is $az_o^2 - pcz_o = (az_o - pc)z_o = -ez_o = -z_o$, which is a unit in B, the matrix is invertible and γ is an isomorphism. Then (B,ν) is a J-Galois algebra.

3. HARRISON COHOMOLOGY

The isomorphism classes of J-Galois algebras are the elements of an abelian group $T(J)$, and the isomorphism classes of associative J-Galois algebras are the elements of a subgroup $A(J)$. If U and P denote the units and Picard functors, respectively, then in terms of the Harrison cohomology of the Hopf algebra J^* there is an exact sequence of abelian groups:

$$0 \to H^1(J^*,U) \to U(J^*) \overset{\delta}{\to} U(J^* \otimes J^*) \to T(J) \to H^1(J^*,P) \to 0$$

[5, Thm. 3.2]. Under the regular representation of the algebra $J^* \otimes J^*$, the units of $J^* \otimes J^*$ correspond to invertible matrices over R of the form displayed in 2.3. But if k_1, k_2, and k_3 are elements of R such that $a_o k_1 = a_1$, $a_o k_2 = a_2$, and $(a_o + pa_1)(a_o + pa_2)k_3 = a_o a_3 - a_1 a_2$, then this matrix may be factored as follows.

$$3.1 \qquad a_o \cdot \begin{vmatrix} 1 & k_1 & 0 & 0 \\ 0 & 1+pk_1 & 0 & 0 \\ 0 & 0 & 1 & k_1 \\ 0 & 0 & 0 & 1+pk_1 \end{vmatrix} \begin{vmatrix} 1 & 0 & k_2 & 0 \\ 0 & 1 & 0 & k_2 \\ 0 & 0 & 1+pk_2 & 0 \\ 0 & 0 & 0 & 1+pk_2 \end{vmatrix} \begin{vmatrix} 1 & 0 & 0 & k_3 \\ 0 & 1 & 0 & pk_3 \\ 0 & 0 & 1 & pk_3 \\ 0 & 0 & 0 & 1+p^2k_3 \end{vmatrix}$$

For any given element t of R, let G(t) be the set of elements a of R such that $1 + ta$ is a unit. If the composition of elements a,a' of G(t) is defined to be $a+a' + taa'$, then G(t) becomes an abelian group, 0 is the identity element of G(t), and the inverse of a is $-a\cdot(1 + ta)^{-1}$. From the factorization 3.1, it now follows readily that $U(J^* \otimes J^*)$ is isomorphic to the product of the groups $U(R) \times G(p) \times G(p) \times G(p^2)$. Similarly, there corresponds to an element $w = a_o\varepsilon + a_1\phi$ of J^* the matrix $\begin{vmatrix} a_o & a_1 \\ 0 & a_o+pa_1 \end{vmatrix}$ under the regular representation of J^*. Then w is a unit in J^* if and only if a_o and $a_o + pa_1$ are units in R. Moreover, if k is an element of R such that $a_o k = a_1$, then $\begin{vmatrix} a_o & a_1 \\ 0 & a_1+pa_1 \end{vmatrix} = a_o \begin{vmatrix} 1 & k \\ 0 & 1+pk \end{vmatrix}$. Therefore $U(J^*)$ is isomorphic to $U(R) \times G(p)$. If $w = a_o\varepsilon + a_1\phi = a_o(\varepsilon + k\phi)$ is a unit in $U(J^*)$, then $w^{-1} = a_o^{-1}(1 + pk)^{-1}((1 + pk)\varepsilon - k\phi)$ and

$$\begin{aligned} \delta(w) &= a_o(1+pk)^{-1}(\varepsilon\otimes\varepsilon+k\phi\otimes\varepsilon)(\varepsilon\otimes\varepsilon+k\varepsilon\otimes\phi)((1+pk)\varepsilon\otimes\varepsilon-k\varepsilon\otimes\phi-k\phi\otimes\varepsilon-qk\phi\otimes\phi) \\ &= a_o(1+pk)^{-1}(\varepsilon\otimes\varepsilon+k\phi\otimes\varepsilon)((1+pk)\varepsilon\otimes\varepsilon-k\phi\otimes\varepsilon-(k^2+pqk^2+qk)\phi\otimes\phi) \\ &= a_o(1+pk)^{-1}((1+pk)\varepsilon\otimes\varepsilon-(1+pk)(k^2+pqk^2+qk)\phi\otimes\phi) \\ &= a_o(\varepsilon\otimes\varepsilon-(-k^2+qk)\phi\otimes\phi) = a_o(\varepsilon\otimes\varepsilon+k(k-q)\phi\otimes\phi). \end{aligned}$$

Thus the restriction of δ to U(R) is the identity map, and δ maps an element k of G(p) to the element $k(k - q)$ of $G(p^2)$. Note that $1 + p^2k(k-q) = 1+ p^2k^2 - p^2qk = 1 + p^2k^2 + 2pk = (1+pk)^2$. If a is an element of R such that $a^2 = qa$, then $(1+pa)^2 = 1 + 2pa + p^2a^2 = 1 - p^2qa + p^2a^2 = 1$ and the set of such elements of R form a subgroup of G(p) which is isomorphic to the kernel of δ. These results are summarized in the following proposition.

3.2 <u>PROPOSITION</u>: (1) $U(J^*) \cong U(R) \times G(p)$, $U(J^* \otimes J^*) \cong U(R) \times G(p) \times G(p) \times G(p^2)$, *and the coboundary map* $\delta: U(J^*) \to U(J^* \otimes J^*)$ *is the identity on* U(R) *and maps an element* k *of* G(p) *to the element* $k(k-q)$ *of* $G(p^2)$

(2) *The set of elements* k *of* R *such that* $k^2 = qk$ *is a group under the composition* $k \cdot k' = k + k' + pkk'$, *and* $H^1(J^*, U)$ *is isomorphic to this group.*

For associative J-Galois algebras there is a left exact sequence of abelian groups: $0 \to H^2(J^*, U) \to A(J) \to H^1(J^*, P)$ [5, Thm. 3.3]. The homomorphism of $A(J)$ into $H^1(J^*, P)$, called the Picard invariant, is the restriction of the homomorphism of $T(J)$ onto $H^1(J^*, P)$, which maps the isomorphism class of a J-Galois algebra (B, μ) to the J^*-module isomorphism class of $B^* = \mathrm{Hom}(B, R)$.

3.3 THEOREM: (1) $H^2(J^*, U)$ *is the cokernel of the mapping* $k \rightsquigarrow k(k-q)$ *from* $G(p)$ *into* $G(p^2)$.

(2) *The Picard invariant is an epimorphism of* $A(J)$ *onto* $H^1(J^*, P)$.

(3) $T(J)$ *is isomorphic to the product* $A(J) \times G(p) \times G(p)$.

Proof: To a unit $u = a_0 \varepsilon \otimes \varepsilon + a_1 \varepsilon \otimes \phi + a_2 \phi \otimes \varepsilon + a_3 \phi \otimes \phi$ of $J^* \otimes J^*$ corresponds a J-Galois algebra (B, μ) with normal basis b_1, b_2. The rules for multiplication in this J-Galois algebra are: $b_1^2 = a_0 b_1$, $b_1 b_2 = a_1 b_1 + (a_0 + pa_1) b_2$, $b_2 b_1 = a_2 b_1 + (a_0 + pa_2) b_2$, and $b_2^2 = a_3 b_1 + (qa_0 - a_1 - a_2) b_2$. The algebra (B, μ) is associative if and only if u is a cocycle in $U(J^* \otimes J^*)$ [5, Prop. 1.8]; and if (B, μ) is associative, then it has an identity element [5, Prop. 1.9]. Now suppose a, a' are elements of R such that $ab_1 + a'b_2$ is an identity element. Then $b_1 = b_1(ab_1 + a'b_2)$ $= (aa_0 + a'a_1) b_1 + a'(a_0 + pa_1) b_2$, and hence $aa_0 + a'a_1 = 1$ and $a'(a_0 + pa_1) = 0$. Since a_0 and $a_0 + pa_1$ are units in R, $a' = 0$ and $a = a_0^{-1}$. Also $b_2 = a_0^{-1} b_1 b_2 = a_0^{-1} a_1 b_1 + a_0^{-1}(a_0 + pa_1) b_2$ and $b_2 = b_2(a_0^{-1} b_1) =$ $a_0^{-1} a_2 b_1 + a_0^{-1}(a_0 + pa_2) b_2$. Therefore $a_1 = o = a_2$. But if $a_1 = o = a_2$, then the rules for multiplication in the algebra (B, μ) become: $b_1^2 = a_0 b_1$, $b_1 b_2 = a_0 b_2 = b_2 b_1$, $b_2 b_2 = a_3 b_1 + qa_0 b_2$; and it may be readily verified that (B, μ) is an associative and commutative algebra. Therefore the cocycles of $U(J^* \otimes J^*)$ correspond to the elements of $U(R) \times G(p^2)$ and $H^2(J^*, U)$ is the cokernel of the mapping $k \rightsquigarrow k(k-q)$ of $G(p)$ into $G(p^2)$.

The Picard invariant is surjective by Corollary 2.6. Therefore there is the following commutative diagram of exact sequences.

$$
\begin{array}{ccccccccc}
U(J^*) & \longrightarrow & U(R) \times G(p^2) & \longrightarrow & A(J) & \longrightarrow & H^1(J^*, p) & \longrightarrow & 0 \\
\| & & \downarrow & & \downarrow & & \| & & \\
U(J^*) & \longrightarrow & U(J^* \otimes J^*) & \longrightarrow & T(J) & \longrightarrow & H^1(J^*, p) & \longrightarrow & 0 \\
 & & \downarrow & & & & & & \\
 & & G(p) \times G(p) & & & & & &
\end{array}
$$

By a diagram chase or a suitable version of the five lemma, the quotient of $T(J)$ by the subgroup $A(J)$ is isomorphic to $G(p) \times G(p)$. Since $U(J^* \otimes J^*)$ is isomorphic to the product of the groups $U(R) \times G(p^2)$ and $G(p) \times G(p)$, the projection of $T(J)$ onto $G(p) \times G(p)$ is split and $T(J)$ is isomorphic to the product $A(J) \times G(p) \times G(p)$.

The preceding argument reproves and extends somewhat results obtained by Lindsay Childs and Andy Magid [4]. This argument may be used also to establish the following result.

3.4 COROLLARY: *A J-Galois algebra* (B,μ) *is associative if and only if it has identity element. Moreover if it has an identity element then* (B,μ) *is commutative*.

Proof: If (B,μ) is associative, then it has an identity element by [5, Thm. 2.3]. Now suppose (B,μ) has an identity element. The algebra (B,μ) is associative and commutative if and only if every localization of (B,μ) at a prime ideal of R is associative and commutative. But if R is a local ring, the J-Galois algebra (B,μ) has a normal basis by [5, Prop. 2.1] and (B,μ) is associative and commutative by the first part of the proof of the preceding theorem.

BIBLIOGRAPHY

1. N. Bourbaki, Algèbre, Éléments de mathématique, Hermann, Paris.

2. N. Bourbaki, Algèbre Commutative, Éléments de mathématique, Hermann, Paris 1961.

3. S. U. Chase and M. E. Sweedler, Hopf Algebras and Galois Theory, Springer Lect. Notes in Math. no. 97 (1969).

4. L.N. Childs and A.R. Magid, The Picard invariant of a principal homogeneous space, Journal of Pure and Applied Algebra 4 (1974), 273-286.

5. T.E. Early and H.F. Kreimer, Galois algebras and Harrison cohomology, Journal of Algebra 58 (1979), 136-147.

6. C. Small, Normal bases for quadratic extensions, Pacific Journal of Mathematics 50 (1974), 601-611.

Florida State University
Tallahassee, Florida 32306

Contemporary Mathematics
Volume 13, 1982

THE LEFT ANTIPODES OF A LEFT HOPF ALGEBRA

by

Warren D. Nichols and Earl J. Taft

To Nathan Jacobson on the occasion of his retirement

1. INTRODUCTION

We shall work over a ground field k, although the basic notions will allow k to be a commutative ring with unit. If (C,Δ,ε) is a coalgebra with comultiplication Δ and counit $\varepsilon:C \to k$, and (A,m,μ) is an algebra with multiplication m and unit $\mu:k \to A$, then $\mathrm{Hom}(C,A)$ is a convolution algebra with product $f*g = m(f \otimes g)\Delta$ and unit $\mu\varepsilon$. A bialgebra $(B,\Delta,\varepsilon,m,\mu)$ is simultaneously a coalgebra and an algebra with Δ and ε algebra morphisms, so that $\mathrm{Hom}(B,B)$ is a convolution algebra. If the identity map Id of B has a left inverse in this convolution algebra, then B is called a <u>left Hopf algebra</u>, and such a left inverse of Id is called a <u>left antipode</u> of B. Using the notation $\Delta x = \Sigma x_1 \otimes x_2$ for $x \in B$ (see [5]), the condition that $S \in \mathrm{Hom}(B,B)$ be a left antipode is that $\Sigma(Sx_1)x_2 = \varepsilon(x)1$ for all $x \in B$. We can similarly define right Hopf algebra and right antipode T, i.e., $\Sigma x_1(Tx_2) = \varepsilon(x)1$ for all $x \in B$. However, if B is a left Hopf algebra with left antipode S and a right Hopf algebra with right antipode T, then $S * \mathrm{Id} = \mu\varepsilon = \mathrm{Id} * T$ implies that $S = T$ is the (unique) antipode for the Hopf algebra B. This is the definition of Hopf algebra (see [5]), and we note here that the antipode of a Hopf algebra is a bialgebra antimorphism ([5], Prop. 4.0.1).

In [2], a left Hopf algebra $H_\ell(C)$ was constructed (Example 21) which is not a Hopf algebra. A left Hopf algebra which is not a Hopf algebra necessarily has an infinite number of left antipodes ([3], p. 89, Exercise 7). In Example 21 of [2], $H_\ell(C)$ has a particular left antipode S_ℓ (called S there) which is a bialgebra antimorphism. The pair $(H_\ell(C),S_\ell)$ is free on the coalgebra $C = M_n(k)^*$, the coalgebra dual to the algebra $M_n(k)$ of n by n matrices over k, in the sense described by Proposition 14 of [2], and indicated by the following diagram:

0271-4132/82/0549/$02.50

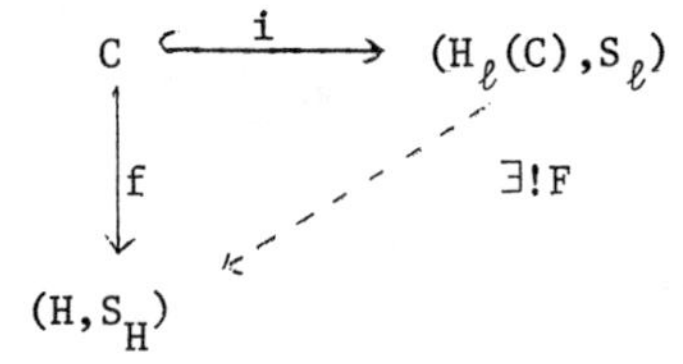

Here H is a left Hopf algebra with left antipode S_H which is a bialgebra antimorphism.

In this paper, we show that a left Hopf algebra can have at most one left antipode which is a bialgebra antimorphism (Corollary 2.2). On the other hand, we also construct a left Hopf algebra for which no left antipode is a bialgebra antimorphism (see Section 3).

2. LEFT ANTIPODES AS ANTIMORPHISMS

PROPOSITION 2.1. _Let H be a left Hopf algebra over k, S and T left antipodes of H. Assume that S is a coalgebra antimorphism and that T is an algebra antimorphism. Then S = T._

Proof: Let $*$ denote the convolution product in $\mathrm{Hom}(H^{coop},H)$, where H^{coop} is the opposite coalgebra of H. Let $x \in H$. Then $(T*TS)(x) = \Sigma(Tx_2)(TSx_1) = T(\Sigma(Sx_1)x_2) = T(\varepsilon(x)1) = \varepsilon(x)1$. Also $(TS*S)(x) = \Sigma(TSx_2)(Sx_1) = \Sigma T((Sx)_1)(Sx)_2 = \varepsilon(Sx)1 = \varepsilon(x)1$. Hence $T * TS = \mu\varepsilon = TS * S$ implies that TS is invertible in $\mathrm{Hom}(H^{coop},H)$ with inverse $S = T$.

COROLLARY 2.2: _A left Hopf algebra can have at most one left antipode which is a bialgebra antimorphism._

3. CONSTRUCTION OF A LEFT HOPF ALGEBRA IN WHICH NO LEFT ANTIPODE IS A BIALGEBRA ANTIMORPHISM

The example we are going to construct has as bialgebra homomorphic image the left Hopf algebra $H_\ell(M_n(k)^*)$ constructed in example 21 of [2]. We recall that example here.

Fix a positive integer $n \geq 2$. $H_\ell = H_\ell(M_n(k)^*)$ has algebra generators $\{X_{ij}^{(r)} | r \geq 0; 1 \leq i,j \leq n\}$ and relations

$$(*) \qquad \sum_{k=1}^{n} X_{ki}^{(r+1)} X_{kj}^{(r)} - \delta_{ij}1 \quad \text{for} \quad r \geq 0; 1 \leq i,j \leq n$$

and

(**) $$\sum_{k=1}^{n} X_{ik}^{(r)}X_{jk}^{(r+1)} - \delta_{ij}1 \text{ for } r \geq 1;\ 1 \leq i,\ j \leq n.$$

The coalgebra structure is described by

$$\Delta X_{ij}^{(r)} = \sum_{k=1}^{n} X_{ik}^{(r)} \otimes X_{kj}^{(r)} \quad \text{and} \quad \varepsilon(X_{ij}^{(r)}) = \delta_{ij}1.$$

H_ℓ has a left antipode S_ℓ (called S in Example 21 of [2]) which is a bialgebra antimorphism satisfying $S_\ell(X_{ij}^{(r)}) = X_{ji}^{(r+1)}$. Note that $r \geq 1$ in the relations (**). In [2], it is shown that the relations (**) with $r = 0$ are not satisfied, so that H_ℓ is a left Hopf algebra which is not a Hopf algebra. If one imposes the additional relations (**) for $r = 0$, one obtains the free Hopf algebra $H(M_n(k)^*)$ discussed in [6] as bialgebra homomorphic image of H_ℓ. Note that by Proposition 2.1, each of the (infinite number) of left antipodes of H_ℓ different from S_ℓ is neither an algebra antimorphism nor a coalgebra antimorphism.

We start with the free k-algebra $k[Y] = k[Y_{ij}^{(r)} | r \geq 0;\ 1 \leq i,\ j \leq n\}$ with coalgebra (bialgebra) structure given by

$$\Delta Y_{ij}^{(r)} = \sum_{k=1}^{n} Y_{ik}^{(r)} \otimes Y_{kj}^{(r)} \quad \text{and} \quad \varepsilon(Y_{ij}^{(r)}) = \delta_{ij}1.$$

We impose only the relations

(*') $$\sum_{k=1}^{n} Y_{ki}^{(r+1)}Y_{kj}^{(r)} - \delta_{ij}1 \quad \text{for} \quad r \geq 0;\ 1 \leq i,\ j \leq n.$$

The ideal of $k[Y]$ generated by relations (*') is a coideal (bi-ideal), so that the resulting algebra H is a bialgebra. This can be seen directly, or by using the proof of Lemma 13 of [2], where x is a $Y_{ij}^{(r)}$, and S is the coalgebra antimorphism of the sub-coalgebra D of $k[Y]$ with basis $\{Y_{ij}^{(r)} | r \geq 0{:}1 \leq i,j \leq n\}$ such that $S(Y_{ij}^{(r)}) = Y_{ji}^{(r+1)}$.

We now obtain a basis for H of irreducible words in the letters $Y_{ij}^{(r)}$, using the techniques of [1], as in [2] and [4]. We order the $Y_{ij}^{(r)}$ lexicographically on the three indices in the order r, i, j. For words w and w', we say $w < w'$ if their lenghts satisfy $\ell(w) < \ell(w')$, or if $\ell(w) = \ell(w')$ and w precedes w' lexicographically. We regard the relations (*') as reduction formulas

(1) $$Y_{ni}^{(r+1)}Y_{nj}^{(r)} \to \delta_{ij}1 - \sum_{k=1}^{n-1} Y_{ki}^{(r+1)}Y_{kj}^{(r)} \quad \text{for} \quad r \geq 0;\ 1 \leq i,\ j \leq n.$$

In order to resolve the overlaps $Y_{ni}^{(r+2)}Y_{nj}^{(r+1)}Y_{nk}^{(r)}$ for $r \geq 0$, we must introduce the additional reduction formulas

$$(2)\qquad Y^{(r+2)}_{ni}Y^{(r+1)}_{n-1,j}Y^{(r)}_{n-1,k} \to \delta_{jk}Y^{(r+2)}_{ni} - \delta_{ij}Y^{(r)}_{nk} + \sum_{a=1}^{n-1} Y^{(r+2)}_{ai}Y^{(r+1)}_{aj}Y^{(r)}_{nk} - \sum_{b=1}^{n-2} Y^{(r+2)}_{ni}Y^{(r+1)}_{bj}Y^{(r)}_{bk}.$$

The only overlaps which must now be considered are $Y^{(r+3)}_{n\ell}Y^{(r+2)}_{ni}Y^{(r+1)}_{n-1,j}Y^{(r)}_{n-1,k}$ for $r \geq 0$, and these can be resolved without any additional reduction formulas, as in [4]. Hence H has a basis of irreducible words, i.e., words containing no subword of the form $Y^{(r+1)}_{ni}Y^{(r)}_{nj}$ nor $Y^{(r+2)}_{ni}Y^{(r+1)}_{n-1,j}Y^{(r)}_{n-1,k}$ for $r \geq 0$.

Let S be the k-linear map of H into H whose action on an irreducible word $w = Y^{(r_1)}_{i_1j_1}Y^{(r_2)}_{i_2j_2}\ldots Y^{(r_t)}_{i_tj_t}$ is given by $S(w) = Y^{(r_t+1)}_{j_ti_t}\ldots Y^{(r_2+1)}_{j_2i_2}Y^{(r_1+1)}_{j_1i_1}$. To see that S is a left antipode of H, we note that if w is an irreducible word, then $\Delta w = \Sigma w_1 \otimes w_2$ where the w_1 are also irreducible words. To see this, note that if a word w is diagonalized, the letters appearing in the left tensor factors retain the same upper indices and the same first lower indices as those in w. Since the form of a reducible word is determined by these two indices, if any w_1 is reducible, then so is w. Since S is explicitly defined on an irreducible word $w = Y^{(r_1)}_{i_1j_1}\ldots Y^{(r_t)}_{i_tj_t}$, we can easily calculate $\Sigma(Sw_1)w_2\mathcal{D} = \delta_{i_1j_1}\ldots\delta_{i_tj_t} = \varepsilon(w)1$, using the relations (*'). Thus S is a left antipode of H.

PROPOSITION 3.1: <u>No left antipode of</u> H <u>is an algebra antimorphism</u>.

<u>Proof</u>: The subcoalgebra D of H with basis $\{Y^{(r)}_{ij} \mid r \geq 0; 1 \leq i,j \leq n\}$ is S-stable, and the restriction of S to D is a coalgebra antimorphism. Let T be a left antipode of H which is an algebra antimorphism. The argument of the proof of Proposition 2.1, applied to the convolution algebra $\mathrm{Hom}(D^{\mathrm{coop}},H)$, shows that the restrictions of S and T to D are equal. Hence applying T to the relation $\sum_{k=1}^{n} Y^{(1)}_{k1}Y^{(0)}_{k2} = 0$, we obtain $0 = \sum_{k=1}^{n} T(Y^{(0)}_{k2})T(Y^{(1)}_{k1}) = \sum_{k=1}^{n} S(Y^{(0)}_{k2})S(Y^{(1)}_{k1}) = \sum_{k=1}^{n} Y^{(1)}_{2k}Y^{(2)}_{1k}$. As the last sum is a sum of distinct irreducible words, this is a contradiction.

PROPOSITION 3.2: <u>No left antipode of</u> H <u>is a coalgebra antimorphism</u>.

<u>Proof</u>: We first note that H_ℓ, described at the beginning of this section is a homomorphic image of H. Since the generators $\{X^{(r)}_{ij} \mid r \geq 0; 1 < i, j \leq n\}$ of H_ℓ satisfy the relations (*), there is an algebra morphism P of H onto H_ℓ for which $P(Y^{(r)}_{ij}) = X^{(r)}_{ij}$ for all $r \geq 0$, $1 \leq i, j \leq n$. P is a

bialgebra morphism. Let $w = Y_{i_1j_1}^{(r_1)} \ldots Y_{i_tj_t}^{(r_t)}$ be an irreducible word in H. Then $PSw = P(Y_{j_ti_t}^{(r_t+1)} \ldots Y_{j_1i_1}^{(r_1+1)}) = X_{j_ti_t}^{(r_t+1)} \ldots X_{j_1i_1}^{(r_1+1)} = S_\ell(X_{i_1j_1}^{(r_1)} \ldots X_{i_tj_t}^{(r_t)}) = S_\ell P(Y_{i_1j_1}^{(r_1)} \ldots Y_{i_tj_t}^{(r_t)}) = S_\ell Pw$. Hence $PS = S_\ell P$, i.e., P is a left Hopf algebra morphism of (H,S) onto (H_ℓ, S_ℓ).

Let * denote the convolution product in $\mathrm{Hom}(H^{coop}, H_\ell)$. Let T be a left antipode of H which is a coalgebra antimorphism. Let $y \in H$. Then

$$(PS*PST)(y) = (S_\ell P*S_\ell PT)(y) = \Sigma(S_\ell Py_2)(S_\ell PTy_1) = S_\ell(\Sigma(PTy_1)(Py_2)\ \mathcal{D}$$
$$= S_\ell P(\Sigma(Ty_1)y_2) = S_\ell P(\varepsilon(y)1) = \varepsilon(y)1.$$

Also

$$(PST*PT)(y) = \Sigma(PSTy_2)(PTy_1) = P(\Sigma(STy_2)(Ty_1)) = P(\Sigma S((Ty)_1)(Ty)_2)$$
$$= P(\varepsilon(Ty)1) = \varepsilon(y)1.$$

Hence PS = PT.

Let $\mu: k \to H$ denote the unit of H, and $tw: H \otimes H \to H \otimes H$ be the twist mapping sending $x \otimes y$ to $y \otimes x$. Let $\pi: H \to k$ be the projection of H on $\mu(k)$ determined by our basis for H of irreducible words, i.e., $\pi(1) = 1$, and $\pi(w) = 0$ for w an irreducible word of positive length. If w is an irreducible word, we have seen that $\Delta w = \Sigma w_1 \otimes w_2$, where the w_1 are irreducible. In fact, the w_1 have the same length as w. It follows that $(\pi \otimes \mathrm{Id})\Delta = \mu\pi$, (identifying $k \otimes H$ and H), since both sides vanish on irreducible words of positive length, and both sides send 1 to 1. Let $\tau = \pi T$. If we apply $\pi \otimes \mathrm{Id}$ to $(T \otimes T)(tw)\Delta = \Delta T$, we obtain $(\tau \otimes T)(tw)\Delta = (\pi \otimes \mathrm{Id})\Delta T = \mu\pi T = \mu\pi$. Applying P (identified with $\mathrm{Id}_k \otimes P$), we obtain $(\tau \otimes PT)(tw)\Delta = P\mu\tau = \mu\tau$. Hence $(\tau \otimes PS)(tw)\Delta = \mu\tau$, i.e., if $y \in H$, then $\tau(y)1 = \Sigma(\tau y_2)(PSy_1)$. We apply this for the irreducible word $y = Y_{11}^{(1)}Y_{11}^{(0)}$. Thus $\tau(Y_{11}^{(1)}Y_{11}^{(0)})1 = (\tau \otimes PS)(tw) \sum_{p,q=1}^{n} Y_{1p}^{(1)}Y_{1q}^{(0)} \otimes Y_{p1}^{(1)}Y_{q1}^{(0)} = \sum_{p,q=1}^{n} \tau(Y_{p1}^{(1)}Y_{q1}^{(0)})X_{q1}^{(1)}X_{p1}^{(2)}$. For $p = q = n$, we use reduction formula (1) with $r = 0$ and $i = j = 1$ to obtain $\tau(Y_{11}^{(1)}Y_{11}^{(0)})1 = \sum_{(p,q)\neq(n,n)} \tau(Y_{p1}^{(1)}Y_{q1}^{(0)})X_{q1}^{(1)}X_{p1}^{(2)} + \tau(1 - \sum_{t=1}^{n-1} Y_{t1}^{(1)}Y_{t1}^{(0)})X_{n1}^{(1)}X_{n1}^{(2)}$. In [2], a basis for H_ℓ of irreducible words in the $\{X_{ij}^{(r)} \mid r \geq 0,\ 1 \leq i, j \leq n\}$ was obtained. All the 2-letter words in the $X_{ij}^{(r)}$ appearing on the right-hand side of the above equation are irreducible. Hence $\tau(Y_{p1}^{(1)}Y_{q1}^{(0)}) = 0$ if $(p,q) \neq (n,n)$; and thus $\tau(1 - \sum_{t=1}^{n-1} Y_{t1}^{(1)}Y_{t1}^{(0)}) = 0$ implies that $\tau(1) = 0$, i.e., $\pi T(1) = 0$. But T is a left antipode of H, so that $T(1) = T(1)1 = \varepsilon(1)1 = 1$. Thus $0 = \pi T(1) = \pi(1) = 1$ and we have arrived at a contradiction.

Clearly either one of Propositions 3.1 and 3.2 will yield

PROPOSITION 3.3: H <u>is a left Hopf algebra such that no left antipode of</u> H <u>is a bialgebra antimorphism</u>.

REFERENCES

[1] G. M. Bergman, The diamond lemma for ring theory, Advances in Math. 29 (1978), 178-218.

[2] J. A. Green, W. D. Nichols and E. J. Taft, Left Hopf algebras, J. Algebra 65 (1980), 399-411.

[3] N. Jacobson, Basic Algebra I, W. H. Freeman, San Francisco, 1974.

[4] W. Nichols, Quotients of Hopf algebras, Comm. Algebra 6 (1978), 1789-1800.

[5] M. Sweedler, Hopf Algebras, Benjamin, New York, 1969.

[6] M. Takeuchi, Free Hopf algebras generated by coalgebras, J. Math. Soc. Japan 23 (1971), 561-582.

W. D. Nichols
Florida State University
Tallahassee, Fla. 32306

E. J. Taft
Rutgers University
New Brunswick, NJ 08903

Contemporary Mathematics
Volume 13, 1982

SIMILARITY OF NILPOTENT, INTEGER MATRICES, OR, FOUR ELEMENTARY CATEGORIES

Daniel Zelinsky

Throughout, R is a principal ideal domain. When we assume that R is local, then, it is a discrete valuation ring. All R-modules considered are finitely generated. The rank of a module is the rank of its torsion-free part.

1. THE FOUR CATEGORIES

Let n be a fixed positive integer. Define

C1: Square matrices T with $T^n = 0$, entries in R. A morphism from T_1 to T_2 is an invertible matrix P with $PT_1P^{-1} = T_2$.

C2: Finitely generated modules over $R[X]/X^n$ which are free over R. Morphisms are $(R[X]/X^n)$-module isomorphisms.

C3: n-step filtrations $\mathcal{F}: F_0 \supset F_1 \supset \dots \supset F_n = 0$ of R-modules with F_0 (hence every F_i) free and finitely generated. Morphisms are isomorphisms of filtrations (defined in the obvious way).

C4: (n-1)-step filtrations $G_0 \supset G_1 \supset \dots \supset G_{n-1} = 0$ of R-modules with G_0 (hence every G_i) finitely generated. Morphisms are isomorphisms of filtrations.

[We are actually only interested in the isomorphism classes of each of these categories; if we were interested in the categories themselves, it would be more natural to enlarge the sets of morphisms in the obvious ways, so that not all morphisms are isomorphisms. The functors below would remain functors.]

Finite direct sums in each category are defined in the obvious way. There are additive functors $\Lambda_{i,i+1}: Ci \to C(i+1)$:

The definition of Λ_{12} is the well known one: $\Lambda_{12}(T)$ is the free R-module on which T acts, made into an $(R[X]/X^n)$-module by letting X act as T. Every module in C2 is isomorphic to $\Lambda_{12}(T)$ for some T. Thus Λ_{12} induces a bijection from the isomorphism classes of C1 to those of C2.

0271-4132/82/0550/$03.50

Define $\Lambda_{23}(M) = \mathcal{F} : F_0 \supset \ldots \supset F_n = 0$ with $F_i = X^{-1}(0) \cap X^i(M)$. As R-module, $M \cong \oplus_i F_i$, the isomorphism depending on the choice of R-module maps $\sigma_i : X^{i+1}M \to X^iM$ which split the epimorphism $X: X^iM \to X^{i+1}M$ (it splits because $X^{i+1}M \subset M$ is R-free). The required isomorphism is deduced from $X^iM = F_i \oplus \operatorname{Im} \sigma_i$ and $\operatorname{Im} \sigma_i \cong X^{i+1}M$, by descending induction on i. It is $m \in M \mapsto \oplus_i (1 - \sigma_i \circ X)X^im$. The inverse isomorphism is defined by additivity from $f_k \in F_k \mapsto \sigma_0\sigma_1 \ldots \sigma_{k-1}(f_k)$.

To make this an $R[X]$-module isomorphism we would have to define the operation of X on $\oplus F_i$ by $\oplus F_i \to M \xrightarrow{X} M \to \oplus F_k$. The result is, for $f_k \in F_k$, $Xf_k =_{DEF} \oplus g_i$ with $g_i = 0$ for $i \geq k$; $g_{k-1} = f_k$; and $g_i = (1 - \sigma_i X)\sigma_{i+1} \ldots \sigma_{k-1}(f_k)$ for $i < k-1$. This shows that <u>every matrix</u> T <u>with</u> $T^n = 0$ <u>is similar to a matrix of</u> (<u>not necessarily square</u>) <u>blocks</u> T_{ij}, $i,j = 1, \ldots, n$ <u>with</u> $T_{ij} = 0$ <u>for</u> $i \geq j$ <u>and</u> $T_{i,i+1}$ <u>a matrix with zero kernel</u> (the matrix of the inclusion $F_{i+1} \to F_i$). See (4.1) for the case $n = 3$.

Another version of most of this calculation is this: We can construct a functor Λ_{32}: C3 $\to$ C2 sending $F_o \supset \ldots \supset F_n = 0$ to $\oplus F_k$ with X acting as the shift; X sends F_{k+1} to F_k by the inclusion map. Then $\Lambda_{23} \circ \Lambda_{32} \cong$ identity, and we have identified C3 with a special subcategory of C2. As a subcategory of C1, C3 is identified with the matrices similar to block matrices as above, but with $T_{ij} = 0$ except when $j = i+1$.

Given $\mathcal{F}$ in C3, define $G_i = F_i/F_{n-1}$, to get an object $\Lambda_{34}(\mathcal{F})$ in C4. Inversely, given any $G_0 \supset \ldots \supset G_{n-1} = 0$ in C4 and a free module F_0 and an epimorphism $\pi : F_0 \twoheadrightarrow G_0$, we can construct a filtration in C3 by defining $F_i = \pi^{-1}(G_i)$. We call this filtration a <u>presentation</u> of the object in C4. In fact C3 is equivalent to the category of diagrams $F_0 \twoheadrightarrow G_0 \supset \ldots \supset G_{n-1} = 0$ with F_0 free; Λ_{34} is just the functor which forgets $F_0 \twoheadrightarrow$. If R is local, so that every R-module G_0 has an essentially unique minimal presentation $F_0 \to G_0$, an isomorphism class in C3 is just a pair: an isomorphism class in C4 together with a nonnegative integer specifying the difference in rank between the F_0 in question and the minimal one.

If $n = 2$, then Λ_{23} and Λ_{32} provide an equivalence of categories C2 $\cong$ C3. Hence the isomorphism classes in C1, C2, C3 and C4 are easy to describe.

For C1, a canonical form is $\begin{pmatrix} 0 & A \\ 0 & 0 \end{pmatrix}$ with A a "diagonal" matrix (possibly with more rows than columns) with diagonal entries $\lambda_i \neq 0$ and with λ_i dividing λ_{i+1} for all i. This results from the preceding description as a supertriangular block matrix, together with this theorem:

(1.1) THEOREM. Two filtrations

$F_0 \supset F_1 \supset F_2 = 0$ and $F_0^* \supset F_1^* \supset F_2^* = 0$ in C3 are isomorphic if and only if $F_0/F_1 \cong F_0^*/F_1^*$ and $F_1 \cong F_1^*$ (i.e. $\mathrm{rk} F_1 = \mathrm{rk} F_1^*$).

Proof. By the "stacked basis theorem" there exist bases of F_0 and F_1 with the latter equal to scalars λ_i times the former; two filtrations are isomorphic if and only if these λ_i agree. The λ_i, of course, are the invariants of the module F_0/F_1.

2. THE CASE OF C3 WHEN n = 3.

Definitions. If $\mathcal{F} : F_0 \supset F_1 \supset F_2 \supset F_3 = 0$, define the factors of $\mathcal{F}$ to be the modules (no longer all free) $G'' = F_0/F_1$, $G' = F_1/F_2$, F_2, and $G = F_0/F_2$. [We should also include F_0 and F_1, but these are determined by their ranks, which are determined by the other factors]. Isomorphic filtrations have isomorphic factors. A filtration $\mathcal{F}$ is naively determined (nd) if the converse is true: any other filtration with factors respectively isomorphic to those of $\mathcal{F}$, is isomorphic to $\mathcal{F}$ in C3.

(2.1) THEOREM . A filtration $F_0 \supset F_1 \supset F_2 \supset F_3 = 0$ in C3 is nd if any one of the following conditions holds:

(i) $F_2 = IF_o$ for some ideal I in R.

(ii) R is local and $\mathrm{rk}\, F_1 \leq 2$.

(iii) R is local and F_1/F_2 is cyclic.

(iv) R is local and F_0/F_1 is free $\oplus$ cyclic.

The hypothesis "R local" cannot be omitted.

(2.2) Example. Let R be the ring $\mathbb{Z}$ of integers,

$F_0 = \mathbb{Z}u \oplus \mathbb{Z}v$, $F_2 = \mathbb{Z}(8u)$, $F_1(a) = \mathbb{Z}(au + 8v) \oplus \mathbb{Z}(8u)$, with a variable but odd. Then $G = F_0/F_2 \cong \mathbb{Z}/8\mathbb{Z} \oplus \mathbb{Z}$, $G'(a) = F_1(a)/F_2 \cong \mathbb{Z}$, $G''(a) = F_0/F_1(a) = \mathrm{coker} \begin{pmatrix} a & 8 \\ 8 & 0 \end{pmatrix} \cong \mathbb{Z}/64\mathbb{Z}$.

Except for the fact that R is not local, (ii), (iii) and (iv) hold, and for any two odd a's the two filtrations have isomorphic facotrs. However, thetwo filtrations are not isomorphic: No automorphism of F_0 leaving F_2 stable can carry $u + 8v$ in $F_1(1)$ into $F_1(3)$.

Example (2.2) also gives two filtrations which are locally isomorphic because local automorphisms of F_0 include $\begin{pmatrix} c & b \\ 0 & d \end{pmatrix}$ with arbitrary local units c and d, and this can carry $F_1(1)$ to $F_1(3)$. This shows that a Hasse principle does not hold in C3; filtrations which are locally isomorphic need not be isomorphic.

Example (2.2) also shows that, when R is not local, filtrations in C3 may not be isomorphic, but after applying Λ_{34} we get isomorphic objects in C4. Here $G \supset G'(1)$ and $G \supset G'(3)$ in C4 are isomorphic by the Localization Lemma, (3.3).

Before we prove (2.1), here is an example of a filtration which is not nd, where each hypothesis (ii), (iii), (iv) just barely fails.

(2.3) Example. R is local with maximal ideal pR,

$$F_0 = Ru_1 \oplus Ru_2 \oplus Ru_3, \quad F_2 = p^4Ru_1 \oplus p^{10}Ru_2 \oplus p^{15}Ru_3, \quad \text{and}$$

$$F_1(a) = Rv(a) \oplus p^6Ru_2 \oplus p^{11}Ru_3 \quad \text{with} \quad v(a) = u_1 + p^au_2 + p^7u_3 \quad \text{and}$$

$a \geq 2$ (to have $F_1 \supset F_2$). Computations show the invariant factors of the inclusion $A(a):F_1(a) \to F_0$ are independent of a. But the isomorphism classes of the filtrations are not. No automorphism of F_0 leaving F_2 stable can carry $v_1(3)$ to a linear combination of $v_1(4)$, p^6u_2 and $p^{11}u_3$.

Now to prove (2.1). Assume $\mathcal{F}$ satisfies (i). If $\mathcal{F}^*$ has the same factors as $\mathcal{F}$, then $F_0^*/F_2^* \cong F_0/F_2 \cong (R/I)^n$, so $F_2^* = IF_0^*$. Hence any isomorphism of F_0 to F_0^* will carry F_2 to F_2^*. To show $\mathcal{F} \cong \mathcal{F}^*$ we need only produce an isomorphism of $F_0 \supset F_1 \supset 0$ to $F_0^* \supset F_1^* \supset 0$, which is possible by (1.1).

The rest of the proof is carried out after the following reduction to the case where G'' is torsion.

(2.4) DECOMPOSITION LEMMA: *Every $\mathcal{F}$: $F_0 \supset F_1 \supset F_2 \supset 0$ in C3 is isomorphic to $\mathcal{G} \oplus \mathcal{H}$ where $\mathcal{G}$: $F_0' \supset F_1 \supset F_2 \supset 0$ with F_0'/F_1 a torsion module, and $\mathcal{H}$: $S \supset 0 = 0 = 0$ with S free. $\mathcal{F}$ is nd if $\mathcal{G}$ is.*

Proof. Let $F_0/F_1 = S \oplus T$ with S free and T torsion. The epimorphism $\pi:F_0 \to F_0/F_1 \to S$ splits, so $F_0 = F_0' \oplus S$ with $F_0' = \operatorname{Ker} \pi \supset F_1 \supset F_2$, and $F_0'/F_1 = T$. The factors of $\mathcal{F}$ then determine the factors, S and 0, of $\mathcal{H}$, hence also the factors of $\mathcal{G}$ (by the cancellation theorem which follows from the structure theorem for R-modules). The isomorphism classes of the factors of H determine the isomorphism class of H; if G is nd then the factors of F determine both G and H and, therefore, F.

Henceforth, we assume F_0/F_1 is a torsion module, that is, $\operatorname{rk} F_0 = \operatorname{rk} F_1$, so the matrix A of the inclusion $F_1 \to F_0$ is square. We shall also use the matrices D and B of the inclusions $F_2 \to F_0$ and $F_2 \to F_1$ (so D = AB). By (1.1) applied to $F_0 \supset F_2 \supset 0$, any filtration with the same factors as F is isomorphic to a filtration $F_0 \supset F_1' \supset F_2 \supset 0$ with the same D as F, but with different A and B. So we hold D fixed, and ask "Where can A be carried by automorphisms of $F_0 \supset F_2$ and isomorphisms of F_1?" The answer is: to PAQ where P and Q are invertible and P leaves the image of D stable. We shall keep D in diagonal form, $D = (\delta_{ij}\lambda_j)$, $i = 1, \ldots, s$, $j = 1, \ldots, m$, $s \geq m$, each $\lambda_j \neq 0$, and λ_j divides λ_{j+1} for all j. The condition on P then becomes: the J,j entry in P is divisible by λ_J/λ_j whenever $J > j$. We shall only use the dilations and transvections of this form (they in fact generate the automorphism group of $F_0 \supset F_2$, but we do not need this fact), so we shall allow the following manipulations on the matrix A:

(2.5a) multiply one row of A by a unit

(2.5b) (add up) add r times the Jth row to the jth $(J > j)$

(2.5c) (add down) add $(\lambda_J/\lambda_j)r$ times the jth row to the Jth $(J > j)$

(2.5d) arbitrary elementary manipulations of columns.

If D is not square $(s \neq m)$ then λ_J/λ_j in (2.5c) is to mean 0 if $J > m > j$ and 1 if $J > j > m$.

It is then clear that F is nd if elementary manipulations (2.5) can convert A to a matrix which is determined by the isomorphism classes of the factors of F, that is, by the sizes of the matrices A and D and the invariant factors of A, B and D.

We use this strategy with the hypothesis (ii) in (2.1). The case $\operatorname{rk} F_1 = 1$ is trivial. So we assume $\operatorname{rk} F_1 = 2$, that is A is 2 by 2. If $\operatorname{rk} F_2 = 0$ we are back in (i), so assume $\operatorname{rk} F_2 = 1$ or 2. In the former case, $D = \begin{pmatrix}\lambda\\0\end{pmatrix}$. Suppose we can bring A into diagonal form $\begin{pmatrix}a & 0\\0 & b\end{pmatrix}$ by manipulations (2.5). Then $B = \begin{pmatrix}\lambda/a\\0\end{pmatrix}$. The invariant factors of D (viz. λ) and of B (viz. λ/a) determine a. The invariant factors of A determine its determinant ab, hence also b.

If $\operatorname{rk} F_2 = 1$ but A cannot be brought into diagonal form, by column manipulations we can always arrange $A = \begin{pmatrix}a & c\\0 & b\end{pmatrix}$. If A cannot be diagonalized, then a does not divide c; <u>we write</u> $c \parallel a$ <u>for this relation</u>; in the valuation ring R this means that the value of c is strictly less than the value of a. Similarly, adding up would make $c = 0$ unless $c \| b$. Use unrestricted manipulations to compute that the invariant factors of A are c and ab/c and that $[c \| b \| b(a/c)$, so$]$ c is the smaller one. Hence the set of invariant factors of A determines c and ab/c; the invariant factors of B and D determine a, so b is also determined.

We also need to know that the invariant factors of A, B and D specify whether or not A is diagonalizable; that is, that the two cases just considered do not overlap. The inequalities on a, b, c do show that in the second case, no invariant factor of A times the invariant factor of B will be λ.

If $\operatorname{rk} F_2 = 2$, then $D = \begin{pmatrix} \lambda_1 & 0 \\ 0 & \lambda_2 \end{pmatrix}$ and a similar argument shows that either A is diagonalizable with diagonal entries determined by the invariant factors of A, B and D; or A can be carried to $\begin{pmatrix} a & c \\ 0 & b \end{pmatrix}$ with $c \| a$, $c \| b$, $b\lambda_1 \| c\lambda_2$.

The invariant factors c and ab/c of A and the invariant factors λ_1/a, λ_2/b of B and the invariant factors λ_1, λ_2 of D determine a,b and c. Furthermore, the cases do not overlap.

To prove (2.1) assuming (iii), focus first on B. Since F_1/F_2 is cyclic, $\operatorname{rk} F_1 - \operatorname{rk} F_2$ is 1 or 0, so B is either m by $m-1$ or m by m. The invariant factors of B are all 1 in the former case, and $1,1,\ldots,b$ in the latter. In either case, the g.c.d. of the $(m-1)$-rowed minors is a unit. Now B can be manipulated by operations which are the duals of (2.5); in particular, arbitrary row operations are allowed. Permute rows to arrange that this invertible $(m-1)$-rowed submatrix Y occurs in the top $m-1$ rows, then left multiply by $\begin{pmatrix} Y^{-1} & 0 \\ 0 & 1 \end{pmatrix}$ to convert B into $\begin{pmatrix} I \\ 0 \end{pmatrix}$. Then $AB = D$ forces $A = (a_{ij})$ with $a_{ii} = \lambda_i$ for $i = 1, \ldots, m-1$, a_{im} possibly not zero; all other entries are zero. We can then manipulate A as in (ii) to arrange those a_{im} which are nonzero satisfy

$$a_{jm} \| a_{Jm} \| a_{mm} \quad (j < J)$$

$$\lambda_j a_{jm} \| \lambda_J a_{jm} \quad (j < J)$$

$$a_j \| \lambda_j$$

We then compute the invariant factors of A and B by using unrestricted manipulations. An induction proves that the invariant factors of A (not necessarily in increasing order) are $\mu_1, \ldots, \mu_m$ with $\mu_J = \lambda_J$ if $a_{Jm} = 0$, $\mu_J = a_{Jm}(\lambda_j/a_{jm})$ if $a_{Jm} \neq 0$ where a_{jm} is the nonzero a_{jm} just preceding a_{Jm}. If we delete the μ's which are equal to λ's, the remaining ones, listed in increasing order, determine the nonzero a_{Jm} and their location in the matrix A. Thus, the μ's and the λ's determine A.

If B is m by m, multiply on the left again by $\begin{pmatrix} Y^{-1} & 0 \\ 0 & 1 \end{pmatrix}$ to get an (m-1)-rowed identity matrix in the top m-1 rows with one column (say the sth) still untamed. Interchanging rows, we can get

$$B = \begin{pmatrix} I_{s-1} & b & 0 \\ c & b_s & d \\ 0 & e & I_{m-s} \end{pmatrix}$$

with $(c\ b_s\ d)$ a single row and $(b\ b_s\ e)$ a single column. By further row operations and adding left we can make $c = d = e = 0$ and arrange that those entries b_i of b which are not zero satisfy $b_J \| b_j \| b_s$ $(s > J > j)$ and $b_j\lambda_j \| b_J\lambda_J$ $(J > j)$. Then $A = DB^{-1}$ is in exactly the same form as in the case when B was not square, but the exceptional column is now the sth instead of the mth. The sth row has entries $a_1, \ldots, a_{s-1}, a_s, 0, \ldots, 0$ with the a's satisfying the same inequalities as before. The rest of the argument is the same.

For (iv) we use precisely the same argument as in (iii) with the roles of A and B reversed.

3. THE CATEGORY C4 FOR n = 3.

Establish this notation (to jibe with section 2): For $G_0 \supset G_1 \supset G_2 = 0$, write $G \supset G' \supset 0$. The object in C4 is just a monomorphism $\alpha: G' \to G$. The <u>factors</u> of α are the modules G, G' and G" = coker α. We say α is <u>naively determined</u> (<u>nd</u>) if the isomorphism class of α is determined by the isomorphism classes of the factors of α.

[This category C4 when n = 3 is equivalent to the cateogry of monomorphisms of R-modules, or to the category of epimorphisms, or to the category of short exact sequences, with isomorphism of morphisms or sequences defined in the obvious ways. The isomorphism classes in C4 are not isomorphism classes of extensions as in $\mathrm{Ext}^1_R(G'', G')$.]

(3.1) THEOREM. <u>An</u> α <u>in</u> C4 <u>is</u> <u>nd</u> <u>if any one of the following holds</u>.

(i) G <u>is free over</u> R/I <u>for some ideal</u> I;

(ii) G <u>can be generated by</u> 2 + rkG" <u>elements</u> (e.g., G" <u>is torsion and</u> G <u>has two generators</u>);

(iii) G' <u>is cyclic</u>;

(iv) G" <u>is cyclic</u> $\oplus$ <u>free</u>.

Proof. As in section 1, every α has a presentation, a filtration $F_0 \supset F_1 \supset F_2 \supset 0$ in C3 with $F_0/F_1 \cong G''$, $F_1/F_2 \cong G'$ and $F_0/F_2 \cong G$. It is clear that

(3.2) α is nd if some presentation of it is nd.

Hence (2.1i) implies (3.1i). For the other parts, we need to reduce (3.1) to the local case (if p is a prime of R we use subscript p to denote $\otimes_R R_p$):

(3.3) LOCALIZATION LEMMA. Suppose $\mathrm{rk}\, G \leq 1$ and α and α^* are objects in C4. If $\alpha_p \cong \alpha^*_p$ for all p, then $\alpha \cong \alpha^*$.

Proof. We are assuming the existence of local isomorphisms

$$\theta p' : G'_p \to G^{*\prime}_p \quad \text{and} \quad \theta p : G_p \to G^*_p \quad \text{with}$$

(3.4) $$(\theta p)\alpha_p = \alpha^*_p(\theta p').$$

We want global isomorphisms $\theta' : G' \to G^{*\prime}$ and $\theta : G \to G^*$ with $\theta\alpha = \alpha^*\theta'$. It suffices to find θ, θ' with $\theta_p = \theta p$ and $\theta'_p = \theta p'$ for all p.

Decompose G' into $T' \oplus S'$ with T' torsion and S' free; and $G = T \oplus S$, and similarly for $G^{*\prime}$ and G^*. Then $\alpha : G' \to G$ can be described by the matrix $\begin{pmatrix} \alpha_T & \alpha_{ST} \\ 0 & \alpha_S \end{pmatrix}$ with $\alpha_T : T' \to T$ $\alpha_{ST} : S' \to T$, $\alpha_S : S' \to S$.

Similarly, $\theta = \begin{pmatrix} \theta_T & \theta_{ST} \\ 0 & \alpha_S \end{pmatrix}$ and the same for θ', θp and $\theta p'$. The diagonal components are monomorphisms, and are isomorphisms if the whole matrix represents an isomorphism. Equation (3.4) translates into

(3.5a) $$\theta p_T \alpha_{Tp} = \alpha^*_{Tp} \quad \theta p'_T$$

(3.5b) $$\theta p_S \, \alpha_{Sp} = \alpha^*_{Sp} \quad \theta p'_S$$

(3.5c) $$\theta p_{ST}\alpha_T + \theta p_S \, \alpha_{ST} = \alpha^*_{ST} \quad \theta p'_T + \alpha^*_S \, \theta p'_{ST}$$

The hypothesis $\mathrm{rk}\, G \leq 1$ implies that all the free modules S, S', S^*, S'^* have rank 1 or 0. Then each item in (3.5b) is a nonzero scalar in R_p, and θp_S is a unit. Since scalars commute with everything, we can multiply all components of θp and $\theta p'$ by θp_S^{-1} to get new θp and $\theta p'$ satisfying (3.5) but now with $\theta p_S = 1$. Since α_S and α^*_S are nonzero elements of R, α_S/α^*_S is an element of the field of fractions of R which, by the new (3.5b), agrees with every $\theta p'_S$. That is, $\theta p'_S$ is independent of p and is a unit in R_p for every p, hence is a unit, θ'_S in R.

Since T and T^* are torsion modules we can write $T = \oplus_p T_p$ and $T^* = \oplus_p T^*_p$ and define an isomorphism $\theta_T: T \to T^*$ to be $\oplus_p \theta p_T$. We get θ'_T, θ_{ST} and θ'_{ST} similarly, and put them together with $\theta_S = 1$ and θ'_S as just constructed, to get the required isomorphisms θ and θ' with $\theta_p = \theta p$ and $\theta'_p = \theta p'$ for all p.

(3.6) <u>Corollary</u>. If $\operatorname{rk} G \leq 1$ and α_p is nd for all p, then α is nd.

(3.7) <u>Lemma</u>. If $\alpha = \alpha_1 \oplus \alpha_2$ is a natural splitting in the sense that every object in C4 with the same factors as α decomposes: $\alpha^* = \alpha_1^* \oplus \alpha_2^*$, with the factors of α_1^* isomorphic to those of α_1; and if α_1 and α_2 are nd, then α is nd.

<u>Proof</u>. The cancellation theorem which follows from the fundamental structure theorem for R-modules asserts that the factors of α_2 will then be isomorphic to those of α_2^*, so if α_1 and α_2 are nd, then $\alpha_1 \cong \alpha_1^*$, $\alpha_2 \cong \alpha_2^*$, and so $\alpha \cong \alpha^*$.

We can now complete the proof of (3.1). Assume (3.1ii). By the construction in the decomposition lemma (2.4) we can write $\alpha = \alpha_1 \oplus \alpha_2$ with $\alpha_1: 0 \to S =$ torsion free part of G'', so the factors of α_1 are determined by G''. Moreover, α_1 is clearly nd, so by (3.7) it suffices to show that α_2 is nd. The rank of the cokernel of α_2 is 0, so by (3.1ii), the rank of the range G_2 of α_2 is at most 2; if it equals 2, then G_2 is free and we are back in case (i). If $\operatorname{rk} G_2 \leq 1$, then α_2 is nd by (3.6), (2.1ii), and (3.2).

If we assume (3.1iii), then, as in (ii) we reduce to the case where G'' is torsion, so $\operatorname{rk} G \leq \operatorname{rk} G' + \operatorname{rk} G'' \leq 1$, and again (2.1iii), (3.2) and (3.6) do the trick.

If we assume (3.1iv), the same decomposition, via (2.4) and (3.7), reduces to the case where G'' is torsion and cyclic. A second decomposition is required to achieve $\operatorname{rk} G \leq 1$. Let $G = S \oplus T$, $G' = S' \oplus T'$ with S and S' free, T and T' torsion. Then α induces a monomorphism $\alpha_S: S' \to S$ whose cokernel S'' is a homomorphic image of G'', hence is cyclic. The stacked basis theorem says that $\alpha_S = \alpha_1 \oplus \alpha_2$ with $\alpha_1: S_1' \to S_1$ an isomorphism and $\operatorname{rk}(\text{range } \alpha_2) = 1$. Lift this decomposition to a decomposition of α by choosing any splitting $\rho': S_1' \to G'$ of the natural epimorphism $G' \to S' \to S_1'$ and defining $\rho: S_1 \to G$ to be $\alpha \rho' \alpha_1^{-1}$, which will be a splitting of $G \to S \to S_1$. Then $\alpha = \alpha_1 \oplus \overline{\alpha}_2$. The cokernel of α_1 is 0; its range and domain are free of rank $\operatorname{rk} G - 1$, so all of these are determined by the factors of α. Then

(3.7), (3.6), (3.2) and (2.1iv) complete the proof.

4. To end where we began, we translate (2.1) into A THEOREM ABOUT Cl.

(4.1) THEOREM. *Every matrix* T *with entries in* R *and with* $T^3 = 0$ *is similar to* $\begin{pmatrix} 0 & A & C \\ 0 & 0 & B \\ 0 & 0 & 0 \end{pmatrix}$ *where* A *and* B *are not necessarily square, but have zero kernels. If we can choose* C = 0, *then the similarity class of* T *is determined by the invariant factors of* A, B *and* AB *provided*

(i) *the invariant factors of* A *are all equal,*

or (ii) A *has* < 2 *columns;*

or (iiia) B *is square and has invariant factors* 1,1, ..., 1, b;

or (iiib) B *has one more row than columns and has all invariant factors* 1;

or (iv) A *has invariant factors* 1,1, ..., 1, a

In these cases a canonical form is obtained by taking A and B in the forms developed in the proof of (2.1) (and C = 0). If R is a field this is the Jordan form with some permutation of rows and columns.

(By a simple extension of a theorem of Gustafson [Linear Algebra and its Applications, 23(1979), 245-251] we can make C = 0 if and only if C = AX + YB for some X and Y.)

NORTHWESTERN UNIVERSITY
EVANSTON, ILLINOIS 60201

Contemporary Mathematics
Volume 13, 1982

NORMAL ANTILINEAR OPERATORS ON A HILBERT SPACE

by

Irving Kaplansky

To Nathan Jacobson

An additive map S on a complex vector space H is _antilinear if_ $S(\lambda x) = (\lambda^*)Sx$ for any $x \in H$ and scalar λ (where λ^* is the complex conjugate of λ). Other names in use are _conjugate linear_ and _semilinear_, the latter applying in greater generality. One comprehensive reference concerning general semilinear transformations is Chapter 3 of [3].

Now let H be a Hilbert space, infinite-dimensional to make it interesting. Assume that S is antilinear and continuous. It has a unique adjoint S^* satisfying $(Sx, y) = (x, S^*y)^*$ for all $x, y \in H$. As usual, S is _normal_ if $SS^* = S^*S$.

The purpose of this note is to prove the following theorem.

THEOREM. _Let S and T be continuous normal antilinear operators on a complex Hilbert space. Then S and T are unitarily equivalent if and only if S^2 and T^2 are unitarily equivalent._

Note that S^2 and T^2 are linear (as well as normal); thus the theorem accomplishes a reduction of the problem to a known one.

In the finite-dimensional case the theorem is due to Jacobson [2, th. 11]. The self-adjoint (finite-dimensional) case was rediscovered by Schur [5] and used in the study of the Bieberbach conjecture [1].

The finite-dimensional methods do not seem to adapt readily to the infinite-dimensional case. I use here a different method: a translation of the problem into real Hilbert space terms.

Let H be demoted to a real Hilbert space by forgetting the action of non-real scalars. However, the complex structure will be kept alive by writing J for what used to be multiplication by i; note that J is a skew operator ($J^* = -J$) satisfying $J^2 = -I$. If S is antilinear on H as a complex Hilbert space then S anticommutes with J : $SJ = -JS$. For later use we note the consequence $S \sim -S$, where we are writing $\sim$ for orthogonal

0271-4132/82/0551/$02.75

equivalence; to see this, observe that $JSJ^* = -S$ and that J is orthogonal ($JJ^* = J^*J = I$).

The theorem above now gets reformulated as follows.

THEOREM'. *Let S and T be normal operators on a real Hilbert space. Let J be a skew operator satisfying $J^2 = -I$. Assume that S and T anti-commute with J and that $S^2 \sim T^2$. Then $S \sim T$ via an operator commuting with J.*

The proof of the modified theorem will be split into a number of lemmas. The first of these lemmas collects some facts, which we shall use several times, about a normal operator on a real Hilbert space. Since these facts are a ready corollary of spectral theory, real style, and since the reader probably has his or her own favorite version of spectral theory, I state the lemma without proof.

LEMMA 1. *Let A be a normal operator on a real Hilbert space H. Then there exists an orthogonal direct sum decomposition $H = N \oplus \Sigma K_i \oplus \Sigma L_j$ with the following properties:*

(a) *N is the null space of A,*

(b) *A is invertible on each K_i and L_j,*

(c) *Each K_i and L_j is invariant under any operator in the double commuter of A (the set of all operators commuting with every operator commuting with A),*

(d) *A is negative definite on each K_i,*

(e) *On each L_j, A has a square root lying in the double commuter of A.*

A recurring theme will be the orthogonal equivalence of skew operators S and T when assumptions are made, above all on S^2 and T^2. The primordial case appears in Lemma 2. I give a proof but note that this proof will be essentially repeated in a more general context in Lemma 9.

LEMMA 2. *Let S and T be skew operators on a real Hilbert space H. Assume that $S^2 = T^2 = -I$. Then $S \sim T$.*

Proof. There is in fact a simple canonical form, say for S. We can find an orthonormal basis for H consisting of sets $\{x_j\}$, $\{y_i\}$ with $Sx_i = y_i$, $Sy_i = -x_i$ for all i. Observe first that $(Sx,x) = 0$ for all x. Take x_1 with $\|x_1\| = 1$ and $y_1 = Sx_1$. Then $\|y_1\| = 1$, x_1 and y_1 are orthogonal, and $Sy_1 = -x_1$. The subspace spanned by x_1 and y_1 is invariant under S; one repeats the procedure in the orthogonal complement of this subspace; and so on ad transfinitum.

When the skew operators of Lemma 2 anticommute with a suitable operator, the desired orthogonal equivalence becomes an exercise in ring theory, with no transfinite argument needed. We present this in two steps.

LEMMA 3. *Let* R *be a ring with unit element and involution* $*$. *Let* e *be a projection (self-adjoint idempotent) in* R. *Let* s *and* t *be skew elements with square* -1. *Assume*

(1) $$es + se = s, \quad et + te = t.$$

Then $s \sim t$ *via an orthogonal element commuting with* e.

Proof. By multiplying (1) on the left and right by e we find

(2) $$ese = ete = 0.$$

Multiply the first of the equations in (1) on the left by s and use $s^2 = -1$. We obtain the first of the equations

(3) $$ses = e - 1, \quad tet = e - 1,$$

the second being of course analogous. Write $p = e - set$. Routine verifications using (2) and (3) show that $ep = pe = e$, $pp^* = p^*p = 1$, and $p^*sp = t$.

LEMMA 4. *Let* R *be a ring with unit element and involution* $*$. *Assume that* 2 *is invertible in* R. *Let* s *and* t *be skew elements with square* -1. *Let* x *be a self-adjoint element with square* 1. *Assume that* s *and* t *anticommute with* x. *Then* $s \sim t$ *via an orthogonal element commuting with* x.

Proof. Set $e = (1 + x)/2$. Then e is a projection and (1) is satisfied. Thus Lemma 3 is applicable.

We return to the context of a real Hilbert space and we shall show in Lemma 6 that the element x of Lemma 4 can be generalized to any normal operator Z. A special argument is needed when Z is skew.

LEMMA 5. *Let* S, T, *and* Z *be skew operators on a real Hilbert space. Assume that* $S^2 = T^2 = -I$, *that* Z *is invertible, and that* S *and* T *anticommute with* Z. *Then* $S \sim T$ *via an orthogonal operator commuting with* Z.

Proof. We work in the weakly closed algebra R generated by S, T, and Z. It follows from our hypotheses that Z^2 and $ST + TS$ are central in R. By taking appropriate direct summands of R we can assume that $ST + TS + 2I$ is either 0 or invertible.

(a) $ST + TS + 2I = 0$. Since $\|S\| = \|T\| = 1$ and $(ST)^* = TS$, this equation can hold only when $ST = TS = -I$, that is $T = -S^{-1} = S$. Then $S \sim T$ can be implemented by I.

(b) $ST + TS + 2I$ invertible. Write $Y = Z(I + ST)$. We find $YY^* = Y^*Y = Z^2(ST + TS + 2I)$ and $SY = YT$. Let Q be the positive square root of YY^*. Then YQ^{-1} is orthogonal, commutes with Z, and implements $S \sim T$.

LEMMA 6. Let Z be a normal operator on a real Hilbert space. Let S and T be skew operators with square $-I$. Assume that S and T anticommute with Z. Then $S \sim T$ via an orthogonal operator commuting with Z.

Proof. We apply Lemma 1 with Z^2 playing the role of A. The problem is reduced to the analogous one on one of the summands N, K_i, L_j. There are thus three cases to consider. If $Z = 0$ we cite Lemma 2. If Z is skew and invertible, Lemma 5 is applicable. Finally, we suppose that Z is invertible and that Z^2 has a square root W lying in the double commuter of Z^2. We work in the commuting algebra R of Z^2. Write $X = ZW^{-1}$. Then $X \in R$, X is normal, and $X^2 = I$ whence X is self-adjoint. Also, S and T lie in R and they anticommute with X since they commute with W. The hypotheses of Lemma 4 are fulfilled, so that we have $S \sim T$ via an orthogonal operator in R commuting with X. Since the orthogonal operator in question lies in R it commutes with Z^2 and hence with W. Thus it commutes with $Z = XW$.

We perform a small transformation on Lemma 6 in order to put it in the form we actually need.

LEMMA 7. Let S and T be orthogonally equivalent normal operators on a real Hilbert space. Let J be a skew operator with square $-I$, anti-commuting with both S and T. Then $S \sim T$ via an orthogonal operator commuting with J.

Proof. Say $P^{-1}TP = S$, P orthogonal. Write $PJP^{-1} = J_1$. The hypotheses of Lemma 6 are fulfilled, with S, T, and Z replaced by J, J_1, and T, respectively. Hence there exists an orthogonal Q with $QJ_1Q^{-1} = J$ and $QT = TQ$. Then $P^{-1}Q^{-1}TQP = S$ and $JQP = QJ_1P = QPJ$. Thus QP is the desired orthogonal operator.

The next lemma is a generalization of Lemma 2 in which the hypothesis $S^2 = T^2 = -I$ is weakened to $S^2 = T^2$.

LEMMA 8. *Let* S *and* T *be skew operators on a real Hilbert space. Then* $S^2 = T^2$ *implies* $S \sim T$.

Proof. By the usual kind of reduction (a portion of Lemma 1) we can assume that S is invertible. Let R be the commuting algebra of S^2 and let Y be the positive square root of $-S^2$. We have that SY^{-1} and TY^{-1} are skew with square $-I$. It will suffice to prove $SY^{-1} \sim TY^{-1}$ via an orthogonal operator in R. So our problem is just that of Lemma 2, with the modification that we must work in the algebra R, rather than in the algebra of all bounded operators. Now R is a real W*-algebra. The fact that R is the commuting algebra of a self-adjoint operator has the further consequence that it is of type I and has the property that if e is any abelian projection in R then $*$ is the identity on eRe. At this point I give an algebraic formulation of the rest of the argument, in the style of [4]; a reader who prefers to avoid this should have no difficulty substituting a direct argument.

LEMMA 9. *Let* R *be a Baer* **-ring satisfying the* EP *and* SR *axioms. Assume that* R *is of Type* I *and that* $*$ *is the identity on* eRe *for any abelian projection* e *in* R. *Let* s *and* t *be skew elements with square* -1. *Then* $s \sim t$ *in* R.

Proof. We repeat the idea in the proof of Lemma 2, showing that s can be put in the form

$$\begin{pmatrix} 0 & 1 \\ -1 & 0 \end{pmatrix}$$

relative to a suitable set of two by two *-matrix units. Since the same is true for t, and such matrix representations of R are unique up to orthogonal equivalence, $s \sim t$ follows.

Take a faithful abelian projection e in R. The skew element ese is 0 since $*$ is the identity on eRe. We have $es \cdot s^*e = e$ and thus $f = s^*e \cdot es$ is a projection equivalent to e. They are orthogonal projections, since $fe = s^*ese = 0$. So in the corner algebra $(e + f)R(e + f)$, the element s has the desired form. Note that s commutes with $e + f$. The process may be repeated in the orthogonal corner $(1 - e - f)R(1 - e - f)$ and continued transfinitely. Lemma 9 is thereby proved, and with it Lemma 8.

LEMMA 10. <u>Let S and T be normal operators on a real Hilbert space.</u> Assume that $S^2 \sim T^2$, $S \sim -S$, and $T \sim -T$. Then $S \sim T$.

<u>Proof</u>. By changing (say T) by the orthogonal operator that sends S^2 into T^2 we can assume that S^2 and T^2 are actually equal. We apply Lemma 1 with $A = S^2$. As in the proof of Lemma 6, there are three cases. In the first case $S = T = 0$ and there is nothing to prove. In the second case S^2 is negative definite; then S and T are skew and Lemma 8 applies. In the remaining case S is invertible and S^2 has a square root W lying in the double commuter of S^2. Let R be the commuting algebra of S^2. Write $S_1 = SW^{-1}$, $T_1 = TW^{-1}$. The orthogonal operator sending S into $-S$ commutes with S^2 and hence commutes with W. Therefore $S_1 \sim -S_1$ in R and likewise $T_1 \sim -T_1$ in R. We have $S_1^2 = T_1^2 = I$. If we prove $S_1 \sim T_1$ via an orthogonal operator in R, then $S \sim T$ will follow since W lies in the center of R. Thus our original problem has been improved by having $S_1^2 = T_1^2 = I$ but we must pay the price of working in R.

The task turns out to be an easy one. As was the case in the proof of Lemma 9, the question is the uniqueness up to orthogonal equivalence of an orthogonal decomposition of I into two equivalent projections. The element $(I + S_1)/2$ is an idempotent and it is self-adjoint since S_1 is self-adjoint (S_1 is normal with square equal to I). Thus $I = (I + S_1)/2 + (I - S_1)/2$ is an orthogonal decomposition into equivalent projections. The same is true with S_1 replaced by T_1. One then knows that $(I + S_1)/2 \sim (I + T_1)/2$ in R, whence $S_1 \sim T_1$ in R, as required.

With this the proof of Lemma 10 is complete, and Theorem' is an immediate corollary of Lemmas 7 and 10.

<u>Remarks</u>. 1. Lemma 10 is valid, as it stands, for operators on a complex Hilbert space. The proof is simpler because skew operators do not need special attention.

2. Lemma 7 can be regarded as yet another structure theorem for normal antilinear operators on a complex Hilbert space. In effect it says that such an operator is determined by its structure on the induced real Hilbert space.

3. This remark is concerned with the finite-dimensional case. There is a parallel theory where the Hilbert space is replaced by a symmetric inner product space. The adjoint is again defined by $(Sx, y) = (x, S^*y)^*$. Suppose that S and T are both self-adjoint, or both skew, or both orthogonal (as well as antilinear). In the first two cases assume further that S and T

are invertible (this cannot be deleted). Then S and T are orthogonally equivalent if and only if S^2 and T^2 are orthogonally equivalent (which is in turn equivalent to the formally weaker statement that S^2 and T^2 are similar).

4. Antilinear operators on a Hilbert space occur with some frequency in the literature of mathematical physics. A sample reference is [6].

REFERENCES

1. P. R. Garabedian, G. G. Ross, and M. M. Schiffer, On the Bieberbach conjecture for even n, J. of Math. and Mechanics 14(1965), 975-989.

2. N. Jacobson, Normal semi-linear transformations, Amer. J. of Math. 61 (1939), 45-58.

3. __________, The Theory of Rings, Math. Surveys no. 2, Amer. Math. Soc. 1943.

4. I. Kaplansky, Rings of Operators, Benjamin, 1968.

5. I. Schur, Ein Satz über quadratische Formen mit komplexen Koeffizienten, Amer. J. of Math. 67 (1945), 472-480.

6. B. Simon, Quantum dynamics: from automorphism to Hamiltonian, pages 327-349 in Studies in Mathematical Physics: Essays in honor of Valentine Bargmann, edited by E. Lieb, B. Simon, and A. Wightman, Princeton, 1976.

University of Chicago
Chicago, IL 60637

P.S. An application of the Jacobson-Schur theorem in physics appears in Neutral Lepton Mass Matrix by Harvey, Ramond, and Reiss, High Energy Physics Conference 1980, pp. 451-4, published by the American Institute of Physics, 1981.

Contemporary Mathematics
Volume 13, 1982

WHY COMMUTATIVE DIAGRAMS COINCIDE WITH EQUIVALENT PROOFS PRESENTED IN HONOR OF NATHAN JACOBSON

by Saunders Mac Lane

1. INTRODUCTION. Proof theory, in the hands of Gerhard Gentzen and his followers, has established a variety of consistency proofs for number theory and other parts of mathematics; the techniques usually involve reducing a formal proof to an equivalent "normal form" and showing that any formal proof can be rewritten so as to eliminate the use of the inference "cut": The inference, from p and from $p \supset q$ to obtain q. In fact, such a "cut elimination" theorem was established in an early paper [6] by Gentzen, and has been reproved in varying contexts by many proof theorists since.

Category theory, on the other hand, has reduced many mathematical statements to visual form as "commutative" diagrams of morphisms. In categories-with-structure, the structure often involves canonical morphisms (such as associativity) and one hopes that "all" diagrams built from such canonical morphisms will be commutative; an early theorem of this sort[10] I called a "coherence" theorem. Then it was Lambek[8] - knowing both logic and category theory - who recognized that the cut-elimination theorem from logic could be employed to good advantage in establishing more difficult coherence theorems for categories with structure. For some time, this overlap between proof theory and category theory seemed partial and therefore mysterious. Recent work by the Soviet logician G. E. Minc, by Miguel LaPlaza, and others has considerably clarified this connection. The present paper is intended to report on this clarification, which rests in part on a considerable correspondence between Minc and Mac Lane. Though the specific topics involved do not lie in the direct line of Nathan Jacobson's research, their inclusion in this conference may seem appropriate because of Jacobson's long interest in the full scope of algebra, as illustrated, for example, in his use of categorical ideas in his treatment of universal algebra in his recent text, Basic Algebra II.

2. MONOIDAL CATEOGRIES. A monoidal category M is a category equipped with a bifunctor $\otimes : M \times M \to M$ which is associative up to a natural isomorphism

$$a = a_{A,B,C} \; : \; A \otimes (B \otimes C) \to (A \otimes B) \otimes C \; .$$

0271-4132/82/0552/$04.75

Moreover, there is an object I which is a (left and right) identify object for $\otimes$, again up to natural isomorphisms

$$b: A \otimes I \to A, \quad b': I \otimes A \to A.$$

There are additional hypotheses. First, the associativity gives two different maps

$$A \otimes (B \otimes (C \otimes D) \rightrightarrows ((A \otimes B) \otimes C) \otimes D; \tag{1}$$

the first is the composite $a_{A \otimes B,C,D} \circ a_{A,B,C \otimes D}$ and the second is the composite $(a \otimes 1) \circ a \circ (1 \otimes a)$. The first hypothesis requires that these two be equal (a commutative pentagonal diagram). Another hypothesis requires a corresponding commutativity for diagrams involving a, b, and b'. These hypotheses are exactly enough to prove the <u>coherence theorem</u> that "all" such diagrams involving a and b are commutative in any monoidal category M.

This statement is not sufficiently accurate, because we must consider a vertex $A \otimes (B \otimes (C \otimes D))$ of a diagram such as (1) not as an <u>object</u> of M, but as a tensor-product <u>formula</u> for such an object, so that two such formulas may count as different even if they happen to represent the same object in some category M. Thus we define a shape S to be an (iterated) tensor product formula in (all different) letters $A, B, C, \ldots$ (and symbols I). Each shape S in n letters determines for each monoidal category M a functor $S_M: M^n \to M$. A natural isomorphism such as a above then involves <u>two</u> shapes $S = A \otimes (B \otimes C)$ and $T = (A' \otimes B') \otimes C'$ in <u>different</u> letters, a bijection $A \mapsto A'$, $B \mapsto B'$ and $C \mapsto C'$, called the <u>graph</u> of a, from the letters of S to the letters of T <u>and</u>, using this graph to identify variables, a natural isomorphism $a_M: S_M \to T_M$ between the functors represented, respectively, by S and by T. We thus have the monoidal category M' with objects all shapes and with arrows $f: S \to T$, each arrow consisting of a graph <u>and</u> a natural isomorphism f_M. We also have the simpler category G with objects all shapes and arrows $f: S \to T$ all graphs; this category G is also a monoidal category. In either of these categories one can now define a <u>canonical arrow</u> to be one obtained from identities and instances of a, b, and b' by tensor products and composition. The desired coherence theorem is then

THEOREM: <u>Two canonical arrows</u> $S \to T$ <u>are equal in all monoidal categories</u> M <u>if and only if they have the same graph</u>.

A <u>symmetric monoidal category</u> is a monoidal category with an added structural natural isomorphism

$$c: A \otimes B \to B \otimes A$$

with three basic commutativities assumed: A condition that $c^2 = 1$, a hexagonal diagram involving both a and c, and the condition that b equals

$$A \otimes I \xrightarrow{c} I \otimes A \xrightarrow{b'} A$$

For canonical arrows (now built up from a, b, and c) in such a category, the same coherence theorem holds:

Two canonical arrows are equal if and only if they have the same graph. This was proved originally in Mac Lane[10]; a fuller exposition appears in §5 of Mac Lane[11].

3. CLOSED CATEGORIES. A symmetric monoidal closed category C is a symmetric monoidal category in which the endofunctor $- \otimes B$ has a right adjoint $[B,-]$, as expressed by the usual natural isomorphism

$$\hom(A \otimes B, C) \cong \hom(A, [B,C]). \tag{2}$$

For example, the category of all vector spaces over a given field F is such a symmetric monoidal closed category, with $A \otimes B$ the usual tensor product and $[B,C] = \mathrm{Hom}(B,C)$ the vector space of all linear transformations $B \to C$, and with the identity object I the ground field F regarded as a vector space.

The adjunction (2) for such a category can better be described in terms of two canonical maps (the unit and the counit of the adjunction)

$$e: [B,C] \otimes B \to C, \qquad d: A \to [B, A \otimes B] \tag{3}$$

which satisfy commutativity conditions; both composites

$$[B,A] \xrightarrow{d} [B,[B,A] \otimes B] \xrightarrow{[1,e]} [B,A] \tag{4}$$

$$A \otimes B \xrightarrow{d \otimes 1} [B, A \otimes B] \otimes B \xrightarrow{e} A \otimes B \tag{5}$$

are the identity. These are simply the usual "triangular" identities for the unit and counit of an adjunction. In the case of vector spaces, e is just the familiar "evaluation" map; for $f: B \to C$ and $b \in B$, take $f \otimes b \to f(b)$.

In such a closed category, shapes are again defined as formulas built up from I and variables, now using both $\otimes$ and $[\ ,\ ]$. Canonical maps are again defined as those constructed using both $\otimes$ and $[\ ,\ ]$ from identities and the basic maps a, b, c, d, and e. Again one can assign to each canonical map $f: S \to T$ between shapes S and T a graph $\Gamma(f)$ - which, however, now must be an involution on the union of the sets of letters in S and T. Thus, in the case of the evaluation map $e:[B,C] \otimes B' \to C'$, this graph links the letter C to C' on the opposite side, where C and C' have the same variance (here both covariant) as before. However, the graph links the letters B and B' both in the domain S, but with opposite variance there (B is contravariant). Such linkage correspond to the facts that the evaluation arrow e is natural in C, in the usual sense, but "dinatural" in the variable B, in the sense that for any arrow $g:B \to D$ in the category one has a commutative diagram

$$\begin{array}{ccc} [D,C] \otimes B & \xrightarrow{1 \otimes g} & [D,C] \otimes D \\ {\scriptstyle [g,1] \otimes 1}\downarrow & & \downarrow{\scriptstyle e} \\ [B,C] \otimes B & \xrightarrow{e} & C. \end{array}$$

In this case, the shapes and graphs again form a symmetric monoidal closed cateogry, but the desired coherence theorem fails: There can be different canonical arrows with the same graph. To see this, observe that $B^* = [B,I]$ in such a category acts as the "dual" or "conjugate" of B, just as in the case of vector spaces. Moreover, the usual canonical map of any vector space B into its double dual B^{**},

$$\varkappa_B: B \to B^{**},$$

can be obtained here by first taking the composite

$$B \otimes [B,I] \xrightarrow{c} [B,I] \otimes B \xrightarrow{e} I$$

and then its transpose $B \to [[B,I],I] = B^{**}$ under the adjunction (2). Now using this canonical map $\varkappa_B$ one may construct a canonical map

$$B^{***} \xrightarrow{(\varkappa_B)^*} B^* \xrightarrow{\varkappa_{(B^*)}} B^{***}$$

which is not the identity of B^{***}, as one may see by specializing to vector spaces and taking B to be a vector space with a denumerable basis. Hence we have here two different canonical maps $B^{***} \to B^{***}$ with the same graph, so that a coherence theorem of the previous type does not hold: A diagram need not commute when the corresponding diagram of graphs commutes. Another striking example of this phenomenon was found by Arens in 1950 in his study of canonical products for Banach algebras[2]. He produced two different canonical maps

$$A^{**} \otimes B^{**} \rightrightarrows (A \otimes B)^{**};$$

they can be readily reconstructed in any symmetric monoidal closed category by using a suitable sequence of transpositions and evaluations which is not symmetric in A and B. In the case considered by Arens, if $A = B$ is a Banach algebra with a product operation $m: A \otimes A \to A$, these two different canonical maps composed with m^{**} give two (usually different) products on A^{**}.

Thus the expected coherence theorem fails for closed categories - and fails for shapes involving I as in the formation of the dual object $B^* = [B,I] = \mathrm{Hom}(B,I)$. Kelly and Mac Lane, in [7], showed that this was the typical and only exception. More exactly, call a shape constant if it involves only I and no (variable) letters $A, B, \ldots$ and call a shape S proper if its construction never involves a subshape $[T,E]$ where T is nonconstant and E constant. Thus, in particular, the dual $B^* = [B,I]$ of a non-constant B can never appear in the formation of a proper shape. With this preparation, Kelly and Mac Lane then proved a partial coherence theorem: Two canonical maps

$$f,g: S \to T$$

between proper shapes for a symmetric monoidal closed category are equal if and only if they have the same graph.

To achieve this theorem, Kelly-Mac Lane also used a method to test, for given shapes S and T, whether there is a canonical map $S \to T$. They first defined a central map to be one built from the monoidals a, b, and c using composition and $\otimes$. They then defined a constructible map $f: S \to T$ by recursion as follows. Every central map is constructible, as are the maps of the following three types (where A, B, C, and D are arbitrary shapes):

$$S \xrightarrow{x} A \otimes B \xrightarrow{f \otimes g} C \otimes D \xrightarrow{y} T \tag{6}$$

where x and y are central, $f: A \to B$ and $g: C \to D$ constructible;

$$S \xrightarrow{\pi f} [B,C] \xrightarrow{y} T \tag{7}$$

where y is central, $f: S \otimes B \to C$ is constructible with transpose πf, and

$$S \xrightarrow{x} [B,C] \otimes A \otimes D \xrightarrow{1 \otimes f \otimes 1} [B,C] \otimes B \otimes D$$
$$\xrightarrow{e \otimes 1} C \otimes D \xrightarrow{g} T \tag{8}$$

where x is central, $f: A \to B$ and $g: C \otimes D \to T$ are constructible and e is the evaluation map as above. They called the latter map a "prepared evaluation," written $g(\langle f \rangle \otimes 1)x$, because it is the evaluation e after "preparation" by the constructible f. Kelly-Mac Lane then proved that every canonical map is constructible. This amounted essentially to a proof (by recursion on the "rank" of canonical maps) that the constructible maps are closed under composition. Technically, for purposes of the induction, this was in fact a proof that a composite

$$T \otimes U \xrightarrow{h \otimes 1} S \otimes U \xrightarrow{k} V \qquad (9)$$

is constructible whenever h and k are such.

This reduction of canonical maps to constructible ones is essential. The canonical maps, as originally described by composition, would include composites $S \to R \to T$ in which a map between two relatively simple shapes S and T could arise through the intermediary of a much more complex shape R. This reduction device is called "cut elimination" in proof theory - and first borrowed for similar purposes in a (flawed) investigation[8] by Lambek.

Our main purpose here is to analyze the confluence between proof theory and category theory provided by this common use of "cut elimination".

4. CUT ELIMINATION AND RELEVANT LOGIC. The slogan "cut elimination" first appeared in the pioneering investigations of Gerhard Gentzen[6] giving a formulation of proof theory designed to establish a Hilbert-style consistency theorem for arithmetic. For example, he analyzed the rules for formal proofs in the predicate calculus including rules of inference such as the famous modus ponens = Schnitt or cut:

$$\frac{p \qquad\qquad p \supset q}{q} \qquad (10)$$

(From p and p implies q, to infer q.) Now if one wishes to establish a consistency theorem, to the effect that no formal proof in number theory will yield a conclusion such as $0 = 1$, then the presence of such a cut[10] in the formal proof will be beyond control, because a proof for a (relatively) simple conclusion such as q might involve a much more complex hypothesis, such as that here denoted by p. His cut elimination theorem then asserted that, in the presence of other rules of inference (to be specified below), any formal proof was equivalent to a different formal proof not involving the use of cut. In this connection, the use of cut in a proof is like the use of composition of canonical morphisms.

Gentzen's original cut elimination applied to the full propositional calculus (classical or intuitionistic). For our purposes, to have an accurate parallel with category theory, we consider a particular fragment of the propositional calculus called relevant logic. This arose first in the study of entailment and strict implication, as in the work of Anderson and Belnap[1] and

Belnap[3]. We use the specific formulation in a paper by Minc[12]. Consider a fragment of the propositional calculus with formulas constructed using only the binary connectives & (for and) and $\supset$ (for implies or entails). A <u>formula</u> is then one built up from I (truth) and proposition letters by the rules that if A and B are formulas, so are both A&B and $A \supset B$. Then, much as in an original calculus of sequents used by Gentzen, a <u>sequent</u> is an expression of the form $\Gamma \to C$, where the succedent C is a formula and the antecent Γ is a possibly empty list of formulas. The intention is that $A,B,D \to C$ is to stand for $A\&B\&D \to C$. Moreover, lists Γ obtained by permuting members are to be identified. Thus in this fragment (following the paper[12] and a communication from G. E. Minc), the axioms are to be the sequents $\to I$ and $C \to C$ for any formula C. The logical rules are the following rules of inference for sequents θ, Γ, Σ and formulas C, D, and E

$$\frac{\theta \longrightarrow C \qquad \Gamma, D \longrightarrow E}{\Gamma, \theta, (C \supset D) \to E\ ,} \tag{11}$$

$$\frac{C, \Gamma \longrightarrow D}{\Gamma \longrightarrow (C \supset D)} \tag{12}$$

In Gentzen's terminology, rule (12) is the introduction of $\supset$, rule (11) its elimination. There are two more logical rules:

$$\frac{\Gamma \longrightarrow C \qquad \Sigma \longrightarrow D}{\Gamma, \Sigma \longrightarrow C \,\&\, D\ ,} \tag{13}$$

$$\frac{\Gamma, C, D \longrightarrow E}{\Gamma, (C \,\&\, D) \longrightarrow E\ .} \tag{14}$$

As above, these are rules for introduction and elimination of &.

There are also two structural rules:

$$\text{Cut} \qquad \frac{\Gamma \longrightarrow C \quad \Sigma, C \longrightarrow D}{\Gamma, \Sigma \longrightarrow D\ ,} \tag{15}$$

$$\text{I-Thinning} \qquad \frac{\Gamma \longrightarrow D}{I, \Gamma \longrightarrow D\ .} \tag{16}$$

(The paper[12] of Minc also involved other connectives, such as disjunction and other rules such as contraction, from $AA \to C$ to infer $A \to C$. These additional rules do not here concern us, and do not affect the conclusions to be stated.)

On this basis, one defines a <u>derivation</u> as a tree starting with axioms and built up with these structural and logical rules. One then defines certain <u>reductions</u> from one derivation to another. These reductions take place in a "cut", replacing it by a cut of lower rank. Thus by an induction, Minc proves (loc. cit. p. 93):

CUT ELIMINATION THEOREM: *Every derivation can be transformed by a finite number of reductions to a cut-free derivation with the same last sequent. If the original derivation contained no thinning, then the same is true of the new one.*

Except for terminology (and the inclusion of other propositional connectives in the original theorem of Minc), this result is essentially identical to the (independently established) cut elimination theorem of Kelly-Mac Lane cited above. In order to make a more detailed connection, one may replace a sequent

$$A_1, A_2, \ldots, A_n \to C$$

by an arrow

$$A_1 \mathbin{\&} A_2 \mathbin{\&} \ldots \mathbin{\&} A_n \to C \tag{17}$$

(One might also have chosen the translation to be

$$A_1 \to (A_2 \supset (A_3 \supset \ldots \supset A_n \supset C) \ \ldots \ .) \tag{18}$$

If one then replaces $\&$ by $\otimes$ and $A \supset B$ by $[A,B]$, each proof-theoretical formula is turned into a shape for a symmetric monoidal closed category and each of the derivations described above becomes the construction of a canonical arrow. Thus rule (11) becomes: from

$$f:T \to C \quad \text{and} \quad g:G \otimes D \to E$$

to construct

$$G \otimes T \otimes [C,D] \to E.$$

Except for the order of factors on the left, the result is exactly the prepared evaluation

$$g(\langle f\rangle \otimes 1)x$$

considered in (8) above. Similarly, (12) simply provides the exponential transpose of a map $C \otimes G \to D$, while (13) describes the tensor product of the constructible maps. Finally, the deduction (14) is simply absorbed by our interpretation (18) of a sequent (and the associative law for $\otimes$). The I-thinning (16) is just composition with the canonical arrow $b':I \otimes A \to A$. Finally, the cut (15) from the premises $f: G \to C$ and $h: S \otimes C \to D$ amounts just to the composite $h(1 \otimes f)$. Except for the order of the factors of the tensor product, this is exactly the form (9) of the composite used in the Kelly-Mac Lane cut elimination theorem.

We conclude that the version of cut elimination for closed categories, as in Kelly-Mac Lane, is, except for form, identical with the Minc version of cut elimination for relevant logic. The conclusion is not surprising; Kelly-Mac Lane's paper followed ideas of Lambek which in turn were based on earlier cut elimination arguments in proof theory. The detailed exposition is different; Minc (like the logicians) removes one cut at a time, by induction on rank; this amounts to considering derivations as trees with various cuts. Kelly-Mac Lane, in effect, consider cut-free derivations and by an induction on a different rank prove them closed under the operation "cut." The difference is superficial; the real idea involved goes all the way back to Gentzen. We emphasize (and will exhibit later) that there are corresponding cut elimination theorems for different fragments of logic, i.e., for canonical arrows in categories with a variety of internal structures.

The proof-theoretic version replaces one derivation by an equivalent one; this amounts to replacing one canonical arrow by an equal one. For example, Minc's "left reduction" in a cut $g(1 \otimes f)$ with $f = h \circ k$ amounts essentially to the equality

$$g(1 \otimes hk) = g(1 \otimes h)(1 \otimes k),$$

while his right reduction is apparently a similar identity. There remains one difference between the proof theory and the category theory: In proof theory, one deals only with the formulas (shapes and derivations); in category theory, one has also models - specific categories with the given structure. This raises additional prospects not readily at hand in proof theory. An example is the notable conjecture: Two canonical arrows are equal for symmetric monoidal closed categories if and only if they are equal in every category of vector spaces (or, of vector spaces over the field $\underset{\sim}{Q}$ of rational numbers).

Note that we also have arrived at a logistic interpretation of the laws for a closed category: Replace $\otimes$ by & and $[A,B]$ by $A \supset B$. Then the basic canonical maps a, b, c, d, and e for closed categories become deductions. For example, $e:[B,C] \otimes B \to C$ becomes: From $B \supset C$ and B, deduce C. The identities are then equivalences between deductions.

5. EVALUATION AND THE λ-CALCULUS. The relation between proof theory and coherence theorems for categories has been further explained in a subsequent paper[13] by Minc. In particular, he provides a notation for the free symmetric

monoidal closed category Γ on a given set of objects as generators. The shapes on these objects, as described above, provide a notation for the objects of this free category. He introduces for each object A a stock of symbols $x^A : A \to A$ for the identity arrows of A. Then, given sequents $f:P \to A$ and $g:G \to (A \supset B)$, he introduces the notation (on p. 92)

$$(g,f): G,P \to B \tag{19}$$

for the result of the evaluation

$$(g,f) = e(g \otimes f): G \otimes P \to [A,B] \otimes A \xrightarrow{e} B \tag{20}$$

(We have changed his notation (b,a) to (g,f) to avoid conflict with the notation above for a as associativity.) Also, for $f:G,A \to B$ he introduces

$$\lambda x^A f:G \to A \otimes B, \tag{21}$$

following the traditional notation for the λ-calculus, in which $f(y,x)$ with $x \in A$ is regarded as a function of x by notation $(\lambda x)f$. Clearly, this λ construction is that of the "exponential" transpose of f, according to the adjunction (2). Thus, the deduction (21) can be represented as the canonical arrow

$$\pi f = [1,f]d:G \xrightarrow{d} [A,G \otimes A] \xrightarrow{[1,f]} [A,B] \tag{22}$$

On the next page 93, Minc defines his equivalence relation on deductions in terms of six rules. The crucial rules are his (1) and (3). This rule (1) states, for $g:G \to (A \supset B)$ the equivalence

$$g \equiv \lambda x^A \; (c,x^A).$$

In categorical notation, with $g:G \to [A,B]$, this amounts to the identity

$$g = [1,e(g \otimes 1)]d = [1,e]d\; g = g;$$

in other words, it is the naturality of d plus the first triangular identity (4). Similarly, his rule (3), for a deduction $h:B \to A$, reads

$$(\lambda x^A, h) = a_{x^A}[h] \tag{23}$$

where the square bracket indicates that h is to be substituted for x^A in a. In terms of the canonical map $b':I \otimes A \to A$, one may write

$$(\lambda x^A, h) = e(\pi b' \otimes h):I \otimes B \to A,$$

where π is transposition, so $\pi b' = [1,b']d$. This gives the commutative diagram (e is natural and ⊗ is a bifunctor):

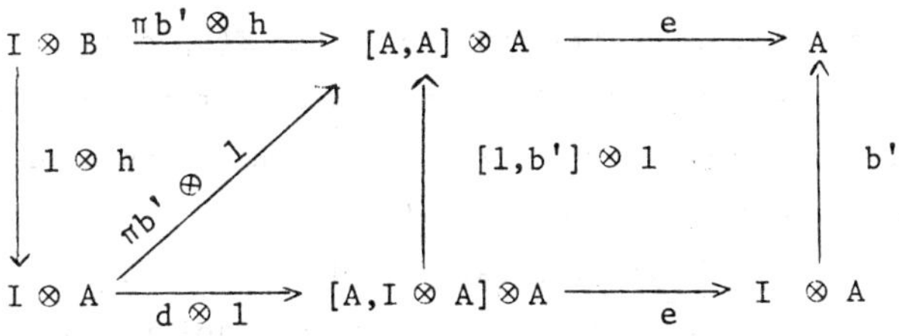

But by the second triangular identity (5), the bottom composite $e(d \otimes 1)$ is just 1. Then $b'(1 \otimes h) = hb' : I \otimes B \to A$. This is the rule (23). In other words, the two basic rules for this equivalence of deductions are just clumsy disguises for the two triangular identities of the basic adjunction. This may indicate that the traditional adherence to the λ-notation of (23) is no longer the effective way of expressing an adjunction.

From this definition of equivalence of deduction, Minc goes on to prove a normal form theorem for deductions: Each deduction is equivalent to (essentially) just one deduction in normal form; from there, he is able to deduce a coherence theorem, and in particular he can obtain the Kelly-Mac Lane coherence theorem for proper shapes. At this point, the proof-theoretic method diverges a little from the categorical one. One may test a diagram of canonical maps for commutativity by reducing each leg of the diagram to normal form; thus, the diagram commutes if the normal forms are the same. The categorical method produces, instead, an invariant (the graph) for each canonical map: then two canonical maps in a diagram are equal if and only if they have the same invariant. Such a use of invariants, when possible, seems a bit simpler. The distinction here is essentially the familiar distinction between invariants and canonical form in classical algebra, as for the case of quadratic form or of linear transformations.

6. OTHER STRUCTURES ON CATEGORIES. Questions of coherence for canonical maps can arise for other structures on categories, and in certain cases these coherence problems correspond to problems of proof theory for various formal calculi. We cite some examples, as follows.

A monoidal closed category is a monoidal category in which the functor $- \otimes B$ has a right adjoint $[B,-]$, exactly as in (2). However, no isomorphism $c : A \otimes B \cong B \otimes A$ of commutativity is assumed. Minc recently observed (private communication) that there was a cut elimination theorem true for the corresponding fragment of the propositional calculus (with & not assumed commutative), and deduced from this a coherence theorem. At about the same time, Laplaza observed that the Kelly-Mac Lane methods could be applied here to yield the same (complete) coherence theorem: Two canonical maps for monoidal closed categories are equal if and only if they have the same graph. Note in particular that no restriction to proper shapes is requisite for this result. The counter-example with the triple dual noted above is not present, because it uses the canonical map $\varkappa: B \to B^{**}$ which can be defined only in the presence of the commutativity c!

A biclosed monoidal category is a monoidal category in which both functors $- \otimes B$ and $A \otimes -$ have right adjoints, not necessarily the same. Such categories exist in nature. If R is a ring, not necessarily commutative, and M the category of bimodules $A = {}_RA_R$ over M, then M is a monoidal category with respect to the usual tensor product $\otimes_R$, and the familiar formulas show that $- \otimes_R B$ has as right adjoint the module $\hom_R(B,-)$ consisting of homomorphism of right-R-modules. On the other hand, $A \otimes_R -$ has a right adjoint ${}_R\hom(A,-)$, consisting of left module homomorphisms. I believe that the standard coherence theorem holds for such biclosed categories, and that it can be derived either by proof theory or by categorical arguments, again with restriction to proper shapes.

A closed category C is one equipped with an internal hom functor, but no corresponding tensor product. Such categories were extensively studied by Eilenberg and Kelly[5], who gave an appropriate system of axioms for the internal hom functor $[\ ,\ ]$, including the composition map, which here appears as the transpose L of the usual composition

$$[B,C] \otimes [A,B] \to [A,C].$$

For a number of years the coherence problem for such categories was formulated but could not be solved. Then at about the same time the problem was solved independently by Minc[13], using deductive terms, and by Laplaza[9], using categorical methods. Both solutions assert that two canonical morphisms are equal for all closed categories if and only if they have the same graph.

Minc[13] also considered the case of a symmetric closed category; that is, a closed category which is also equipped with a natural isomorphism

$$[A,[B,C]] \to [B,[A,C]]$$

satisfying suitable identities. In such a category (an object I is also present), one can introduce the usual canonical map $\varkappa: A \to A^{**}$ to the double dual, with the corresponding non-commutative diagrams noted above. For these categories, using equivalence of deductions, Minc proved an analog of the Kelly-Mac Lane theorem: Two canonical maps between proper shapes are equal if and only if they have the same graph.

A cartesian closed category is a monoidal closed category in which the tensor product $\otimes$ is the categorical product $\times$; such a category is also necessarily symmetric. It can also be described as a category with a terminal object and finite products $A \times B$ in which the product functor $- \times B$ has a right adjoint. In such a category, the product is equipped with the usual canonical projections

$$A \leftarrow A \times B \rightarrow B.$$

They correspond to deductions $A \& B \rightarrow A$ and $A \& B \rightarrow B$, familiar properties of the usual propositional connective &. Associates of Minc are reported to have established an appropriate coherence theorem in this case. Scott, in lectures, has observed that the study of cartesian closed categories is substantially equivalent to the classical typed λ-calculus.

7. NEW CRITERIA FOR COMMUTATIVITY. For a symmetric monoidal closed category, the Kelly-Mac Lane coherence theorem is not a complete test for the commutativity of diagrams of canonical maps. It asserts that canonical maps $S \rightrightarrows T$ with the same graph are equal only when the shapes S and T are proper (have no subshapes of the form [non-constant, constant]). Voreadou, in an elaborate investigation[14], using graphs which also involved the constant I in shapes, did obtain a complete criterion for commutativity in such a category. She constructed a particular symmetric monoidal closed category K which serves as a test category: Two canonical maps $h, h':S \rightarrow T$ (between possibly improper shapes) are equal if and only if they have the same components for all realizations in K.

The critical difficulty in this study is the possibility that a given canonical map may be expressed in two different ways as a prepared evaluation as in (8). For example, the two different canonical maps $B^{***} \rightarrow B^{***}$ are both prepared evaluations but in different ways. In the Kelly-Mac Lane argument, the hypothesis of propriety was used essentially to surmount this difficulty.

Voreadou's work involves a more subtle treatment of this point. First, because there is a canonical isomorphism $[I,A] \cong A$, it is possible to replace each shape by an isomorphic shape which is simple, in the sense that it does not involve any subshape of the form [constant,-]. Voreadou then proves (a version of) the following

LEMMA: *If a canonical map $f:S \rightarrow T$ between simple shapes can be expressed (modulo central maps) in two ways $g(\langle h\rangle \otimes 1)$ and $m(\langle n\rangle \otimes 1)$ as a prepared evaluation, where neither of the preparation maps h and n is itself a prepared evaluation, then either $h = n$, up to central maps, or else g can be*

<u>written</u> (modulo central maps) <u>as a prepared evaluation</u> $g = k(\langle n\rangle \otimes 1)$ <u>using</u> n.

Recently (and not yet published), Dole[4] has found a new and more perspicuous proof for this lemma and has used it to obtain a straightforward commutativity criterion. He calls two canonical maps $f,g:S \to T$ <u>similar</u> if there are central maps x and y such that $xf = gy$. A shape S is prime if it is a variable or a shape $[U,V]$; the collection of all prime subshapes of S is then a poset $P(S)$. To each canonical map $f:S \to T$ we can then assign a poset $P(f)$ which is the disjoint union of $P(S)$ and $P(T)$. If h is a canonical map used in a construction of f, there is then a map $z:P(h) \to P(f)$ of posets which specifies the position of h in (the construction of) f. In a prepared evaluation map $f = g(\langle h\rangle \otimes 1)$, call the preparation map h <u>elementary</u> if it is not itself a prepared evaluation. Then let $E(f)$ be the set of all positions (h,z) of elementary preparation maps h used in (any) construction of f, calling two such equal when they are similar. Dole's final theorem now states that two canonical maps $f,f' : S \to T$ between simple shapes are equal if and only if they have the same graph and $E(f) = E(f')$. This theorem is completed by a prescription for calculating $E(f)$ from a construction of f.

Communication from Minc indicates that some of his associates have used proof-theoretic methods to obtain the results of Voreadou.

In reporting on this progress in coherence theorems, I would like to acknowledge the assistance of Kreisel (who put me in touch with Minc), of Minc (for his careful letters explaining his methods), and of Laplaza (for additional analysis and translations of papers by Minc).

References

[1] Anderson, A. R. and N. D. Belnap. The pure calculus of entailment, J. Symbolic Logic 27 (1962), 19-52.

[2] Arens, R. F., Operations induced in conjugate spaces, Proc. Internat. Congr. of Math. (Cambridge, Mass., 1950), vol. 1, Amer. Math. Soc., Providence, R. I., 1952, 532-533.

[3] Belnap, N. D. Intensional models for first-degree formulas, J. Symbolic Logic, 32 (1967), 1-24.

[4] Dole, K. A coherence theorem for closed categories. Doctoral thesis, University of Chicago, 1981.

[5] Eilenberg, S. and G. M. Kelly. A generalization of the functorial calculus, J. Algebra 3 (1966), 366-375.

[6] Gentzen, G. Untersuchungen über das logische Schliessen I, II, Math. Z 39 (1934-35), 176-210 and 405-431.

[7] Kelly, G. M. and S. Mac Lane. Coherence in closed categories, J. Pure Appl. Algebra 1 (1971), no. 1, 97-140; erratum, ibid. no. 2, 219

[8] Lambek, J. Deductive systems and categories I, Syntactic calculus and residuated categories, Math. Systems Theory 2 (1968), 287-318; Deductive systems and categories II, Standard constructions and closed categories, Lecture Notes in Math., vol. 86, Springer-Verlag, Berlin, 1969. pp. 76. 76-122.

[9] LaPlaza, M. Coherence in non-monoidal closed categories, Trans. Amer. Math. Soc. 230 (1977), 293-311.

[10] Mac Lane, S. Natural associativity and commutativity, Rice University Studies 49 (1963), 28-46.

[11] __________, Topology and logic as a source of algebra (retiring Presidential address), Bull Amer. Math. Soc. 82 (Jan. 1976), no. 1, 1-40.

[12] Minc, G. E. A cut elimination theorem for relevant logics (Russian, English summary), Investigations in constructive mathematics and mathematical logic V. Zap. Naučn Sem. Leningrad Otdel Mat. Inst. Steklov (LOMI) 32 (1972), 90-97; 156.

[13] __________, Closed categories and the theory of proofs (Russian, English summary), Zap Naučn. Sem. Leningrad Otdel Mat. Inst. Steklov (LOMI) 68 (1977), 83-114; 145. (Trans. circulated as preprint).

[14] Voreadou, R. Coherence and non-commutative diagrams in closed categories, Mem. Amer. Math. Soc. No. 182 (1977).

THE UNIVERSITY OF CHICAGO
CHICAGO, ILLINOIS 60637

Contemporary Mathematics
Volume 13, 1982

AXIOMATIC GAME THEORY

By David J. Winter

Department of Mathematics
University of Michigan
Ann Arbor, Michigan

The concept of "game" has been formalized in a wide variety of different ways, one of the most interesting and remarkable being the formulation of John Conway [1]: "If L and R are two sets of games, there is a game (L,R). All games constructed in this way." Such <u>Conway games</u> form a subclass CONG of the class ZFC of sets in Zermelo-Fraenkel Set Theory with Choice. A ZFC set z is a 1-person game, the player being <u>Membership</u> $\in$ and its move being the elements of z. A Conway game (L,R) is a 2-person game with players <u>Left</u> and <u>Right</u>, the moves for Left (respectively Right) being the games in L (respectively R). The sequence of play (e.g. alternate play) is stipulated as a side condition. A player who is to move, yet has no move, loses. A game in CONG is represented as the base of a tree with branches labelled "Left" or "Right" and nodes labelled by games. Right, for example, makes a move by climbing from the tree base along some "Right" branch to the next game node, bringing Left along, thereby establishing a new tree base and new game. The theory in Conway [1] is a fascinating development of 2-person games and numbers (strengths of positions in 2-person games), the real closed field NO (class of standard and nonstandard Conway numbers), ON (class of Conway ordinal numbers), ON_2 (the field into which turns the class ON with Nim addition and multiplication) and properties of specific games.

In this paper, we extend horizons by axiomatically introducing games with unlimited numbers of players. Games are represented as bases of trees whose branches are labelled by players y_i and whose nodes are labelled by games x_i: "If the x_i are games and the y_i are players, then there is a game $z = \cup x_i y_i$. All games are constructed in this way." The resulting Game Theory G_P has models in the Set Theory ZFC and in turn, contains models of ZFC and CONG.

0271-4132/82/0553/$02.75

The underlying language is comprised of a countable collection G of move and game variables $v_0, v_1, \ldots$, a countable collection P of player variables $w_0, w_1, \ldots$, symbols $\neg, \vee, \wedge, \Rightarrow, \Leftrightarrow, \forall, \exists$, (,), and formulas formed as follows (with appropriate additional use of parenthesis):

(1) for x,z in G and y in P, xyz is a formula read "x is a move for the player y in the game z;"

(2) for any formulas p,q and variable x, p, $p \vee q$, $p \wedge q$, $p \Rightarrow q$, $p \Leftrightarrow q$, $(\forall x)q$ are formulas.

Game Theory G_P consists of the above language together with appropriate Logical Axioms and Rules of Inference, and the Game Theory Axioms introduced below. We let G_G represent Game Theory G_P in the most interesting variation G = P where there is only one collection G of variables interchangeably called moves, games, players. The discussion below applies to this variation G_G as well as to the variation G_P with G and P disjoint.

A subgame of a game z' is a game z such that all moves for any player y in z are moves for y in z' : $z \subset z'$ iff $xyz \Rightarrow xyz'$ for all x in G, y in P. Games z,z' are equal if they are subgames of each other: $z = z'$ if $z \subset z' \wedge z' \subset z$.

There is an endgame e in which no player has a move: $\neg x\,y\,e$ for all x in G, y in P (Endgame Axiom).

For any game x and player y, there is a productgame xy wherein the only move is the move x by the player y : $ab(xy) \Leftrightarrow (a = x \wedge b = y)$ (Simple Game Axiom). If xyz, we also call xy a move x by y in z.

For any game x and player y, there exists a successor game x^y in which all moves are as in x except that, also, x is a move for y : $ab(x^y) \Leftrightarrow (abx \vee ab(xy))$ (Successor Game Axiom).

For every player y, there exists a y-inductive game z, that is, a game z such that e y z and $(x^y)y\,z$ for all moves x for y in z (Inductive Game Axiom).

For no game x is there an infinite succession of moves x_1y_1z, $x_2y_2x_1$,.. (Game Termination Axiom). In the variation G_G, this axiom is strengthened: "For no game x is there an infinite succession of moves x_1y_1z, $x_2y_2z_2$,... with $z_2 = x_1$ or y_1, $z_3 = x_2$ or y_2, ..."

For any formula p and game z, there exists a game z'' denoted $z'' = \bigcup_{xyz \wedge p(z,y,z')} z'$ such that $xyz'' \Leftrightarrow (\exists z' \text{ in } G)\ (\exists x \text{ in } G,\ y \text{ in } P)$ $(xyz \wedge p(x,y,z') \wedge xyz')$ (<u>Game Union Axiom</u>). It follows that for any formula q and game z, there exists a game z''' denotes $z''' = \bigcup_{x \subset z \wedge q(x,z')} z'$ such that $xyz''' \Leftrightarrow (\exists z' \text{ in } G)\ (\exists x \text{ in } G)\ (x \subset z \wedge q(x,z') \wedge xyz')$.

For every player y and every game z, there exists a <u>move choice function</u> f (as defined below), which chooses a move $f(x)$ for y in every subgame $x \subset z$ in which y has some move b (<u>Choice Axiom</u>). Letting z be the game $\bigcup_{x \subset z \wedge (\exists b) b\ y\ x} xy$ whose y-moves are such x, and letting z' be the game $\bigcup_{x \subset z \wedge b\ y\ x} by$ whose y-moves are those games b which are y-moves of some such x, f is a function from z to z' such that $f(x) y\ x$ for all $x\ y\ z$. In the variation G_G, this axiom is strengthened so that for every move x and every game z, there also exists a <u>player choice function</u> g which, for every game $y \subset z$ in which x is a move for some player, chooses a player $g(y)$ having x as move in y : $xg(y)y$ for every $y\ x\ z''$ where $x'' = \bigcup_{y \subset z \wedge (\exists b) x\ b\ y} yx$.

Consequences of these axioms and definitions are as follows. For any games x,z and player y, $x\ y\ z \Leftrightarrow xy \subset z$. Accordingly, whenever $x\ y\ z$ we call the product $x\ y$ <u>a move x by y of z</u>. Any game z is union $z = \bigcup_{x\ y\ z} x\ y$ of its moves x by y. In particular, the products $x\ y$ are the <u>simple games</u> z, that is, those games $x \neq e$ such that e and z are the only subgames of z.

For any games z,z' there is a <u>union game</u> $z \cup z'$ such that $xy(z \cup z') \Leftrightarrow x\ y\ z \vee x\ y\ z'$ for all games x and players y, as one verifies using the Game Union, Endgame and Successor Game Axioms. Note that the successor x^y is $x^y = x \cup x\ y$.

For any formula p and game z, there is the <u>subgame</u> $\bigcup_{x\ y\ z \wedge p(x,y)} xy$ of z <u>defined</u> by p, which we denote $\{x\ y\ z \mid p(x,y)\}$, whose moves are those moves x by y of z such that $p(x,y)$. In particular, any formula p defines the <u>intersection game</u> $\bigcap_{p(z)} z = \{x\ y\ z' \mid (\forall z)\ p(z) \Rightarrow x\ y\ z\}$ where z' is any game for which $p(z')$. The intersection is e if $\neg p(z)$ for all z. Note that $xy\left(\bigcap_{p(z)} z\right) \Leftrightarrow (\forall z)\ (p(z) \Rightarrow x\ y\ z)$ for all x in G, y in P.

For any game z and player y, there is a <u>power game</u> $P_y(z) = \bigcup_{x \subset z} xy$ whose moves are those moves x by y such that $x \subset z$.

Letting the <u>y-pair</u> $(a,b)_y$ be the game $(ay)y \cup (ay \cup by)y$ for games a,b and player y, we have $(a,b)_y = (c,d)_y \Leftrightarrow a = c$ and $b = d$. We define the <u>y-Cartesian product</u> of games z and z' by $z \times_y z' = \bigcup_{a\,y\,z \wedge b\,y\,z'} (a,b)_y y$. The <u>Cartesian Product</u> of games z and z' is $z \times z' = \bigcup z \times_y z'$ (union over all y for which $a\ y\ z$ for some a). A <u>relation</u> is a subset f of some $z \times z'$. A <u>function</u> from z to z' is a relation $f \subset z \times z'$ such that for all x,y, $x\ y\ z$ implies that there exists a unique $f(x)$ that $f(x)yz'$ and $(x,f(x))_y yf$. We have the <u>Replacement Property</u> that if z is a game, y a player and $p(a,b)$ a formula which determines a unique game b for every move a for y in z, there exists a game z' and a function f from z to z' such that $a\ y\ z \Rightarrow p(a,f(a))$ for all a in G; namely, $f = \bigcup_{a\,y\,z \wedge p(a,b)} (a,b)_y y$ with $z' = \bigcup_{a\,y\,z \wedge f(a) = b} by$.

Following John Conway's definition [1] of addition for 2-person games, we define sums $z + z'$ as $z + z' = \bigcup_{a\,b\,z} (a+z')b \cup \bigcup_{a\,b\,z'} (z+a)b$.

We henceforth consider the Game Theory variation G_G where $G = P$. A review of the axioms shows that there is a <u>Move-Player Duality</u>. That is, G_G with $x\ y\ z$ (x, y, z in G) has, as model, G_G with $y\ x\ z$ (x, y, z in G), denoted G_G^*: every theorem in G_G has a dual theorem obtained by interchanging the roles of moves and players. Clearly $G_G^{**} = G_G$.

In G_G, we refer to the simple games xy as <u>Conway games</u> in G_G. Thus, any game z is the union $z = \bigcup_{x\,y\,z} xy$ of its Conway subgames, and we have: "If x and y are unions of Conway games, then xy is a Conway game. All Conway games are constructed in this way." We say that a Conway game ab <u>is a move for Left</u> (respectively, <u>Right</u>) in the Conway game xy if a is a move for b in x (respectively, in y). Thus, the class $(G_G)^2$ of Conway games in G_G is closely related to the class CONG of Conway games in ZFC. In fact, $(G_G)^2$ is essentially, CONG in the model G_G of G_G in ZFC which we give shortly.

Note that every game z is a subgame of the game $\bar{z} = z^0 \cup z^1 \cup z^2 \cup \ldots$ (countable union) where $z^0 = z$, $z^1 = \bigcup_{(\exists y)xyz} x \cup \bigcup_{(\exists x)xyz} y, \ldots, z^{n+1} = (z^n)^1, \ldots$. If $z = \bar{z}$, z is transitive. Note that z is transitive iff $xyz \Rightarrow x \subset z \wedge y \subset z$ for all x,y in G. In particular, $\bar{z}$ is transitive and, accordingly is called the transitive closure of z.

We now prove the powerful Principal of Induction for Games: If P is a formula such that for all games z, $P(z)$ is true if $P(x) \wedge P(y)$ is true for all x,y in G such that xyz, then $P(z)$ is true for all z. To prove this, we suppose otherwise, letting z be a game with $P(z)$ false. We let f be a choice e-function for $P_e(\bar{z})$, which assigns an e-move to every subgame $s \subset P_e(\bar{z})$ except $s = e$. We now generate moves $x,y,z,x_2,y_2,z_2,\ldots$ as follows. We let $s_1 = \bigcup_{xyz \wedge \neg(p(x) \wedge p(y))} (xy)e$, we note that $s_1 \subset P_e(\bar{z})$, we define $x_1y_1 = f(s_1)$ and we let $z_2 = x_1$ if $\neg P(x_1)$, otherwise $z_2 = y_1$. Letting $s_2 = \bigcup_{xyz_2 \wedge \neg(P(x) \wedge P(y))} (xy)e$, we similarly note that $s_2 \subset P_e(\bar{z})$, and define $x_2y_2 = f(s_2)$ and $z_3 = x_2$ if $P(x_2)$, otherwise $z_3 = y_2$. Since $\neg P(z)$, $\neg P(z_2)$, $\neg P(z_3),\ldots$, the move succession $x_1y_1z,x_2y_2z_2,\ldots$ never terminates, which contradicts the Game Termination Axiom. Thus, there is no game z with $P(z)$ false.

Next, we consider the class S_y of y - sets z, where z is a y-set if z is a subgame of a transitive y set z', that is, of a game z' such that $abz' \Rightarrow a \subset z' \wedge b = y$ for all a,b in G. This class S_y together with y as a Membership is a model of Set Theory ZFC, as one verifies by checking the ZFC axioms using the game properties discussed above.

In particular, we imbed ZFC in G_G as the set theory model S_e with Membership e. We henceforth refer to e-sets as sets and e-functions as functions. And we adopt the notation $\{z\} = ze$, $\{z_1,\ldots,z_n\} = z_1e \cup \ldots \cup z_ne$, $(x,y) = (x,y)_e$. Note that $xez \Leftrightarrow xe \subset z \Leftrightarrow \{x\} \subset z$, $(x,y) = (xe)e \cup (xe \cup ye)e = \{\{x\},\{x,y\}\}$ and $x^e = x \cup \{x\}$. Now G_G encompasses set theory S_e and its classes CONG, NO, ON, VON (class of Von Neumann ordinal numbers $0 = e$, $1 = 0^e$, $2 = 1^e,\ldots$).

For any game g or any set or class g of sets, we now let G_g denote the class of g-games z. These are defined as the subgames z of transitive g-games z', where z' is a transitive g-game if $xyz' \Rightarrow x \subset z' \wedge y \; e \; g$ for all x,y in G. Note that G_g has the following properties:

(1) e is in G_g, z is in G_g for any $z \subset z'$ with z' in G_g, G_g is closed under intersections and unions, xy and x^y are in G_g for all x in G_g and $y \in g$;

(2) for any x in G_g and any x,y in G such that xyz, x is in G_g and $y \in g$.

For any ordinal number $n \in$ VON, we refer to G_n as n-<u>Person Game Theory</u>. Note that $G_1 = S_e$ is set theory ZFC and that $G_2 = G_{\{0,1\}}$ is closely related to $(G_G)^2$ and CONG.

For any bijective function $f: g \to h$ from a set g to a set h, denoted $f(y) = \hat{y}$, there is a <u>unique one-to-one inclusion perserving corrrespondence</u> $x \to \hat{x}$ <u>between</u> G_g <u>and</u> G_h <u>such that</u> $\hat{e} = e$ <u>and</u> $\widehat{xy} = \hat{x}\hat{y}$ <u>for all</u> x <u>in</u> G_g <u>and</u> $y \in g$. This is proved using the Principle of Induction. This <u>product isomorphism</u> $\widehat{xy} = \hat{x}\hat{y}$ from G_g to G_h is, equivalently, <u>move preservation</u> in the sense that $xyz \Leftrightarrow \hat{x}\hat{y}\hat{z}$, for x,z in G_g and $y \in g$. More specifically, if $x \to \hat{x}$ is any one-to-one mover preserving correspondence, then:

(1) $x \subset z \Leftrightarrow \hat{x} \subset \hat{z}$ since, for example, $(abx \Rightarrow abz) \Rightarrow (\hat{a}\hat{b}\hat{x} \Rightarrow \hat{a}\hat{b}\hat{z})$; and

(2) for $z = xy$, $\hat{z} = \hat{x}\hat{y}$ by (1), since $xy \subset z \Rightarrow xyz \Rightarrow \hat{x}\hat{y}\hat{z} \Rightarrow \hat{x}\hat{y} \subset \hat{z} \Rightarrow \hat{x}\hat{y} = \hat{z}$ by the simplicity of z and $\hat{z}$.

The correspondence $\wedge: x \to \hat{x}$ also preserves unions and intersections.

We now describe models for the Game Theories G_P and G_G in ZFC Set Theory S_ϵ. For any sets x,y,z in S_ϵ, we define xyz iff $(x,y) \in z$ where (x,y) is the Kuratowski ordered pair $\{\{x\},\{x,y\}\}$. For any nonempty set or class g of sets, a <u>transitive</u> g-<u>game</u> is a relation z such that xyz implies $x \subset z$ and $y \in g$ for all x,y. A g-<u>game</u> is a subset z of a transitive g-game. The class $\mathcal{G}_g$ of g-games in S_ϵ is a model for G_P with moves and games in $\mathcal{G}_g$, players in g and xyz true or false for x,z in $\mathcal{G}_g$ and y in g. A <u>transitive game</u> is a relation z in S_ϵ such that xyz implies $x \subset z \wedge y \subset z$. A <u>game</u> is a subset of a transitive game. The class $\mathcal{G}$ of games is a model for G_G, with xyz true or false for all x,y,z in $\mathcal{G}$. Note that $\mathcal{G}_{\mathcal{G}} = \mathcal{G}$.

In the models $\mathcal{G}_g$ and $\mathcal{G}_{\mathcal{G}}$ in S, the product is $xy = \{(x,y)\}$ and $\subset, \cup, \cap$ are the same for games as for sets. In the model $\mathcal{G}_{\mathcal{G}}$ of G_G, the Conway games in G_G are represented by the products $xy = \{(x,y)\}$ $(x,y \in \mathcal{G}_{\mathcal{G}})$, which satisfy: "If x and y are unions of products, then xy is a product. All products are constructed in this way." It follows that the class $(\mathcal{G}_{\mathcal{G}})^2$ of products is $(\mathcal{G}_{\mathcal{G}})^2 = \{ab | (a,b) \in \mathrm{CONG}\}$. That is, $(\mathcal{G}_{\mathcal{G}})^2$ is, essentially, the class of CONG of Conway games in ZFC. From this, it follows, in turn, that $\mathcal{G}_{\mathcal{G}}$ is the class $\mathcal{G}_{\mathcal{G}} = 2^{\mathrm{CONG}}$ of sets of Conway games.

REFERENCES

[1] Conway, John H., On Numbers and Games, London Mathematical Society Monographs No. 6, Academic Press, London, 1976.

ABCDEFGHIJ–AMS–898765432